Managing Engineering and Technology

PRENTICE HALL INTERNATIONAL SERIES IN
INDUSTRIAL AND SYSTEMS ENGINEERING
W. J. Fabrycky and J. H. Mize, Editors

Managing Engineering and Technology
An Introduction to Management for Engineers

Second Edition

DANIEL L. BABCOCK

Professor Emeritus
University of Missouri–Rolla

Prentice Hall, Upper Saddle River, New Jersey 07458

LIBRARY OF CONGRESS CATALOGING-IN-PUBLICATION DATA

Babcock, Daniel L.
   Managing engineering and technology: an introduction to
management for engineers/Daniel L. Babcock.–2nd ed.
     p. cm.–(Prentice hall International series in industrial
and systems engineering)
   Includes index.
   ISBN 0-13-141392-9
   1. Engineering–Management.   I. Title. II. Series.
TA190.B32   1996
658.4'0024'62–dc20   95-12378
                                        CIP

> This book is dedicated to the current generation of
> practicing engineers, especially my children Jean, Bethany, and Sterling,
> my stepson Jim, and now *his* son Chris; the engineering students
> who will join them soon; and my late wife Bettie, whose unfailing love
> and encouragement made it all possible.

Acquisitions Editor: Alice Dworkin
Editorial-Production Service: Electronic Publishing Services
Manufacturing Buyer: Donna Sullivan
Cover Design: Karen Salzbach

PRENTICE HALL INTERNATIONAL SERIES IN INDUSTRIAL AND SYSTEMS ENGINEERING
W.J. Fabrycky and J. H. Mize, Editors

 © 1996, 1991 by Prentice-Hall, Inc.
Simon & Schuster/A Viacom Company
Upper Saddle River, New Jersey 07458

The author and publisher of this book have used their best efforts in preparing this book. These
efforts include the development, research, and testing of the theories and programs to determine
their effectiveness. The author and publisher shall not be liable in any event for incidental or con-
sequential damages in connection with, or arising out of, the furnishing, performance, or use of
these programs.

Printed in the United States of America

10 9 8 7 6 5 4 3 2 1

ISBN 0-13-141392-9

Prentice-Hall International (UK) Limited, *London*
Prentice-Hall of Australia Pty. Limited, *Sydney*
Prentice-Hall Canada, Inc., *Toronto*
Prentice-Hall Hispanoamericana, S.A., *Mexico*
Prentice-Hall of India Private Limited, *New Delhi*
Prentice-Hall of Japan, Inc., *Tokyo*
Simon & Schuster Asia Pte. Ltd., *Singapore*
Editora Prentice-Hall do Brasil Ltda., *Rio de Janeiro*

Contents

Preface

For twenty years after I began teaching engineering management in 1970 I searched for a textbook to use in teaching management principles to young engineers, both upper level undergraduates and young professionals in graduate degree programs or noncredit courses. Certainly there was (and is) no shortage of excellent introductory textbooks in business management. I had several shelves full—generally well written, printed attractively in multiple colors, and accompanied by stacks of supplemental instructors' aids. However, I had been looking for something different—a book that would help introduce the engineer and applied scientist to the ways in which management principles are applied in the kinds of work they are most likely to encounter. And I'd found that young engineering and science students and beginning technical professionals have little understanding of the environment and the profession they are entering. For a limited group of engineers there were solutions. There were several books on construction management suitable for civil engineers interested in that aspect of management. For manufacturing engineers, or for mechanical engineers interested in manufacturing, there were several suitable books on manufacturing management. And a variety of books on project management were available for use in credit and noncredit classes and personal study in this area. Finally, there were several books on various aspects of engineering management, but none seemed to have the breadth and content I was looking for.

In the late 1980s I began to prepare handout materials that supplemented the management textbook I was then using. I incorporated them into an outline of the book I really wished were available. Prentice-Hall liked the outline, so I fleshed it out, and the first edition, appearing in late 1990, was the result. I've been pleased by the reception the book received from my colleagues in engineering management education and related fields. It was adopted for use in more than fifty U.S. institutions, as well as in several other countries (especially in Australia).

However, we live in a fast-moving world. American industry, portrayed in the first edition as being in serious difficulty, is now much more competitive worldwide. Our better companies have been "reengineered" by reducing layers of middle management, motivating and empowering employee teams to continually improve quality and productivity, emphasizing customer satisfaction, and eliminating activities that are not essential. New methods and tools such as concurrent engineering, strategic management of technology, activity-based costing, and total productive maintenance have seen increasing use. International trade is increasingly important in the era of NAFTA, the expansion of GATT, and the collapse of the Soviet empire. These considerations, plus updating of statistical information and increasing end-of-chapter discussion questions, have been incorporated into this second edition.

As in the first edition, I begin this book with a chapter on the profession of engineering, the art and science of management, and their integration. This is followed by a chapter on the history of management from the engineer's perspective. Chapters 3 through 8 concern themselves with the management functions of planning, organizing, motivating, and controlling as they might appear in a more conventional management text, but somewhat condensed and written with an emphasis on the management of technology. The next chapters then apply these basics of management to the working environments the engineer is most likely to see, organized into chapters on research, design, production, and technical sales and service. Then two chapters on project management treat the application of management principles to this common type of engineering practice. Chapters 16 and 17 consider the early career development of the engineer—in getting off to the right start, maintaining technical competence, and considering the transition to management. Chapter 18 closes with supporting topics of interest to the engineer—the position of women and minorities in engineering, using time effectively, and the importance of professional ethics.

Instructors adopting this book may wish to tailor the content to their specific needs. For industrial and manufacturing engineering classes, Chapters 11 and 12 may be redundant to similar material covered in more detail in other classes. Chapter 16, dealing with the beginning of the engineering career, may be unnecessary for graduate classes of practicing engineers. Discussion questions at the end of each chapter include some that ask working engineers for application of chapter content to their own professional experience, and others more appropriate to undergraduates. Adopters have confirmed my experience that the book has more than enough breadth of material, even when selectively pruned in this manner, for use in a 45-contact-hour, 3-semester-hour program.

Suggestions for improvement or identification of errors are welcomed.

ACKNOWLEDGMENTS

The contributions of my colleagues Drs. Henry Metzner, Donald Myers, Yildirim "Bill" Omurtag, and Henry Sineath in providing portions of Chapters 9, 12, and 13 are acknowledged with sincere appreciation. Thanks also to Professor Benjamin Blanchard and series editors Professors Wolter Fabrycky and Joe Mize for their careful review of the manuscript, to many other colleagues for their suggestions for improvement of the first edition, and to Prentice Hall editor Alice Dworkin and Electronic Publishing Services Production Editor Anthony Calcara for their patient help with this second edition.

Daniel L. Babcock, PhD, P.E.
Rolla, Missouri

PART 1

Introduction to Engineering Management

CHAPTER 1

Engineering and Management

PREVIEW

We begin this chapter with a discussion of the origins of engineering practice and education, the nature of the engineering profession, and the types of engineers, their work, and their employers. Next, we define *management* and characterize managerial jobs and functions. Finally, these topics are synthesized by defining *engineering management* and discussing the expectation of managerial responsibilities in the typical engineering career.

ENGINEERING

Origins of Engineering

The words *engineer* and *ingenious* both stem from the Latin *ingenium,* which meant a talent, natural capacity, or clever invention. Early applications of "clever inventions" often were military ones, and *ingeniarius* became one of several words applied to builders of such "ingenious" military machines.

Heritage of the engineer. By whatever name, the roots of engineering lie much earlier than the time of the Romans, and the engineer today stands on the shoulders of giants. William Wickenden said this well in 1947:

> Engineering was an art for long centuries before it became a science. Its origins go back to utmost antiquity. The young engineer can say with truth and pride, "I am the heir of the ages.

Tubal Cain, whom Genesis places seven generations after Adam and describes as the instructor of every artificer in brass and iron, is the legendary father of my technical skills. The primitive smelters of iron and copper; the ancient workers in bronze and forgers of steel; the discoverers of the lever, the wheel, and the screw; the daring builders who first used the column, the arch, the beam, the dome, and the truss; the military pioneers who contrived the battering ram and the catapult; the early Egyptians who channeled water to irrigate the land; the Romans who built great roads, bridges, and aqueducts; the craftsmen who reared the Gothic cathedrals; all these are my forbears. Nor are they all nameless. There are: Hero of Alexandria; Archimedes of Syracuse; Roger Bacon, the monk of Oxford; Leonardo da Vinci, a many-sided genius; Galileo, the father of mechanics; Volta, the physician; the versatile Franklin. Also, there are the self-taught geniuses of the industrial revolution: Newcomen, the ironmonger; Smeaton and Watt, the instrument makers; Telford, the stone mason; and Stephenson, the mine foreman; Faraday and Gramme; Perronet, Baker, and Roebling; Siemens and Bessemer; Lenoir and Lavassor; Otto and Diesel; Edison, Westinghouse, and Steinmetz; the Wright brothers, and Ford. These are representative of the trail blazers in whose footsteps I follow."[1]

Beginnings of engineering education. Florman contrasts the French and British traditions of engineering education in his "Engineering and the Concept of the Elite,"[2] and the following stems both from that and from my own writings. In 1716 the French government, under Louis XV, formed a civilian engineering corps, the *Corps des Ponts et Chausées,* to oversee design and construction of roads and bridges, and in 1747 founded the *École des Ponts et Chausées* to train members of the corps. This was the first engineering school, in which the study of mathematics and physics was applied not only to roads and bridges, but also to canals, water supply, mines, fortifications, and manufacturing. The French followed by opening other technical schools, most notably the renowned *École Polytechnique* under the revolutionary government in 1794. In England, on the other hand, gentlemen studied the classics, and it was not until 1890 that Cambridge added a program in "mechanical science" and 1909 when Oxford established a chair in "engineering science." True, the Industrial Revolution began in England, but "Knowledge was gained pragmatically, in the workshop and on construction sites, and engineers learned their craft—and such science as seemed useful, by apprenticeship."

America was heir to both traditions. Harvard and other early colleges followed the British classical tradition, and in the Revolutionary War we had to borrow engineers from France and elsewhere to help us build (and destroy) military roads, bridges, and fortifications. "In the early days of the United States, there were so few engineers—less than 30 in the entire nation when the Erie Canal was begun in 1817—that America had no choice but to adopt the British apprenticeship model. The canals and shops—and later the railroads and factories—were the 'schools' where surveyors and mechanics were developed into engineers. As late as the time of World War I, half of America's engineers were receiving their training 'on the job.'"[3]

The U.S. Military Academy was established in 1802, at the urging of Thomas Jefferson and others, as a school for engineer officers, but they did not distinguish themselves in the War of 1812. Sylvanus Thayer, who taught mathematics at the academy, was sent to Europe to study the *École Polytechnique* and other European schools; on his return in 1817 as superintendent of the Academy, he introduced a four-year course in civil engineering, and he hired the best instructors he could find. As other engineering schools opened, they followed this curriculum and employed Academy graduates to teach from textbooks authored by Academy faculty. Florman continues:

> Perhaps the most crucial event in the social history of American engineering was the passage by Congress of the Morrill Act—the so-called "land grants" act—in 1862. This law authorized federal aid to the states for establishing colleges of agriculture and the so-called "mechanic arts." The founding legislation mentioned "education of the industrial classes in their several pursuits and professions in life." With engineering linked to the "mechanic arts," and with engineers expected to come from the "industrial classes," the die was cast. American engineers would not be elite polytechnicians. They would not be gentlemen attending professional school after graduation from college [as law and medicine became]. . . . Engineering was to be studied in a four-year undergraduate curriculum.

Engineering as a Profession

Given the preceding, is engineering *really* a profession? "Profession" has been defined (in *Webster's Third International Dictionary*) as:

> a calling requiring *specialized knowledge* and often long and *intensive preparation,* including instruction in skills and methods as well as in the scientific, historical or scholarly principles underlying such skills and methods, maintaining by force of *organization* or concerted opinion *high standards of achievement and conduct,* and committing its members to *continued study* and to a kind of work which has for its prime purpose the rendering of a *public service.* [emphasis added]

The first issue (1866) of the English journal *Engineering* began with a description of

> the profession of the engineer, as defined in the charter which Telford obtained [in 1818 for the Institute of Civil Engineers], for himself and his associates, from [King] George the Fourth— "the art of directing the great sources of power in nature, for the use and convenience of man."[4]

A more modern and complete definition was created in 1979 by American engineering societies, acting together through the Engineers' Council for Professional Development (ECPD, the precursor to ABET, the Accrediting Board for Engineering and Technology). ECPD defined *engineering* as

> the profession in which a knowledge of the mathematical and natural sciences gained by study, experience, and practice is applied with judgement to develop ways to utilize, economically, the materials and forces of nature for the benefit of mankind.[5]

Certainly, engineering meets all the criteria of a proud profession. Engineering undergraduates recognize the need for "intensive preparation" to master the specialized knowledge of their chosen profession, and practicing engineers understand the need for lifelong learning to keep up with the march of technology. In Part 5 of this book we look at engineering societies and their ethical responsibilities in maintaining standards of conduct. Finally, engineers provide a public service not only in the goods and services they create for the betterment of society, but also by placing the safety of the public high on their list of design criteria. Each generation of engineers has the opportunity and obligation to preserve and enhance by its actions the reputation established for this profession by its earlier members.

What Engineers Do

Engineering work. The U.S. Bureau of Labor Statistics defines the nature and variety of engineering work as follows:

> Engineers apply the theories and principles of science and mathematics to the economical solution of practical technical problems. Often their work is the link between a scientific discovery and its application. Engineers design machinery, products, systems, and processes for efficient and economical performance. They design industrial machinery and equipment for manufacturing goods; design defense and weapons systems for the Armed Forces; and design, plan, and supervise the construction of buildings, highways, and rapid transit systems. They also design and develop consumer products and systems for control and automation of manufacturing, business, and management process.[6]

Numbers of engineers. To identify the number of engineers in the United States at any specific date, one must first specify carefully what is meant by an engineer. For example, the U.S. Bureau of Labor Statistics (BLS) included in their totals of "employed civilians" taken from January *Employment and Earnings* data a total of 1,572,000 engineers for 1983 and 1,716,000 for 1993 (Table 1-1). However, BLS also lists from *Monthly Labor Review* data 1,353,800 engineers and 666,200 "computer systems analysts, engineers, and scientists," for a total of 2.02 million in 1992 (Table 1-2); the comparable table for 1990 data,[7] using earlier category definitions, listed 1,519,000 engineers but only 463,200 "system analysts, EDP," for a comparable total of 1.98 million. Moreover, the National Science Foundation (NSF) reported the U.S. engineering population in the labor force in *1986* as 2.47 million![8] Only a small portion of the excess reported by NSF can be accounted for by engineers in military uniform, excluded from the BLS total. A National Research Council study sums up the difficulty:

> There is considerable diversity of opinion among individuals, groups, and organizations on what constitutes engineering. . . . In fact, the engineering community contains significant numbers of
>
> - individuals with baccalaureate or higher degrees in science or mathematics who have acquired status as engineers;
> - people whose highest degree is an associate engineering [technician] or technologist degree and who have acquired status through experience as engineers or as engineering technicians or technologists; and
> - individuals who over years of experience and/or noncollegiate training have acquired the skills and knowledge to do bona fide engineering work.[9]
>
> There also appears to be a large "strategic reserve" represented by those who have qualified as members of the engineering community in the past but who either left the profession entirely or, more likely, are currently engaged in managerial and engineering support activities.[10]

TABLE 1-1 U.S. Civilian Employment of Engineers by Type

Type of Engineer	Number in Thousands (percent)			
	1983		1993	
Electrical & electronic	450	(28.6)	533	(31.1)
Mechanical	259	(16.5)	296	(17.2)
Civil	211	(13.4)	221	(12.9)
Industrial	210	(13.3)	201	(11.7)
Aerospace	80	(5.1)	83	(4.8)
Chemical	67	(4.3)	58	(3.4)
Other types of engineers	295	(18.8)	324	(18.9)
Total Engineers	1572	(100.0)	1716	(100.0)

Source: *Statistical Abstract of the United States 1994*, 114th ed., U.S. Department of Commerce, Bureau of the Census, Washington, DC, 1994, Table 637, taken from U.S. Bureau of Labor Statistics, *Employment and Earnings*, monthly, January issues.

Types of engineers. Since the first engineers were military, the term *civil engineer* was first used to refer to all other engineers. With the age of machinery came the *mechanical* engineer, and today there are many kinds of engineers. The relative number of engineers of the most common types in the United States in 1983 and 1993 is shown in Table 1-1 (taken from the BLS *Employment and Earnings* data previously cited). About 31% of all (1.716 million) engineers by 1993 were electrical (and electronic) engineers, and another 17% mechanical. This difference may decrease: from 1987 to 1992, U.S. undergraduate enrollment decreased from 104,000 to 72,000 for electrical engineers, while increasing slightly from 65,000 to 67,000 for mechanical engineers, perhaps due to changes in perceived demand.[11] The large proportion (19%) of engineers in categories "other" than the six listed shows the continuing diversification of the engineering profession and the formation of new and rapidly growing specialties. In developing countries where the major engineering emphasis is on building infrastructure, civil engineers make up a much greater proportion than the 13% shown for the United States.

Employers of engineers. In 1992, engineers not working primarily on computer systems held 1,353,800 civilian jobs. As shown in Table 1-2, almost half (640,700, or 47.3%) of them worked in manufacturing industries—mostly in electrical and electronic equipment, aircraft and parts, machinery, scientific instruments, chemicals, motor vehicles, fabricated metal products, and primary metals industries. They were not evenly distributed as a percentage of total employment across industry. According to the National Research Council[12], in 1980 about 25.2% of engineering services, 15.7% of aircraft (aerospace), and 12.7% of commercial R&D personnel were engineers. Engineers then made up 9.6% of electronic computer, 7.1% of electrical machinery, and 4.8% of nonelectric machinery employment, but only about 4% in chemicals, communications, and motor vehicles and roughly 2.2% in primary and fabricated metals industries.

Another 302,200 (22.3%) of civilian engineers in 1992 worked in services, especially in engineering and architectural services and business and management consulting services, where firms designed construction projects or did other engineering work on a con-

TABLE 1–2 U.S. Civilian Employment of Scientists, Engineers, and Technicians, by Occupation and Industry: 1992
[In thousands. Based on sample and subject to sampling error. For details, see source]

Occupation	Total[1]	Wage and Salary Workers								Self-employed
		Mining[2]	Construction	Manufacturing	Transportation[3]	Trade	FIRE[4]	Services	Government	
Scientists and engineers	**2,673.7**	**41.6**	**27.1**	**913.5**	**108.4**	**86.8**	**99.3**	**761.1**	**461.6**	**163.0**
Scientists	1,319.8	19.5	1.2	272.8	33.1	30.8	88.0	458.9	271.7	134.0
Physical scientists	197.5	14.8	0.1	59.4	3.2	2.5	0.6	65.3	45.2	6.0
Life scientists	182.1	0.1	—	27.6	0.9	1.7	(NA)	57.7	72.9	12.0
Mathematical scientists	16.3	—	—	1.5	0.7	—	1.6	7.4	5.1	—
Social scientists	257.8	0.3	—	—	1.3	—	7.7	99.0	65.1	84.0
Computer systems analysts, engineers and scientists	666.2	4.2	1.1	184.4	26.9	26.5	78.1	229.5	83.3	32.0
Engineers[5]	1,353.8	22.2	25.9	640.7	75.4	56.0	11.3	302.2	189.9	29.0
Civil engineers	172.8	1.1	10.9	8.3	6.2	0.5	0.9	66.2	73.7	5.0
Electrical/electronics	369.9	0.9	5.8	167.6	34.5	35.5	1.1	78.7	39.8	6.0
Mechanical engineers	227.0	1.7	4.0	136.6	5.1	7.1	1.3	52.3	13.8	5.0
Engineering and science technicians	**1,253.1**	**19.2**	**25.5**	**449.5**	**74.0**	**84.4**	**5.1**	**400.3**	**160.8**	**26.0**
Electrical/electronics technicians	322.8	1.6	5.1	128.0	26.0	62.1	1.8	67.3	26.7	4.0
Engineering technicians	372.1	5.4	4.5	132.0	26.6	11.4	0.3	94.7	90.4	6.0
Drafters	314.2	2.7	15.9	100.1	17.5	6.6	1.1	144.9	11.3	12.0
Science technicians	244.1	9.5	—	89.3	3.9	4.3	1.9	93.4	32.4	4.0
Surveyors	**99.3**	**2.6**	**3.3**	**—**	**3.3**	**—**	**0.6**	**55.9**	**23.0**	**10.0**
Computer programmers	**554.7**	**3.4**	**1.3**	**80.8**	**24.1**	**57.0**	**75.4**	**246.1**	**50.1**	**16.0**

— Represents or rounds to zero. NA Not available. [1] Includes agriculture, forestry, and fishing not shown separately. [2] Includes oil and gas extraction. [3] Includes communications and public utilities. [4] Finance, insurance, and real estate. [5] Includes kinds of engineers and technicians not shown separately.

Source: *Statistical Abstract of the United States 1994,* 114th ed., U.S. Department of Commerce, Bureau of the Census, Washington, DC, 1994, Table 979, taken from U.S. Bureau of Labor Statistics, *Monthly Labor Review,* November 1993. (Data collected biennially.)

tract basis for organizations in other parts of the economy. Federal, State, and local governments employed another 189,900 (14.0%). Over half of these were in the Federal Government, mainly in the Departments of Defense, Transportation, Agriculture, Interior, and Energy, and in the National Aeronautics and Space Administration (NASA). Most engineers in state and local government agencies worked in highway and public works departments.

Another 75,400 (5.6%) of (especially electrical) engineers worked in communications, public utilities, or transportation. The remaining 144,400 (10.7%) were employed in various commercial occupations, construction, or mining, or were self-employed consultants.

Engineering jobs in a company. Manufacturing organizations offer many types of jobs for engineers. Figure 1-1 gives an example.

Many of the engineering positions in this hypothetical manufacturing company would fall under the vice president for research and engineering. Positions in engineering research, engineering design, and related design support activities such as reliability and maintainability engineering are discussed in Chapters 9 and 10. As many or more engineers might report to the vice presidents for manufacturing and (if separate) quality control. These industrial, plant, maintenance, manufacturing, and quality engineering functions are discussed in Chapters 11 and 12. The more technically complex the product, the more engineers will be involved in technical sales, field service engineering, and logistics support, as discussed in Chapter 13. A smaller number of engineers will find temporary positions or permanent careers in areas such as purchasing (of technically complex parts and services) and recruiting (of technical personnel). Finally, in today's age of technical complexity, many general management positions are held by engineers.

MANAGEMENT

Management Defined

The Australian Edmund Young, in supplementary notes used in teaching from the original edition of this chapter, wrote that

> "Management" has been one of the most ubiquitous and misused words in the 20th century English language. It has been a "fad" word as well. Civil engineers discuss river basin management and coastal management, doctors discuss disease management and AIDS management, garbage collectors are now "waste management experts," and the glamorous Hollywood actress Jane Fonda even has a "school of figure management."[13]

McFarland traces the meaning of the words "manage" and "management" as follows:

> The word *manage* seems to have come into English usage directly from the Italian *maneggiare,* meaning "to handle," especially to handle or train horses. It traces back to the Latin word *manus,* "hand." In the early sixteenth century *manage* was gradually extended to the operations of war and used in the general sense of taking control, taking charge, or directing. . . . *Management* was originally a noun used to indicate the process for managing, training, or directing. It was first applied to sports, then to housekeeping, and only later to government and business.[14]

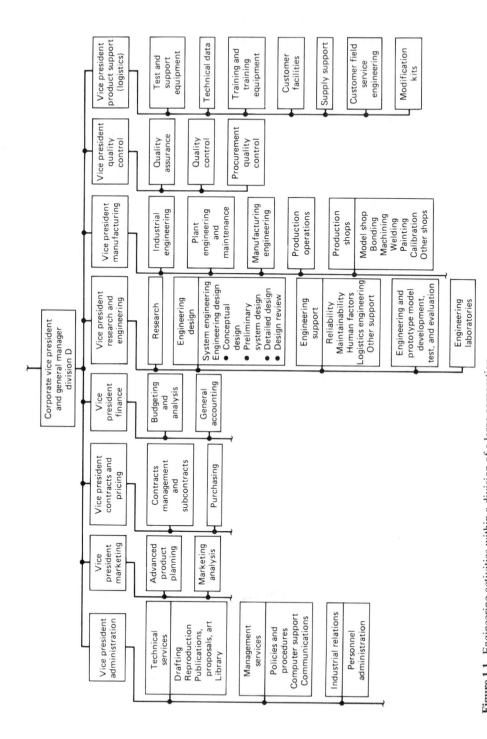

Figure 1-1. Engineering activities within a division of a large corporation. (From Benjamin S. Blanchard, *Engineering Organization and Management*, ©1976, Figure 10-3, p. 280. Reprinted by permission of Prentice Hall, Inc., Englewood Cliffs, NJ.)

McFarland continues by identifying "four important uses of the word *management,* as (1) an organizational or administrative process; (2) a science, discipline, or art; (3) the group of people running an organization; and (4) an occupational career." Sentences illustrating each of these in turn might be (1) "He practices good management"; (2) "She is a management student"; (3) "Management *doesn't really believe* in quality"; and (4) (heard from innumerable college freshmen) "I wanna get inta management." Of these four, most authors of management textbooks are referring to the first meaning (the *process*) when they define "management." According to some of these authors, management is

- The work of creating and maintaining environments in which people can accomplish goals efficiently and effectively (Albanese[15])
- The process of achieving desired results through efficient utilization of human and material resources (Bedeian[16])
- The process of reaching organizational goals by working with and through people and other organizational resources (Certo[17])
- A set of activities (including planning and decision making, organizing, leading, and controlling) directed at an organization's resources (human, financial, physical, and information) with the aim of achieving organizational goals in an efficient and effective manner (Griffin[18])
- The process by which managers create, direct, maintain, and operate purposive organizations through coordinated, cooperative human effort (McFarland[19])
- The process of acquiring and combining human, financial, informational, and physical resources to attain the organization's primary goal of producing a product or service desired by some segment of society (Pringle, Jennings, and Longnecker[20])

Albanese[21] provides a set of definitions of the word "management" suggested by a sample of business executives:

- Being a respected and responsible representative of the company to your subordinates
- The ability to achieve willing and effective accomplishments from others toward a common business objective
- Organizing and coordinating a profitable effort through good decision making and people motivation
- Getting things done through people
- The means by which an organization grows or dies
- The overall planning, evaluating, and enforcement that goes into bringing about "the name of the game"—profit
- Keeping your customers happy by delivering a quality product at a reasonable cost
- Directing the actions of a group to accomplish a desired goal or objective in the most efficient manner

Management Levels

Ensign or admiral, college president or department chair, maintenance foreman, plant manager, or company president: all are managers. What skills must they have, what roles do they play, what functions do they carry out, and how are these affected by the level at which they operate? Let us look at each of these questions in order.

Management is normally classified into three levels: first-line, middle, and top management. Managers at these three levels need many of the same skills, but they use them in different proportions. The higher the management level, the further into the future a manager's decisions reach and the more resources are placed at risk.

First-line managers directly supervise nonmanagers. They hold titles such as foreman, supervisor, or section chief. Generally, they are responsible for carrying out the plans and objectives of higher management, using the personnel and other resources assigned to them. They make short-range operating plans governing what will be done tomorrow or next week, assign tasks to their workers, supervise the work that is done, and evaluate the performance of individual workers. First-line managers may only recently have been appointed from among the ranks of people they are now supervising. They may feel caught in the middle between their former fellows and upper management, each of which feels the supervisor should be representing them. Indeed, they must provide the "linking pin" between upper management and the working level, representing the needs and goals of each to the other.

Many engineers who go into a production or construction environment quickly find themselves assigned as a foreman or supervisor. Many find such an assignment a satisfying chance to "make things happen" through their own actions and decisions. Doing so effectively, while according the workers newly assigned to you the courtesy and respect merited by their years of experience, requires tact and judgment. If you can achieve this balance, however, you may be surprised to find how willing your team members are to accord you the same respect, and to help you learn *your* job.

Middle managers carry titles such as plant manager, division head, chief engineer, or operations manager. Although there are more first-line managers than any other in most organizations, most of the *levels* in any large organization are those of middle management. Even the lowest middle manager (the second-line manager, who directly supervises first-line managers) is an *indirect manager,* and has the fundamentally different job of managing through other managers. Middle managers make plans of intermediate range to achieve the long-range goals set by top management, establish departmental policies, and evaluate the performance of subordinate work units and their managers. They also provide an integrating and coordinating function so that the short-range decisions and activities of first-line supervisory groups can be orchestrated toward achievement of the long-range goals of the enterprise. A major management movement of the 1990s, driven by the need to become more competitive, has been drastic reduction in the number of middle managers—often leading to elimination of half the management levels between supervisor and top manager. This has become possible in part because modern computer-based management information systems bring decision-making information directly to upper management that previously had to be summarized in turn by each level of middle management, and in part because nonmanagers are now better educated and are often organized into empowered teams that can make some of the decisions previously reserved for lower management.

Top managers bear titles such as chairman of the board, president, or executive vice president; one of these will normally be designated "chief executive officer" (CEO). In government the top manager may be the administrator (of NASA) or secretary (of State or Commerce) or governor or mayor. While they may report to some policy-making group (the board of directors, legislature, or council), they have no full-time manager above them.

Top managers are responsible for defining the character, mission, and objectives of the enterprise. They must establish criteria for and review long-range plans. They evaluate the performance of major departments, and they evaluate leading management personnel to gauge their readiness for promotion to key executive positions. Bedeian[22] paints a picture of the typical top manager: a college graduate (85%), probably with some postgraduate work (58%) and often a graduate degree (40%); usually from a middle-class background, often born to fathers in business or a profession; age 50 to 65, with work experience concentrated in one, two, or three companies; and with a work week of 55 to 65 hours. Most CEOs have previously specialized in finance, banking, administration, or marketing (13 to 15% each); about 11% each come from technical, production/operations, or legal careers. One often finds a household products company led by a marketer, an electric utility by a lawyer, and an electronics firm led by an engineer (who has mastered the art of management). Often an organization will look for a top manager with particular strength in the functional area in which the enterprise is currently facing a challenge.

Managerial Skills

Katz[23] suggests that managers need three types of skills: technical, interpersonal, and conceptual. *Technical skills* are skills (such as engineering, accounting, machining, or word processing) practiced by the group supervised. Figure 1-2 shows that the lowest level of manager has the greatest need for technical skills, since they are directly supervising the people who are doing the technical work, but even top managers must understand the underlying technology on which their industry is based. *Interpersonal skills,* on the other hand, are important at every management level, since every manager achieves results through the efforts of other people. *Conceptual skills* represent the ability to "see the forest in spite of the trees"—to discern the critical factors that will determine an organization's success or failure. This ability is essential to the top manager's responsibility for setting long-term objectives for the enterprise, although it is needed at every level.

Managerial Roles—What Managers Do

Henry Mintzberg[24] gives us another way to view the manager's job by examining the varied *roles* a manager plays in the enterprise. He divides them into three types: *interpersonal, informational,* and *decisional* roles.

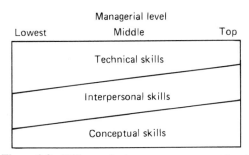

Figure 1-2. Skills required versus management level.

Interpersonal roles are further divided into three types, depending on the direction of the relationship:

- The *figurehead* role involves the ceremonial or legal actions of the symbolic head of an organization in welcoming dignitaries and signing official documents, largely *outward* relationships. Many such events lose significance if they are delegated.
- The *leader* role is the widely recognized *downward* relationship of selecting, guiding, and motivating subordinates. This role is considered in detail in Chapter 7.
- The *liaison* role consists primarily of the *horizontal* relationships with peers and people in other organizations that are built and nurtured for mutual assistance. The modern term *networking* is much the same.

Informational roles are also of three types, depending on the direction of information flow:

- The *monitor* role involves *collecting* information about both internal operations and external events. This is done by reviewing activities and reading reports internally, attending professional meetings or trade shows, and reading the professional and trade literature to monitor the external environment and understand the trends that will affect the future of the enterprise. The researcher (often a supervisor) who performs this function is known as a *gatekeeper*.
- The *disseminator* role involves the *transmission* of information internally to subordinates, superiors, and peers so that everyone has the information they need to know to do their job. The manager here acts as a sort of "telephone switchboard" in transmitting information. This role as the source of information, if carefully handled, can strengthen a manager's formal authority.
- The *spokesman* [or *spokesperson*] role, normally carried out by higher management, involves speaking for the organization to the press, the public, and other *external* groups. In an *internal* version of this role, which might be called *advocate,* successful supervisors "carry the ball" for their subordinates to get the resources they need or the rewards they have earned.

Decisional roles are of four kinds in this typology:

- The *entrepreneurial* role of initiating change, assuming risk, and transforming ideas into useful products.
- The *disturbance handler* role of dealing with unforeseen problems or crises and resolving them. The use of penalties is only one—and often the least effective—mechanism for handling disturbances.
- The *resource allocator* role of distributing the (normally scarce) resources of money, labor, materials, and equipment where they will provide greatest benefit to the organization.
- The *negotiator* role of bargaining with suppliers or customers, or subordinates, peers, or superiors, to obtain agreements favorable to the enterprise (or at least the portion of it for which the manager is responsible).

Functions of Managers

Henri Fayol, the famous French mining engineer and executive, divided managerial activities into five "elements": planning, organizing, command, coordination, and control.[25] These elements, now called *"functions of managers,"* have proven remarkably useful and durable over the decades. Although each management author has his or her favored set of functions, almost all include planning, organizing, and controlling on their list. "Command" became too authoritative a word in today's participative society and has been replaced by "leading," "motivating," or "actuating," and few authors treat coordinating as a separate function. Nonetheless, as the late management author Harold Koontz concluded, "There have been no new ideas, research findings, or techniques that cannot readily be placed in these classifications."[26] Koontz chose and (with co-author Heinz Weihrich) defined his favored list of the functions of managers as follows:

- *Planning* involves selecting missions and objectives and the actions to achieve them; it requires decision making, that is, choosing future courses of action from among alternatives.
- *Organizing* . . . is that part of managing that involves establishing an intentional structure of roles for people to fill in an enterprise.
- *Staffing* [included by most authors with Organizing] involves filling, and keeping filled, the positions in the organization structure.
- *Leading* is influencing people to strive willingly and enthusiastically toward the achievement of organization and group goals. It has to do predominantly with the interpersonal aspect of managing.
- *Controlling* is the measuring and correcting of activities of subordinates to ensure that events conform to plans.

Engineering managers need to understand the body of knowledge that has been developed by management theorists and practitioners and organized under this framework, and this is the purpose of Part 2 of this book. Planning and the associated subfunction of decision making is treated in Chapters 3 and 4; organizing in Chapter 5; staffing, motivating, and leading in Chapters 6 and 7; and controlling in Chapter 8. Wherever possible, the particular implications of these functions for the technical employee and the technology-affected organization is emphasized.

The engineering manager also needs to understand the particular problems involved in managing research, development, design, production/operations, projects, and related technical environments. In Parts 3 and 4 we treat the application of these management functions to the specific environments in which most engineers and engineering managers will work.

Management: Art or Science?

Earlier in this chapter the characteristics of a profession were discussed, and engineering was shown to meet all the criteria of a profession. Management also has a body of *specialized knowledge,* which we introduce in Part 2. Here the justification of management as a profession largely ends. Many managers will have first completed bachelor's or master's degree programs in business administration, public administration, or engineering management, but as I have observed elsewhere:

> The knowledge need not be obtained only in such formal programs. It may be acquired by personal study, in-house employee education programs, seminars by all kinds of consultant entrepreneurs, or programs of many professional societies. Sometimes this formal or informal education is obtained before promotion [into] the management hierarchy, but often it occurs after promotion.
>
> A very small proportion of the broad range of managers belong to management-specific organizations such as the American Management Association or the Academy of Management or (for engineers) the American Society for Engineering Management. They are more likely (especially in technical areas) to belong to management divisions or institutes within discipline-oriented professional societies. Considerations of standards, ethics, certification, and the like become those of the parent societies, not the management subset.[27]

ENGINEERING MANAGEMENT: A SYNTHESIS

What Is Engineering Management?

Some writers would use a narrow definition of "engineering management," confining it to the direct supervision of engineers or of engineering functions. This would include, for example, supervision of engineering research or design activities. Others would add an activity we might consider the "engineering of management"—the application of quantitative methods and techniques to the practice of management (often called "management science"). However, these narrow definitions fail to include many of the management activities engineers actually perform in modern enterprises.

If engineering management is broadly defined to include the general management responsibilities engineers can grow into, one might well ask how it differs from "ordinary" management. I have tried to answer this question elsewhere:

> The engineering manager is distinguished from other managers because he [or she] possesses both an ability to apply engineering principles and a skill in organizing and directing people and projects. He is uniquely qualified for two types of jobs: the management of *technical functions* (such as design or production) in almost any enterprise, or the management of broader functions (such as marketing or top management) in a *high-technology enterprise.*[28]

Need for Engineers in Management

Herbert Hoover, a very successful mining engineer and manager, recognized the importance of the American engineering manager in an address to engineers the year he was elected President of the United States:

> Three great forces contributed to the development of the engineering profession. The first was the era of intense development of minerals, metallurgy, and transportation in our great West. . . . Moreover, the skill of our engineers of that period owes a great debt to American educators. The leaders of our universities were the first of all the educators of the world to recognize that upon them rested the responsibility to provide fundamental training in the application of science to engineering under the broadening influence and cultivation of university life. They were the first to realize that engineering must be transformed into a practice in the highest sense, not only in the

> training and character but that the essential quality of a profession is the installation of ethics. . . . A third distinction that grew in American engineering was the transformation from solely a technical profession to a profession of administrators—the business manager with technical training.[29]

There are several reasons engineers can be especially effective in the general management of technically oriented organizations. High-technology enterprises make a business of doing things that have never been done before. Extensive planning is therefore needed to make sure that everything is done right the first time—there may not be a second chance. Planning must emphasize recognizing and resolving the uncertainties that determine whether the desired product or outcome is feasible. Since these critical factors are often technical, the engineer is best capable of recognizing them and managing their resolution. In staffing a technically based enterprise, engineering managers can best evaluate the capability of technical personnel when they apply for positions and rate their later performance. Further, they will better understand the nature and motivation of the technical specialist and can more easily gain their respect, confidence, and loyalty. George H. Heilmeier, President and CEO of Bellcore (and an electrical engineer), makes clear the advantages of an understanding of technology in top management:

> Competition is global, and the ability to compete successfully on this scale is fostered by corporate leaders who:
>
> - Really understand the business;
> - Understand both the technology that is driving the business today and the technology that will change the business in the future;
> - Treat research and development as an investment to be nurtured, rather than an expense to be minimized;
> - Spend more time on strategic thinking about the future as they rise higher in the corporation;
> - Are dedicated to solving a customer's problem or satisfying a need, which is how I would define true marketing as opposed to sales; and
> - Place a premium on innovation.[30]

Management and the Engineering Career

A National Engineers Registry Survey conducted in 1969 analyzed the extent to which engineers were employed in management.[31] This survey revealed that about 18% of engineers had no regular supervisory responsibility and another 18% provided only indirect or staff supervision. The remainder (almost two-thirds) were acting as managers: 12% over a team or unit, 22% over a project or section, 20% over a major department, division, or program, and 10% in the general (top) management of an organization. This survey is now 25 years old, and in the last decade many companies have reduced the numbers and levels of management positions and given more decision-making authority to teams at the working level. Nevertheless, most engineers can expect a transition to management responsibilities at some point in their professional careers. Despite this, undergraduate engineering education offers little preparation for such a possibility. To meet this need, many engineering schools now provide degree programs in engineering management, which blends business and engineering as shown in Figure 1-3.

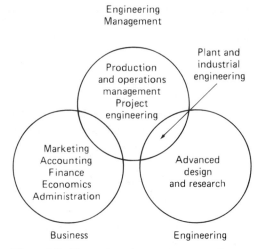

Figure 1-3. The engineering management education program.
(From Daniel L. Babcock, "B.S. and M.S. Programs in Engineering Management," *Engineering Education,* November 1973, p. 102.)

This book provides some insight into the nature of management and the environments in which the engineer is most likely to encounter the need for an understanding of management as his or her career progresses. In the last three chapters, we sum up the career implications for the engineer. In Chapter 16, we discuss some significant concerns in the development of the engineer as a technical professional; in Chapter 17, managerial and international opportunities for the engineer; and in Chapter 18, several special topics of interest to engineering and management.

DISCUSSION QUESTIONS

1-1. The precursors of today's engineers listed in the quotation from Wickenden had no classes and few or no books from which to learn scientific principles. How can you explain their success?

1-2. Compare how well *engineering* and *management* satisfy the several parts of the definition of a profession.

1-3. Why is it so difficult to answer the simple question "How many engineers are there in the United States?" Is the question "How many physicians are there in the United States?" any easier?

1-4. Comment on the sensitivity of U.S. engineering employment to a major change in the Department of Defense budget. What types of engineers would be especially affected?

1-5. What are the similarities in the definitions of *management* quoted from authors of management textbooks? Do the definitions provided by business executives differ in any way? Synthesize your own definition of management.

1-6. How does the job of supervisor or first-line manager differ from that of a higher-level manager?

1-7. How does the job of a top manager differ from those of the several levels of middle management?

1-8. Identify the three types of skills needed by an effective manager as conceived by Robert L. Katz, and describe how the relative need for them might vary with the level of management.

1-9. From the 10 managerial roles provided by Mintzberg, choose three or four that you consider most important for the first-line manager and explain your selection. Repeat for middle-level and top managers.

1-10. How would you distinguish between engineering management and management in general?

NOTES

1. Drafted by William E. Wickenden before his 1947 death, later edited and collated by G. Ross Henninger as *A Professional Guide for Young Engineers,* rev. ed. (New York: Accreditation Board for Engineering and Technology, 1981), p. 7.

2. Samuel C. Florman, "Engineering and the Concept of the Elite," *THE BENT of Tau Beta Pi,* Fall 1992, p. 19.

3. Florman, "Engineering and the Concept of the Elite," p. 19.

4. "Breaking Ground," *Engineering* [London], 1:1, January 5, 1866, p. 1.

5. *The Engineering Team* (New York: Engineers' Council for Professional Development (now Accreditation Board for Engineering and Technology), 1979).

6. *Occupational Outlook Handbook,* 1992–93 ed. (Washington, DC: U.S. Bureau of Labor Statistics Bulletin 2400, May 1992), p. 64.

7. *Statistical Abstract of the United States 1993,* 113th ed. (Washington DC: U.S. Bureau of the Census, 1993), Table 644, p. 405.

8. *Statistical Abstract of the United States 1989,* 109th ed. (Washington, DC: U.S. Bureau of the Census, 1989), Table 983, p. 582.

9. National Research Council, Committee on the Education and Utilization of the Engineer, *Engineering Education and Practice in the United States: Engineering Infrastructure Diagramming and Modeling* (Washington, DC: National Academy Press, 1986), p. 3.

10. National Research Council, *Engineering Infrastructure,* p. 5.

11. American Association of Engineering Societies, Engineering Workforce Commission data in *ASEE Prism,* September 1994, p. 12.

12. National Research Council, Committee on the Education and Utilization of the Engineer, *Engineering Education and Practice in the United States: Engineering Employment Characteristics* (Washington, DC: National Academy Press, 1985), p. 23, from Bureau of Census data.

13. Edmund J. Young, personal communication, August 1988.

14. Dalton E. McFarland, *Management: Foundations and Practices,* 5th ed. (New York: Macmillan Publishing Company, 1979), pp. 4–5.

15. Robert Albanese, *Managing: Toward Accountability for Performance,* 3rd ed. (Homewood, IL: Richard D. Irwin, Inc., 1981), p. 5.

16. Arthur G. Bedeian, *Management,* 2nd ed. (New York: Holt, Rinehart and Winston, 1989), p. 6.

17. Samuel C. Certo, *Modern Management: Diversity, Quality, Ethics, and the Global Environment,* 6th ed. (Needham Heights, MA: Allyn and Bacon, 1994), p. 6.

18. Ricky W. Griffin, *Management,* 4th ed. (Boston: Houghton Mifflin Company, 1993), pp. 5–6.

19. McFarland, *Management,* p. 5.

20. Charles D. Pringle, Daniel F. Jennings, and Justin G. Longnecker, *Managing Organizations: Functions and Behaviors* (Columbus, OH: Merrill Publishing Company, 1988), p. 4.

21. Robert Albanese, *Management: Toward Accountability for Performance* (Homewood, IL: Richard D. Irwin, Inc., 1975), p. 28.

22. Bedeian, *Management,* p. 15.

23. Robert L. Katz, "Skills of an Effective Administrator," *Harvard Business Review,* 52:5, September–October 1974, pp. 90–112.

24. Selected excerpts from Henry Mintzberg, *The Nature of Managerial Work,* Chapter 4. Copyright© 1973 by the author. Reprinted by permission of HarperCollins Publishers, Inc.

25. Henri Fayol, *Administration Industrielle et Générale,* Constance Storrs, trans. (London: Sir Isaac Pitman & Sons Ltd., 1949).

26. Heinz Weihrich and Harold Koontz, *Management: A Global Perspective,* 10th ed. (New York: McGraw-Hill Book Company, 1993), pp. 20–21.

27. Daniel L. Babcock and Bernard R. Sarchet, "Is Engineering Management a Profession?" *IEEE Trans. on Engineering Management,* November 1981, pp. 107–109.

28. Daniel L. Babcock, "Is the Engineering Manager Different?" *Machine Design,* March 9, 1978, pp. 82–85.

29. Herbert C. Hoover, "The Engineer's Contribution to Modern Life," an address to the American Institute of Mining and Metallurgical Engineers on receiving their Saunders Mining Medal at their 1928 annual meeting, reprinted in Dugald C. Jackson Jr. and W. Paul Jones, eds., *The Profession of Engineering* (New York: John Wiley & Sons, Inc., 1929), pp. 119–120.

30. George H. Heilmeier, "Room for Whom at the Top?: Promoting Technical Literacy in the Executive Suite," *THE BENT of Tau Beta Pi,* Spring 1994.

31. *Engineering Manpower Bulletin 25* (New York: Engineers' Joint Council, September 1973).

CHAPTER ● 2

Historical Development of Engineering Management

PREVIEW

The story of the development of management thought and of our ability to organize and control complex activities has already been documented. Two excellent books on this subject are those by George[1] and Wren.[2] In this chapter we introduce only a small part of this history, concentrating on people and situations of most significance and interest to the engineer in management. We begin by considering the great construction projects of ancient civilizations, and then the medieval production facility that was the Arsenal of Venice. We will see how the Industrial Revolution changed not only manufacturing but society as well, first in England and then in America.

As the nineteenth century ended and the twentieth began, the United States led the world in finding better, more efficient ways to do things in a movement that became known as "scientific management," while Europeans such as Weber and Fayol were developing philosophies of management at the top level. About 1930, a series of experiments at the Hawthorne Works near Chicago led to studies, which continue, on the impact of individual and group behavior on the effectiveness of managing. Engineering management continues to evolve, with the development in the last half of this century of methods for managing large projects such as the Apollo program, quantitative methods for analysis of operations, and the revolution in our lives that computer technology is creating.

ORIGINS

Ancient Civilizations

Even the earliest civilizations required management skills wherever groups of people shared a common purpose: tribal activities, estates of the rich, military ventures, governments, or organized religion. Indeed, the prototypes of civil engineering and construction management became necessary as soon as "plants and animals were domesticated and people began living in communities. By 6000 B.C. these communities sometimes contained over 1000 people and Jericho is known to have had a wall and defensive towers."[3] By 4500 B.C. the first canals diverted water from the River Gangir in eastern Iraq for crop irrigation. As canals proliferated, it became possible to store crops for commerce, and written records as well as management organization became necessary:

> In ancient Mesopotamia, lying just north and west of Babylon, the temples developed an early concept of a "corporation," or a group of temples under a common body of management. Flourishing as early as 3000 B.C., temple management operated under a dual control system: one high priest was responsible for ceremonial and religious activities, while an administrative high priest coordinated the secular activities of the organization. Records were kept on clay tablets, plans made, labor divided, and work supervised by a hierarchy of officials.[4]

Many ancient civilizations left behind great stone structures that leave us wondering how they could have been created with the few tools then available. Examples include the great wall of China, the monoliths on Easter Island, Mayan temples in South America, and Stonehenge in England. Especially impressive are the pyramids of Egypt. The great pyramid of Cheops, built about 4500 years ago, covers 13 acres and contains 2,300,000 stone blocks weighing an average 5000 pounds apiece. Estimates are that it took 100,000 men from 20 to 30 years to complete the pyramid—about the same effort in worker-years as it later took the United States to put a man on the moon. The only construction tools available were levers, rollers, and immense earthen ramps. Yet the difference in height of opposite corners of the base is only ½ inch![5]

Hammurabi (2123–2081 B.C.) of Babylon "issued a unique code of 282 laws which governed business dealings . . . and a host of other societal matters."[6] One law which should interest the civil engineer:

> If a builder builds a house for a man and does not make its construction firm, and the house which he has built collapses, and causes the death of the owner of the house, that builder shall be put to death.[7]

Today's engineer should be thankful that, while penalties for faulty design can be expensive and damaging to one's career, they are not terminal!

Problems of controlling military operations and dispersed empires have made necessary the development of new management methods since ancient times. Alexander the Great (336–323 B.C.) is generally credited with the first documented (European) use of the staff system. He developed an informal council whose members were each entrusted with a specific function (supply, provost marshal, and engineer).

Imperial Rome governed an estimated 50 million people spread from England to Syria and from Europe to North Africa by dividing the empire in turn into four major regions, 13 dioceses, and 110 provinces for civil government, with a separate structure for the military forces garrisoned throughout the provinces to maintain control. The great Roman roads that made it possible to move messages and Roman legions quickly from place to place were an impressive engineering achievement that helped the empire survive as long as it did.

It should not be inferred that early management skills were confined to Western civilization as it developed around the Mediterranean Sea. George[8] describes the consistent use of advisory staff by Chinese emperors as early as 2350 B.C. and "ancient records of Mencius and Chow (1100 to about 500 B.C.) [that] indicate that the Chinese were aware of certain principles bearing on organizing, planning, directing, and controlling." In India one Brahman Kautilya described in *Arthasastra* in 321 B.C. a wide range of topics on government, commerce, and customs. Because he analyzed objectively rather than morally the political practices that brought success in the past, his name "has become synonymous with sinister and unscrupulous management" in India (just as has Niccolo Machiavelli's name for his similar analysis in *The Prince* in the early seventeenth century in Italy).[9]

The Arsenal of Venice

George abstracts from Lane[10] a fascinating story of "what was perhaps the largest industrial plant of the [medieval] world." As Venice's maritime power grew, the city needed an armed fleet to protect her trade, and by 1436 it was operating its own government shipyard, the Arsenal. The Arsenal "had a threefold task: (1) the *manufacture* of galleys, arms, and equipment; (2) the *storage* of the equipment until needed; and (3) the *assembly and refitting* of the ships on reserve."[11]

Most impressive was the assembly line used to outfit ships. A Spanish traveler, one Pero Tafur, wrote in 1436:

> And as one enters the gate there is a great street on either hand with the sea in the middle, and on the one side are windows opening out of the houses of the Arsenal, and the same on the other side, and out came the galley towed by a boat, and from the windows they handed out to them from one the cordage, from another the bread, from another the arms, and from another the balistas and mortars, and so from all sides everything which was required, and when the galley had reached the end of the street, all the men required were on board, together with the complement of oars, and she was equipped from end to end. In this manner there came out ten galleys, fully armed, between the hours of three and nine.[12]

George identifies several other industrial management practices of the Arsenal that were ahead of their time:

1. Systematic warehousing and inventory control of the hundreds of masts, spars, and rudders and thousands of benches, footbraces, and oars needed to make the assembly line above work.
2. Well-developed personnel policies, including piecework pay for some work (making oars) and day wages for both menial labor and artisans (the latter with semiannual merit reviews and raises).

3. Standardization, so that any rudder would meet any sternpost and all ships handled the same way.
4. Meticulous accounting in two journals and one ledger, with annual auditing.
5. Cost control. As an example, one accountant discovered that lumber was stored casually in piles, and the process of searching through the piles to find a suitable log was costing three times as much as it did to buy the log in the first place; as a result of this early industrial engineering study an orderly lumberyard was established, which not only saved time and money but permitted accurate inventory of lumber on hand.

An important innovation developed in Venice during this period was *double-entry bookkeeping*. Luca Pacioli published an instruction manual (*Summa de arithmetica, geometria, proportioni et proportionalia*) in 1494 describing the system then in use and recommending it. His discussion of the use of memorandum, journal, and ledger, supporting documents, and internal checks through periodic audits were so modern that "Many excerpts from Pacioli's writing could be inserted into our current accounting textbooks with virtually no change in wording."[13] Pacioli's work was translated into English about 50 years later and was in widespread use by the early eighteenth century.

THE INDUSTRIAL REVOLUTION

End of Cottage Industry

Before the late eighteenth century, farm families would spin cotton, wool, or flax to yarn or thread on a spinning wheel, weave it on a hand loom, wet the goods with mild alkali, and spread them on the ground for months to bleach in the sun before selling these "gray goods" at a local fair for whatever price they could get. Even when under the "putting out" system, where merchants at the fairs would provide the family with materials and buy their output at a negotiated rate, the work could be done in the farm cottage as well as anywhere else.

In the last third of the eighteenth century, a series of eight inventions (six English and two French) changed society irretrievably. Summarized from Amrine et al.,[14] they were:

1. The *spinning jenny,* invented by James Hargreaves in 1764, which could spin eight threads of yarn (later, 80) at once instead of one.
2. The *water frame,* a spinning machine driven by water power, patented by Richard Arkwright and incorporated by him in 1771 in the first of many successful mills.
3. The *mule,* a combination of the spinning jenny and water frame invented by Samuel Crompton in 1779, which enormously increased productivity and eliminated hand spinning.
4. The *power loom,* a weaving machine patented in 1785 by Edmund Cartwright, which with time and improvements destroyed the ancient system of making cloth in the home.
5. *Chlorine bleach,* discovered in 1785 by the French chemist Claude Louis Berthollet (and bleaching powder in 1798 by Charles Tennant), which provided quick bleaching without the need for large open areas or inconstant sun.

6. The *steam engine,* patented by James Watt in 1769 and used in place of water power in factories beginning about 1785.

7. The *screw-cutting lathe,* developed in 1797 by Henry Maudslay, which made possible more durable metal (rather than wood) machines.

8. *Interchangeable manufacture,* commonly attributed to the American Eli Whitney in carrying out a 1798 contract for 10,000 muskets, but perhaps adopted by him as a result of a letter dated May 30, 1785, from Thomas Jefferson (while in France) to John Jay, describing the approach of one Leblanc at the *manufacture de Versailles:*

> An improvement is made here in the Construction of muskets, which it may be interesting to Congress to know. . . . It consists in the making of every part of them so exactly alike, that what belongs to any one, may be used for every other musket in the magazine. . . . [Leblanc] presented me the parts of fifty locks, taken to pieces, and arranged in compartments. I put several together myself, taking pieces at hazard as they came to hand, and they fitted in the most perfect manner. The advantage of this when arms are out of repair [is] evident.[15]

Problems of the Factory System

The innovations of the late eighteenth century just described caused major upheavals in England's society as well as its economy. Cottage industry could not compete with factories powered first by water and then by steam. Underground coal mines provided fuel for the steam engine, and the steam engine powered the pumps that removed water seepage from the mines. A mass movement of workers from the farms and villages to the new industrial centers was required.

The new factory managers had problems of recruiting workers, training the largely illiterate work force, and providing discipline and motivation to workers who had never developed the "habits of industry." Wren quotes Powell[16]: "If a person can get sufficient [income] in four days to support himself for seven days, he will keep holiday for the other three" and adds "Some workers took a weekly holiday they called 'Saint Monday' which meant either not working or working very slowly at the beginning of the week." In the United States today, urban plant managers who hire the "hard-core unemployed" can face exactly this same problem with workers who have neither personal experience nor family tradition with the "habits of industry," such as regular attendance and punctuality; similarly, automobiles assembled on Monday have in the past been statistically less reliable than those assembled in midweek because of plant absenteeism.

Explosive growth of the English mill towns led to filthy, overcrowded living conditions, widespread child labor, crime, and brutality. Falling wages, rampant unemployment, and rising food prices led to a rash of smashing of textile machinery by the "Luddites," peaking in 1811–12. This movement "soon died for lack of leadership" by dint of hanging Luddites in at least four cities.[17]

England's agrarian history provided no source of professional managers. Supervisors often were illiterate workers who rose from the ranks and were paid little more than the workers they supervised, and there was no common body of knowledge about how to manage. Upper management often consisted of the sons and relatives of the founders, a condition that persists today in many developing countries. Gradually, the forerunners of modern factory management began to develop. One early firm was Boulton and Watt, founded by Matthew Boulton and James Watt to manufacture Watt's steam engine. By 1800 their sons, inheriting the firm, instituted innovations at their Soho Engineering Foundry such as factory layout planning, inventory control, production planning, work flow study, sophisticated analysis of piecework rates, and paid overtime.

Another pioneer was Robert Owen, part owner of a mill complex in New Lanark, Scotland. Owen was ahead of his time in proposing that as much attention be paid to vital "human machines" as to inanimate ones. He told a group of factory owners:

> Your living machines may be easily trained and directed to procure a large increase of pecuniary gain. Money spent on employees might give a 50 to 100 per cent return as opposed to a 15 per cent return on machinery. The economy of living machinery is to keep it neat and clean, treat it with kindness that its mental movements might not experience too much irritating friction.[18]

Owen reshaped the whole village of New Lanark, improving housing, streets, sanitation, and education. Although he continued to employ children, he lobbied for legislation that ultimately forbade employing children under the age of 9, limited the workday to $10\frac{1}{2}$ hours, and forbade night work for children.

Industrial Development in America

England regarded her colonies as markets for English factories; as early as 1663, all manufactured goods were required to be purchased in England (even if made elsewhere in Europe), and "the 1750 English Iron Act made it illegal to set up in the Colonies mills and furnaces for the manufacture of finished products."[19] Although emigration of skilled labor to America was prohibited after the American Revolution, an experienced textile machinery builder and mechanic named Samuel Slater emigrated from England as a "farmer" and joined with three prosperous Rhode Island merchants to build the first technically advanced American textile mill at Pawtucket, RI, in 1790; by 1810 the census listed 269 mills in operation.[20] Although growth of American industry was accelerated by the War of 1812 with England, most American firms before 1835 were small, family owned, and water powered. Only 36 firms employed more than 250 workers: 31 textile firms, three in iron, and two in nails and axes. The greatest sophistication in manufacturing was at the government-owned Springfield (MA) Armory, and this knowledge "provided the basis for the later manufacture of axes, shovels, sewing machines, clocks, locks, watches, steam engines, reapers, and other products" in the 1840s and 1850s.[21]

Canals provided the first construction challenge for the new nation. Although the Middlesex Canal Company obtained the rights to build a canal from Boston to Lowell, MA, in 1793, they experienced great difficulty until they called in an immigrant engineer, William Weston, who had worked under a canal builder in England. Weston "went on to provide 'know-how' for all the major projects of that period in New England. . . . This experience was available when the Erie Canal was built, 363 miles from Albany to Buffalo, NY" between 1816 and 1825, a project that provided the training for many of our early "civil" engineers.[22]

Railroads and steel were the high-technology growth industries of the nineteenth century. Colonel John Stevens, dubbed the "father of American engineering," built the first rail line—the 23-mile Camden and Amboy Railroad—in 1830; by 1850 there were 9000 miles of track extending west to Ohio. Morse's first experimental telegraph line was built in 1844; by 1860 there were 50,000 miles of telegraph line—much of it along railroad right of way and used in part to facilitate rail shipment.

The railroads presented management problems of a dimension not seen before, and the men who mastered these challenges became the leaders of the American industrial explosion. One such was Andrew Carnegie (1835–1919), at age 24 made superintendent of the largest division of the nation's largest railroad (the Pennsylvania). In 1872, attracted by Sir Henry Bessemer's new process, he moved into steelmaking—integrating operations, increasing volume, and selling aggressively. In 1868 the United States produced 8500 tons of steel and England 110,000; by 1902, thanks to Carnegie and others, America produced 9,100,000 tons to England's 1,800,000![23]

The large industrial firms of the nineteenth century were precursors of the industrial giants of the twentieth century, headquartered primarily in Europe, the United States, or Japan but manufacturing and selling all over the world. The nature of these *multinational corporations* and the opportunities they offer engineers are discussed in Chapter 17.

Development of Engineering Education

Most engineering skill through the eighteenth century was gained through apprenticeship to a practitioner. According to Mihalasky:

> The first engineering school was probably established in France in 1747 when [Jean Rodolphe] Perronet, engineer to King Louis XV, set up his staff as a school. This group was later chartered in 1775 under the official name École des Ponts et Chaussées [School of Bridges and Roads]. Other early schools were the Bergakadamie at Freiburg in Saxony (1766), École Polytechnic in Paris (1794), Polytechnic Institute in Vienna (1815), Royal Polytechnic of Berlin (1821), and University College of London (1840).[24]

When the American colonies revolted in 1776, they did not have the engineering resources needed to build (or destroy) fortifications, roads, and bridges, and they had to rely on French, Prussian, and Polish assistance. At the urging of Thomas Jefferson and others, the new nation quickly established the United States Military Academy at West Point, NY, in 1802 to provide training in, among other things, practical science. Graduates did not acquit themselves as well as hoped in the War of 1812 with England, and so Sylvanus Thayer, assistant professor of mathematics at the Academy 1810–12, and Lt. Colonel William McRee were sent to Europe in 1815 to examine curricula at École Polytechnic, the most famous scientific military school in the world; unfortunately it was closed following the Battle of Waterloo, and they had to wait for it to reopen in 1816![25] On their return in 1817, Thayer was appointed Superintendent at West Point, collected the best teachers of physics, engineering, and mathematics available, and set up a four-year civil engineering program. Ross emphasizes the importance of this program:

> The influence of the Academy extended far beyond the institution's cadets. "Every engineering school in the United States founded during the nineteenth century copied West Point, and most found their first professors and president among academy graduates."[26] Many of the great

canals, railroads, and bridges constructed during the nineteenth century were built by West Point graduates. The faculty, recruited by Thayer, wrote textbooks that dominated the subjects of mathematics, chemistry, and engineering during the 1800s.[27]

For example, Mihalasky reports[28] that Captain Partridge, an early Academy superintendent, founded in 1819 the first civilian engineering school in the country, which later became known as Norwich (CT) University, followed at Rensselear (NY) Polytechnic Institute in 1823 with "a practical school of science," and 12 years later a school of civil engineering. Other early engineering schools were Union College (1845), and Harvard, Yale, and Michigan (1847). Mihalasky reports that only these six engineering schools existed in the United States when the Civil War opened, although Reynolds and Seely "have identified at least 50 institutions that at one time or another offered instruction in engineering before 1860"[29] (although not necessarily as full curricula).

As I have reported elsewhere,[30] "the event that had the greatest influence on engineering education was passage of the Morrill Land Grant Act in 1862. This act gave federal land (ultimately [totaling] 13,000,000 acres, an area 46% greater than Taiwan) to each state to support 'at least one college where the leading object shall be . . . scientific and classical studies . . . agriculture and mechanic arts.' This made education in the 'mechanic arts' (which became engineering) available and affordable throughout the country. By 1928 president-elect Herbert Hoover (himself a distinguished engineer and manager) could say:

> The leaders of our universities were the first of all the educators of the world to . . . provide fundamental training in the application of science to engineering under the broadening influence and cultivation of university life . . . [Another] dimension that grew in American engineering was the transformation from solely a technical profession to a profession of administrators—the business manager with technical training."[31]

An International Engineering Congress, with one division of the meetings on engineering education, was held as part of the 1893 Columbian Exposition in Chicago. Since there were then more than one hundred engineering schools in the country, the engineering education sessions were well attended, and interest there led to the 1893 formation of the Society for the Promotion of Engineering Education, which became the American Society for Engineering Education. In the century since, the meetings, journals, and studies of the ASEE have represented another major factor in the quality of American engineering education.

SCIENTIFIC MANAGEMENT

Charles Babbage

Babbage (1792–1871) lived in England during the Industrial Revolution. His work was far ahead of its time: Wren calls him both "patron saint of operations research and management science" and "grandfather of scientific management"; and so I have chosen to discuss him here, out of chronological sequence. Wren continues by describing the work for which Babbage is popularly known:

> He demonstrated the world's first practical mechanical calculator, his "difference engine," in 1822.

Ninety-one years later its basic principles were being employed in Burroughs' accounting machines. Babbage had governmental support in his work on the difference engine but his irascibility cost him the support of government bureaucrats for his "analytical engine," a versatile computer that would follow instructions automatically. In concept, Babbage's computer had all the elements of a more modern version. It had a store or memory device, a mill or arithmetic unit, a punch card input system, an external memory store, and conditional transfer [the modern "if statement"].[32]

Babbage's "engines" never became a commercial reality, largely because of the difficulty of producing parts to the necessary precision and reliability. This frustration led him to visit a wide variety of English factories, and his fascination with what he observed there led to publication of his very successful book *On the Economy of Machinery and Manufactures,* in 1832. In this he described at length his ideas on division of labor, his "method of observing manufactures," methods of optimizing factory size and location, and proposed a profit-sharing scheme. He showed a sophisticated understanding of effective time-study methods when he said:

> If the observer stands with his watch in his hand before a person heading a pin, the workman will almost certainly increase his speed, and the estimate will be too large. A much better average will result from inquiring what quantity is considered a fair day's work. When this cannot be ascertained, the number of operations performed in a given time may frequently be counted when the workman is quite unconscious that any person is observing him. Thus the sound made by . . . a loom may enable the observer to count the number of strokes per minute . . . though he is outside the building.[33]

Henry Towne and the ASME

The science of management made little progress over the centuries, largely because almost no one considered management as a legitimate subject for study and discussion. Although engineers frequently became enterprise managers, the first American engineering societies (the American Society of Civil Engineers, founded 1852, and the American Institute of Mining Engineers, founded 1871) were not interested in machine shop operation and management. Wren[34] believes that the first American forum for those interested in factory management was the *American Machinist,* an "illustrated journal of practical mechanics and engineering" founded in 1877, which soon began including a series of letters to the editor from one James Waring See on machine shop management. The *Machinist* was instrumental in the formation of the American Society of Mechanical Engineers (ASME), which elected its first officers April 7, 1880, at the Stevens Institute of Technology in Hoboken, NJ; ASME was formed to address itself to "those issues of factory operation and management that the other groups had neglected." Speaking in this vein before the May 1886 ASME meeting in Chicago was an engineer named Henry R. Towne, who was co-founder of Yale Lock Company and president of Yale & Towne Manufacturing Company. Towne began his famous paper "The Engineer as Economist":

> The monogram of our national initials, which is the symbol of our monetary unit, the dollar, is almost as frequently conjoined to the figures of an engineer's calculations as are the symbols indicating feet, minutes, pounds, or gallons. The final issue of his work, in probably a majority of cases, resolves itself into an issue of dollars and cents, of relative or absolute values.[35]

Towne then observed that although engineering had

become a well-defined science, with a large and growing literature of its own, . . . the matter of shop management is of equal importance with that of engineering, as affecting the successful conduct of most, if not all, of our great industrial establishments, and that the *management of works* has become a matter of such great and far reaching importance as perhaps to justify its classification also as one of the modern arts.[36]

Towne cited the need for a medium for the interchange of management experience "by the publication of papers and reports, and by meetings for the discussion of papers and interchange of opinions" and called for a new section of the ASME to carry this out. Although such a Management Section was not organized until 1920, consideration of matters of shop management became part of ASME meetings, and the ASME Management Division dates its official history from Towne's 1886 paper.

Frederick W. Taylor

Frederick Winslow Taylor (1856–1915), called the "father of scientific management," was born in 1856 to a well-to-do family in Germantown (Philadelphia), PA, and completed a four-year apprenticeship as a machinist. In 1878 he joined Midvale Steel Company as a laborer, and he was promoted to time clerk and then foreman of a machine shop.

As foreman, he was frustrated because his machinists were producing only about a third of what Taylor knew (as a machinist himself) and demonstrated they should be producing. Even on piecework pay, production did not improve, because the workers knew that as soon as they increased production, the rate paid per piece would be decreased, and they would be no better off. With permission of the president of Midvale Steel, Taylor began a series of experiments in which work was broken down into its "elements" and the elements timed to establish what represented a "fair day's work."

During this period he was a mechanical engineering student at Stevens Institute (where the ASME held its first meeting), graduating in 1883. Next year, at the age of 28, Taylor became chief engineer at Midvale Steel; a year later he joined the ASME and in May 1886 attended its meeting in Chicago; biographers report that Taylor was encouraged there to continue his studies of work methods and shop management by Henry Towne's paper (described above). Another paper at that meeting was by Captain Henry Metcalf, describing a "shop-order system of accounts" he established at the Frankford (PA) and Watervliet (NY) Arsenals in 1881 that helped management determine direct and indirect costs of work activity. In the extensive discussion that followed, Taylor reported on a similar system Midvale had been using for 10 years. For the first time in recorded history, engineers now had a medium for sharing their management problems and solutions. Taylor contributed further to this interchange with papers presented to the ASME in 1895 ("A Piece Rate System") and in 1903 ("Shop Management") and became president of ASME in 1906. Today, most larger engineering societies have active management divisions, and one society (the American Society for Engineering Management) is totally devoted to such concerns.

Taylor's "piece rate system" involved breaking a job into elementary motions, discarding unnecessary motions, examining the remaining motions (usually through stopwatch studies) to find the most efficient method and sequence of elements, and teaching the resulting method to workers. The worker would be paid according to the quantity of work produced. Taylor went further in his "differential piecework" method in establishing one piece rate if the worker produced the standard number of pieces, and a higher rate for all work if the worker produced more. For example, if three pieces were deemed a standard day's work and the two rates were 50 and 60 cents per piece, the worker would earn $1.50 for making three pieces a day, but $2.40 for four.

The best-known examples of Taylor's studies occurred after he became a consultant to Bethlehem Iron (later Steel) Company in 1898. One such was a study of a crew of pig-iron handlers: workers who picked up 92-pound "pigs" of iron, carried them up an inclined plank, and loaded them on railroad flat cars. By developing a method that involved frequent rest periods to combat the cumulative fatigue resulting from such drudgery, Taylor was able to increase the long tons loaded by a worker in a day from 12.5 to 47.5.

In another example, Taylor examined the work of shoveling at Bethlehem:

> Operation of the three blast furnaces and seven large open-hearth furnaces required a steady intake of raw materials—sand, limestone, coke, rice coal, iron ore, and so forth. Depending on the season, 400 to 600 men were employed as shovelers in the 2-mile-long and a half-mile wide Bethlehem yard. Taylor noted that the shovelers were organized into work gangs of 50 to 60 men under the direction of a single foreman. Each owned his own shovel and used it to shovel whatever he was assigned. . . . Taylor's analysis revealed that a shovel-load (depending on the shovel and the substance shoveled) varied in weight from 3.5 to 38.0 pounds, and that a shovel-load of 21.5 pounds yielded the maximum day's work. As a result, instead of permitting workers to use the same shovel regardless of the material they were handling, Taylor designed new shovels so that for each substance being shoveled the load would equal 21.5 pounds.[37]

In the latter example, the average amount shoveled per day increased from 16 to 59 tons. In both these cases the savings produced were shared. Workers' earnings increased from $1.15 to about $1.85 a day, while management's cost per ton handled was reduced by 55% or more.

Taylor summarized his methods in his 1911 book *Principles of Scientific Management* as a combination of four principles:

> *First.* Develop a science for each element of a man's work, which replaces the old rule-of-thumb method.
>
> *Second.* Scientifically select, then train, teach, and develop the workmen, whereas in the past he chose his own work and trained himself as best he could.
>
> *Third.* Heartily cooperate with the men so as to insure all of the work being done in accordance with the principles of the science which has been developed.
>
> *Fourth.* There is an almost equal division of the work and the responsibility between the management and the workmen. The management take over all work for which they are better fitted than the workmen [defining *how* work is to be done], while in the past almost all of the work and the greater part of the responsibility were thrown upon the men.[38]

The Gilbreths

Frank B. Gilbreth (1868–1924) passed the entrance exams for the Massachusetts Institute of Technology, but he chose instead to apprentice himself as a bricklayer. Trying to learn the trade, he found that bricklayers used three sets of motions: one when working deliberately but slowly, another when working rapidly, and a third when trying to teach their helpers! Gilbreth resolved to find the "one best way." As he described it later in testimony before the U.S. Interstate Commerce Commission:

> Bricks have been laid the same way for 4,000 years. The first thing a man does is to bend down and pick up a brick. Taylor pointed out that the average brick weighs ten pounds, the average weight of a man above his waist is 100 pounds. Instead of bending down and raising this double load, the bricklayer could have an adjustable shelf built so that the bricks would be ready to his hand. A boy could keep these shelves at the right height. When the man gets the brick in his hand, he tests it with his trowel. If anything, this is more stupid than stooping to pick up his material. If the brick is bad he discards it, but in the process it has been carried up perhaps six stories, and must be carted down again. Moreover, it consumes the time of a $5-a-day man when a $6-a-week boy could do the testing on the ground. The next thing the bricklayer does is to turn it over to get its face. More waste: more work for the $6 boy. Next what does the bricklayer do? He puts the brick down on the mortar and begins to tap it with his trowel. What does his tapping do? It gives the brick a little additional weight so it will sink into the mortar. If anything this is more stupid then any of the others. For we know the weight of the brick and it would be a simple matter in industrial physics to have the mortar mixed so that just that weight will press it down into the right layer. And the result? Instead of having eighteen motions in the laying of a brick, we have only six. And the men put to work to try it lay 2,700 with no more effort than they laid a thousand before.[39]

By 1895 Gilbreth had his own construction firm based on "speed work." He analyzed each job to eliminate unnecessary motions, devising a system of classifying hand motions into 17 basic divisions (which he called *therbligs* from his last name) such as "search," "select," "transport loaded," "position," and "hold." He soon became one of the best-known building contractors in the world, but by 1912 had given up the construction business and was devoting full time to management consulting.

Lillian Moller Gilbreth (1878–1972) earned bachelor's and master's degrees in English at the University of California (and qualified for Phi Beta Kappa, although as a woman she was not included on the official list of recipients). She interrupted Ph.D. studies for a trip to Europe by way of the port of Boston, where she met Frank Gilbreth on the outgoing leg and married him on her return. As Lyndall Urwick rhapsodizes in a discussion of scientific management pioneers:

> —there was the amazing fact that one of them, Frank Bunker Gilbreth, happened to fall in love with a girl who was a psychologist by education, a teacher by profession, and a mother by vocation [see the book and subsequent movie *Cheaper by the Dozen*]. I know of no occurrence more worthy of the epithet "providential" than that fact. Here were three engineers—Taylor, Gantt, and Gilbreth—struggling to realize the wider implications of their technique, in travail with a "mental revolution", their great danger that they might not appreciate the difference between applying scientific thinking to material things and to human beings, and one of them married Lillian Moller, a women who by training, by instinct, and by experience was deeply aware of human beings, the perfect mental complement in the work to which they had set their hands.[40]

Lillian quickly became interested in Frank's work and assisted him in preparation of six books published between 1908 and 1917 (*Field System, Concrete System, Bricklaying System, Motion Study, Fatigue Study,* and *Applied Motion Study*). Meanwhile, she continued work on her Ph.D. thesis, "The Psychology of Management," one of the earliest contributions to understanding the human factor in industry, submitting it in 1912. The work was serialized in *Industrial Engineering Magazine* and finally published as a book (the latter with the proviso that the author be listed as "L. M. Gilbreth" without identifying her as a woman, so that it might have some credibility!).

Frank prepared an invited paper for the 1925 International Management Conference in Prague, but he dropped dead of a heart attack June 14, 1924. Lillian presented the paper in his place, then continued Frank's work and established a strong reputation of her own as one of the creators of industrial psychology. She was the first woman admitted to the Society of Industrial Engineers and the ASME, the first woman professor of management at an engineering school (Purdue University and later the Newark College of Engineering), and the only woman to date to be awarded the Gilbreth Medal, the Gantt Gold Medal, or the CIOS Gold Medal. She has understandably been called the "first lady of management."[41] Her life was so long (she outlived Frank by 48 years) and distinctive that many contemporary women engineers speak of her as an early inspiration that led them into engineering work.

Growth and Implications of Scientific Management

Taylor's work attracted many disciples who propagated the scientific management method. *Carl Barth,* a mathematics teacher, was recruited by Taylor to help *Henry Laurence Gantt* solve the speed and feed problems in metal-cutting studies conducted at Bethlehem. Barth later helped Taylor apply scientific management to the problems of Link Belt, Fairbanks Scale, and Yale & Towne companies and then helped George D. Babcock install scientific management at the Franklin Motor Car Company (1908–12).

Gantt (1861–1919) earned degrees from Johns Hopkins University and from Stevens Institute of Technology (in mechanical engineering in 1884, a year after Taylor). He joined Taylor at Midvale Steel in 1887, followed him to Simond's Rolling Machine Company and then to Bethlehem Steel, and became an independent consulting industrial engineer in 1901. Gantt modified Taylor's "differential piece rate" by providing a standard day rate regardless of performance, which provided security to workers during training and when delayed because materials were not available; workers who accomplished the specified daily production received an additional bonus, as did their foreman. Gantt was also noted for his work in developing charts that graphed some function of performance against time; their application to project management is discussed in Part 4.

Another protégé of Taylor was *Morris L. Cooke* (1872–1960), a mechanical engineer (Lehigh University, 1895), who began "applying a questioning method to the wastes of industry long before he met or heard of Taylor,"[42] then started reading Taylor's writings, and met him. Taylor funded Cooke to study the administrative effectiveness of ASME, sent him to perform an "economic study" of administration in educational organizations for the Carnegie Foundation for the Advancement of Teaching, and then (1911–15) sent him to help the newly elected reform mayor of Philadelphia improve the efficiency and effectiveness of municipal government. Cooke later advised the president of the American Federation of

Labor (Samuel Gompers) and co-authored a book with the president of the Congress of Industrial Organizations (Phillip Murray), emphasizing that labor was as important for production as management, and then he headed the Rural Electrification Administration, which brought inexpensive electric power to rural America.

Taylor's system received extensive publicity in the 1911 Eastern rate case. The Eastern railroads petitioned the Interstate Commerce Commission for an increase in rates, but Boston lawyer (and later Supreme Court Justice) Louis D. Brandeis took up the cause of shippers with the theme that no increase would be necessary if railroads would only apply "scientific management" (the name adopted instead of the "Taylor system" in a meeting of Brandeis, Gilbreth, Gantt, and others in preparation for this case). The parade of witnesses supporting this view included Taylor, Gilbreth (as quoted above), Henry Towne, and others. *Harrington Emerson* (1853–1931), who had been very successful as a troubleshooter on the Burlington Railroad and then a consultant to the Santa Fe Railroad, testified that preventable labor and material waste was costing the railroad industry "a million dollars a day."

Scientific management spread rapidly because media and institutions for the sharing of knowledge and experience were becoming available in an unprecedented way. Many of the practitioners were active in ASME, presented and critiqued papers at their meetings, and two became ASME presidents. Industrial and popular journals were increasing in number, and they reported on progress in scientific management and even serialized books by Taylor and Lillian Gilbreth. Most of the major participants authored several books each, many of which were widely read. Universities increasingly decided management was, after all, worthy of study: Taylor was persuaded to lecture at what would become the Harvard Business School, Lillian Gilbreth and Carl Barth each lectured at two universities, and Henry Gantt lectured at four. Bachelor's degree programs that combined engineering and business were founded at Stevens in 1902, Yale in 1911, and MIT in 1913. The discipline of industrial engineering (and the Institute of Industrial Engineers) originated from the work of scientific management, and the newer discipline of engineering management owes a great debt to it as well.

The effectiveness of American support of the Allies in World War I spread interest in scientific management throughout Europe, even to newly communist Russia. Lenin is quoted from a June 1919 address on "Scientific Management and the Dictatorship of the Proletariat" as saying:

> The Russian is a poor worker in comparison with the workers of the advanced nations, and this could not be otherwise under the regime of the Czar and other remnants of feudalism. To learn how to work—this problem the Soviet authority should present to the people in all its comprehensiveness. . . . We must introduce in Russia the study and the teaching of the Taylor system and its systematic trial and adaptation.[43]

Scientific management did, however, have some negative impact, which still affects us today. Taylor divided work into planning and training (a management responsibility) and rote execution (by the uneducated laborer of the day). Only in the last decade have executives in mass production industries such as General Motors realized how much they were losing by "only hiring workers from the neck down" rather than encouraging them to participate in improving work methods. Konosuke Matsushita, founder of Matsushita Electric Industrial Company (Osaka, Japan), believes his country will win the manufacturing war with the United States as a result. He explains:

You cannot do anything about it because your failure is an internal disease. Your companies are based on Taylor's principles. Worse, your heads are Taylorized too. You firmly believe that sound management means executives on one side and workers on the other, on one side men who think and on the other men who can only work. For you, management is the art of smoothly transferring the executives' ideas to the workers hands.

We have passed the Taylor stage. We are aware that business has become terribly complex. Survival is very uncertain in an environment increasingly filled with risk, the unexpected, and competition. Therefore, a company must have the constant commitment of the minds of all of its employees to survive. For us, management is the entire work force's intellectual commitment at the service of the company . . . without self-imposed functional or class barriers.

We have measured—better than you—the new technological and economic challenges. We know that the intelligence of a few technocrats—even very bright ones—has become totally inadequate to face these challenges. Only the intellects of all employees can permit a company to live with the ups and downs and the requirements of its new environment. Yes, we will win and you will lose. For you are not able to rid your minds of the obsolete Taylorisms that we never had.[44]

Matsushita lays down a challenge to American engineers working in production and operations management. In Chapter 7 we look at some of the theories of human motivation, and in Chapter 12, their application to production operations using techniques of Total Quality Management and empowered teams.

ADMINISTRATIVE MANAGEMENT

As we have seen, initial American management study emphasized management at the production-shop level. In the meantime two Europeans, Henri Fayol and Max Weber, were making significant contributions to general management theory.

Henri Fayol

Fayol (1841–1925) was a 1860 graduate of the National School of Mines at St. Etienne, France. His distinguished career is described in Urwick's foreword to the 1949 English translation of his most noted work, *Administration Industrielle et Générale* (General and Industrial Management).[45] He believed that the activities of industrial undertakings could be divided into six groups: technical (production), commercial (marketing), financial, security, accounting, and administrative activities. The first five he considered well known, but the last, administrative (French has no exact equivalent of the word "management"), he considered most important above the first two levels of management, yet least understood. Fayol divided administration into planning/forecasting (*prevoyance*), organization, command, coordination, and control. He decried the absence of management teaching in technical schools, but stated that without a body of theory no teaching is possible, and then proceeded to develop a set of 14 "general principles of administration," most of which have meaning to this day.

Today's critics of engineering education would agree with Fayol that

Our young engineers are, for the most part, incapable of turning the technical knowledge received to good account because of their inability to set forth their ideas in clear, well-written reports, so compiled as to permit a clear grasp of the results of their research or the conclusions to which their observations have led them.[46]

Engineering educators today would be less comfortable with his observation that "Long personal experience has taught me that the use of higher mathematics counts for nothing in managing businesses and that engineers, mining or metallurgical, scarcely ever refer to them."[47] However, every engineering student should consider his "Advice to Future Engineers":

> You are not ready to take over the management of a business, even a small one. College has given you no conceptions of management, nor of commerce, nor of accounting, which are requisite for a manager. Even if it had given you them, you would still be lacking in what is known as practical experience, and which is acquired only by contact with men and with things. . . . Your future will rest much on your technical ability, but much more on your managerial ability. Even for a beginner, knowledge of how to plan, organize, and control is the indispensable complement of technical knowledge. You will be judged not on what you know but on what you do and the engineer accomplishes but little without other people's assistance, even when he starts out. To know how to handle men is a pressing necessity.[48]

Max Weber and Bureaucracy

A contemporary of Fayol, the German sociologist Max Weber (1864–1920) influenced classical organization theory more than any other person. Weber developed a model for a rational and efficient large organization, which he termed a bureaucracy. Weber described the following as characteristics of "legal authority with a bureaucratic administrative staff":

- The basic organizational unit is the *office* or position, which is designated a specific set of functions (based on division of labor), with clearly defined authority and responsibility.
- Members of the organization owe loyalty to the office, not (as with traditional authority or charismatic authority) to the individual.
- Candidates for offices are selected and appointed (not elected) based on their technical capability.
- Offices are organized in a clearly defined hierarchy: each lower office is under the control and supervision of a higher office.
- Officials (office holders) are "subject to strict and systematic discipline and control in the conduct of the office," and subordinates have a right of appeal.
- Administrative acts, decisions, and rules must be reduced to writing.
- The office is the primary occupation of the incumbent, who is reimbursed by a fixed salary.
- Promotion is based on the judgment of superiors.
- Officials are not the owners of the organization.[49]

The term "bureaucracy" need not imply an organization that is mired in red tape, delay, and inefficiency, with no concern for the human dimension. Most of Weber's elements are necessary in any large organization to assure consistent and reasonably efficient operation. The U.S. Postal Service or Internal Revenue Service must have the same rules of operation at every local office; an army must have common procedures so that replacement officers and men can function quickly on assuming new positions; General Motors, or a large university or hospital, or the Boy Scouts of America must have fairly uniform structures and rules among their divisions to function smoothly. The challenge of a large organization is to incorporate into this necessary structure some flexibility to handle exceptions and an ability to recognize and reward individual contributions.

Russell Robb

Robb (1864–1927) was an American electrical engineer and manager whose original contributions on organization theory have not received the attention they deserve. After graduating from MIT, Robb spent most of his career as an executive in the Stone and Webster Engineering Corporation. He expressed his views on organization in three lectures presented to the Harvard University Graduate School of Business Administration in 1909 and later published.[50] Young summarizes their import:

> These three lectures . . . contain more practical observations on organizations and concepts of organization theory than Weber. He was a practising engineer manager, whereas Weber was a sociologist. . . . His penetrating observation of organizations as "only a means to ends—it provides a method" and analysis of principles and concepts make him more a "pioneer of organization theory" than Weber.[51]

Lyndall Urwick

Urwick was an Englishman who majored in history at Oxford. His contribution lay not in creating concepts of management, but in being the first to try to develop a unified body of knowledge. Using Fayol's management functions as a framework, he analyzed the writings of Fayol, Taylor, Mary Parker Follett, James Mooney, and others, and attempted to correlate them with some of his own views into a consistent system of management thought. His 1943 book, *The Elements of Administration,* can therefore be viewed as the first general textbook on, as opposed to personal observations about, management. Toward the end of his long career he summarized his observations on the contribution of engineers to management:

> The study of management, as we all know, started with engineers. It was the sciences underlying engineering practice—mathematics, physics, mechanics, and so on—which were first applied by Frederick Winslow Taylor to analyzing and measuring the tasks assigned to individuals. That is where the science of management started.[52]

BEHAVIORAL THEORIES: THE HAWTHORNE STUDIES

What is arguably the single biggest departure in management thinking grew out of a series of studies conducted at the Hawthorne Works (near Cicero, IL) of the Western Electric Company (now AT&T Technologies) in the 1920s and early 1930s. The first phase of the studies, known as the Illumination Experiments, were conducted between 1924 and 1927 under the direction of Vannevar Bush, a famous electrical engineer from MIT who later developed systems that made the modern computer possible. The original intent was to find the level of illumination that made the work of female coil winders, relay assemblers, and small parts inspectors most efficient. Workers were divided into test and control groups, and lighting for the test group was increased from 24 to 46 to 70 foot-candles. Production of the test group increased as expected, *but production of the control group increased roughly the same amount.* Again, when lighting for the test group was decreased to 10 and then 3 foot-candles, their output *increased,* as did that of the control group. Production did not drop appreciably until illumination was lowered to that of moonlight (0.06 foot-candle).

To try to understand these unexpected results, Australian-born Harvard professor Elton Mayo and his colleague Fritz Roethlisberger conducted a second phase (1927–32), known as the Relay Assembly Test Room Experiments. A large number of women were employed in assembling about 40 parts into the mechanical relays that were needed for telephone switching in the days before solid state electronics. Six women whose prior production rates were known were moved from the large assembly room to a special test room to test the effects of changes in length and frequency of rest periods and hours worked. The women were given regular physical examinations (with free ice cream), their sleep each night and food eaten were carefully recorded, and room temperature and humidity were controlled. The room had an observer who recorded events as they happened and maintained a friendly atmosphere. The women had no supervisor, but they increasingly assumed responsibility for their own work and were allowed to share in decisions about changes in their work [a precursor of today's emphasis on "empowered teams"]. Birthdays were regularly celebrated at work, and the women became fast friends after-hours as well. Incentive pay had been used in the main workroom based on overall production of a large number of workers, but in the test room incentive pay was based just on production of the group of six.

After production rates had been stabilized in the new room, rest periods were added and maintained for periods of four or five weeks each at levels of (1) two 5-minute periods, (2) two 10-minute periods, (3) six 5-minute periods, and (4) two 10- or 15-minute periods with light snacks. Shorter workdays and elimination of Saturday work were also tried. Throughout this period daily production continued to increase, as it also did in a subsequent 12-week period when *rest periods, refreshments, and shortened workdays were eliminated,* and again when they were reinstated. Absenteeism among the six was only a third of that in the main room.

A third phase of study (the Bank Wiring Observation Room Experiment of 1931–32) involved a group of 11 wiremen, soldermen, and inspectors who assembled terminal banks used in telephone exchanges. It became clear that the men formed a complex social group, had established their own standard of a fair day's work, and despite the piecework pay that existed, ridiculed and abused any worker who tried to work faster (or slower) than the group norm.

Mayo and others have attributed the surprising results in the first two phases to the pride of the women in being part of something important, the esprit de corps developed in the work group, and the satisfaction of having some control over their own destiny, and the behavior of the men to the need for affection from the group (and the fear that management would lower pay rates if productivity improved). Later analysts such as Rice have criticized the studies as lacking the rigorous controls now demanded in scientific experiments,[53] but the criticism is moot. These studies focused the attention of an army of behavioral scientists— psychologists, sociologists, even anthropologists (who turned their attention from the culture of remote tribes to the culture of General Motors and IBM)—on the behavior of workers individually and in groups; their work over the ensuing decades has added immeasurably to our knowledge of the art of management. The contributions of some of these social scientists to our knowledge of leadership and human motivation are discussed in Chapter 7.

OTHER CONTRIBUTIONS

Successful Executives

Management theory owes a great deal to practical executives who took the time to set down the wisdom they had accumulated in a successful management career. Henri Fayol, discussed previously, was such a man, as were *Chester Barnard,* who summarized his findings about people in organizations in *The Functions of the Executive,*[54] and *Alfred P. Sloan,* who documented his development of the decentralized organization with central control in *My Years with General Motors.*[55] Just as often, these contributions are related secondhand by management writers. For example, Peters and Waterman have highlighted in their *In Search of Excellence*[56] the following: the wisdom of Walt Disney in treating theme park customers as "guests"; the emphasis of *Thomas Watson Jr.* of IBM on service and customer satisfaction; the revolution in the U.S. Navy by *Admiral Zumwalt* "based on the simple belief that people will respond well to being treated as grownups"; and the success of MBWA ("management by walking around") by Bill Hewlett and Dave Packard at HP.

Quantitative Methods and Systems Theory

In the last 50 years mathematicians, economists, and a variety of quantitative specialists have made their own contributions to management theory. The decision theory school of thought, for example, views decision making as the central focus of management, and concentrates on study of the decision process and the organizational group making the decision. The operations research school (also known as "management science") attempts to express management problems and occasions for decision in terms of mathematical symbols and relationships. In Chapter 4 we provide some insight into these two approaches. Finally, the systems theorist characterizes organizations as interacting subsystems and as "open systems" that influence and are influenced by the external environment in which they are located.

Project Management

Many of the most difficult management challenges of recent decades have been to design, develop, and produce very complex systems of a type that has never been created before. Examples include establishment of vast petroleum production systems in the waters of the North Sea or the deserts of Saudi Arabia, the collaboration of 400,000 people in the Apollo program to place two of them on the moon, and development of very complex jet aircraft such as the F-15 fighter or the Boeing 767 airliner. To create these systems with performance capabilities not previously available and within schedule and budget limitations has required development of new forms of planning, organizing, and controlling events; since engineers are especially likely to spend significant parts of their careers in this type of environment, two chapters (14 and 15) are devoted to discussing project management.

DISCUSSION QUESTIONS

2-1. What was the lasting significance of the Sumerian priests' need to maintain a permanent record of temple property (which they did on clay tablets)?

2-2. Stones for the pyramids were quarried far to the south (upstream on the Nile River) and were brought downstream on rafts only during the spring flood of the Nile. Discuss some of the planning and organizational implications of this immense logistic effort.

2-3. Why was the Venetian development of double-entry bookkeeping so important to the development of management?

2-4. The development of cotton and woolen mills in the mill cities of England, and later New England, caused tremendous sociological change as potential workers (especially women) swarmed from rural areas to the growing industrial cities. Cite examples of similar occurrences in more recent times in developing countries.

2-5. Summarize the contribution of the American Society of Mechanical Engineers to the dissemination of better methods of (production) shop management. What does this say about *your* need as an engineer to be active in at least one professional society?

2-6. Matsushita emphasized the residual disadvantages to the United States of the teachings of Frederick Taylor. Discuss the *positive* contributions Taylor and his contemporaries in the scientific management movement made.

2-7. What was the positive value of Max Weber's model of "bureaucracy"?

2-8. The essence of the Relay Assembly Test Room Experiments at the Hawthorne Works was that expected correlations between productivity and physical factors such as rest periods were not demonstrated. What other factors could explain the regular productivity increases observed in these experiments?

2-9. Read at least part of *In Search of Excellence* and elaborate on one significant finding of Peters and Waterman.

2-10. As made clear in this chapter, engineers and engineer managers have made strong contributions to management theory and practice. List the engineers and engineer managers identified in this chapter together with their contributions, and add any others you may know of.

NOTES

1. Claude S. George Jr., *The History of Management Thought,* 2nd ed. (Englewood Cliffs, NJ: Prentice-Hall, Inc., 1972).

2. Daniel A. Wren, *The Evolution of Management Thought,* 3rd ed. (New York: John Wiley & Sons, Inc., copyright © 1987; reprinted by their permission).

3. Christopher J. Davey, "Engineering and Civilization," *Professional Development Handbook* (Parkville, Victoria, Australia: The Institution of Engineers, Australia, September 1992), p. 21.

4. Wren, *Management Thought,* p. 13.

5. Arthur G. Bedeian, *Management* (Hinsdale, IL: The Dryden Press, 1986), p. 29.

6. Wren, *Management Thought,* p. 13.

7. Wren, *Management Thought,* p. 13.

8. George, *Management Thought,* pp. 11–13.

9. George, *Management Thought,* p. 19.

10. Frederick C. Lane, *Venetian Ships and Shipbuilders of the Renaissance* (Baltimore, MD: Johns Hopkins Press, 1934), abstracted in George, *Management Thought,* pp. 35–41.

11. Lane, *Venetian Ships,* p. 164.

12. Lane, *Venetian Ships,* p. 172.

13. George, *Management Thought,* p. 32.

14. Harold T. Amrine, John A. Ritchey, and Colin L. Moodie, *Manufacturing Organization and Management,* 5th ed. (Englewood Cliffs, NJ: Prentice-Hall, Inc., 1982), pp. 15–16.

15. W. F. Durfee, "The History and Modern Development of the Art of Interchangeable Construction in Mechanisms," *Journal of the Franklin Institute,* 137:2, February 1894, quoted in George, *Management Thought,* pp. 63–64.

16. J. Powell, *A View of Real Grievances* (publisher unknown, 1772), quoted in Asa Briggs, ed., *How They Lived,* Vol. 3 (Oxford: Basil Blackwell Publisher Ltd., 1969), p. 184.

17. Wren, *Management Thought,* p. 41.

18. Harwood F. Merrill, ed., *Classics in Management* (New York: American Management Associations, Inc., 1960), p. 13, quoted in George, *Management Thought,* p. 63.

19. Denis L. Johnston, "Engineering Contributions to the Evolution of Management Practice," *IEEE Trans. on Engineering Management,* 36:2, May 1989, p. 106.

20. R. L. Breedon, *Those Inventive Americans* (Washington, DC: National Geographic Society, 1971), p. 48.

21. Wren, *Management Thought,* pp. 68–73.

22. Johnston, "Engineering Contributions," pp. 106–107.

23. Wren, *Management Thought,* pp. 73–85.

24. John Mihalasky, *The Role of Professional and Engineering Education Societies in the Development of the Undergraduate Industrial Engineering Curriculum,* Ed.D. dissertation, Columbia University, 1973 (draft copy), Chapter 2.

25. W. L. Ross, "Early Influences of the U.S. Military Academy on Engineering Technology and Engineering Graphics Education in the United States," *Proceedings of the 1991 Annual Conference, American Society for Engineering Education,* p. 1604.

26. Stephen E. Ambrose, *Duty, Honor, Country: A History of West Point* (Baltimore, MD: Johns Hopkins Press, 1966)

27. Ross, "Early Influences," p. 1605.

28. Mihalasky, *The Role of . . .*

29. Terry Reynolds and Bruce Seely, "Reinventing the Wheel?" *ASEE Prism,* October 1992, p. 42.

30. Daniel L. Babcock and Brian E. Lloyd, "Educating Engineers to Manage Technology," *Proceedings of the 1992 International Engineering Management Conference, IEEE Engineering Management Society and American Society for Engineering Management,* p. 248.

31. Herbert C. Hoover, "The Engineer's Contribution to Modern Life," reprinted in *The Profession of the Engineer* (New York: Wiley, 1929), p. 119 ff.

32. Wren, *Management Thought,* p. 59.

33. Charles Babbage, *On the Economy of Machinery and Manufactures* (London: Charles Knight Ltd., 1832; reprinted New York: Augustus M. Kelley, Publishers, 1963), p. 132, quoted in George, *Management Thought,* pp. 76–77.

34. Wren, *Management Thought,* pp. 86–87.

35. Henry R. Towne, "The Engineer as Economist," *American Society of Mechanical Engineers Trans.,* No. 207, 1886, reprinted in Charles M. Merrick, ed., *ASME Management Division History 1886–1980* (New York: American Society of Mechanical Engineers, 1984), p. 71.

36. Towne, "Engineer as Economist," p. 71.

37. Arthur G. Bedeian, *Management,* 2nd ed. (New York: Holt, Rinehart and Winston, 1989) p. 40.

38. Frederick Winslow Taylor, *Principles of Scientific Management* (New York: Harper & Brothers, 1911), pp. 36-37.

39. Frank B. Gilbreth, quoted in Arthur G. Bedeian, "Finding The One Best Way," *Conference Board*

Record, 13, June 1976, pp. 37–38. In later work, Gilbreth reduced these 6 motions further to 4.5 per face brick and only 2 motions per interior brick, where excess mortar need not be scraped off.

40. Lyndall Urwick, "Management's Debt to Engineers," *Advanced Management,* December 1952, p. 7.

41. Wren, *Management Thought,* p. 141.

42. Wren, *Management Thought,* p. 152.

43. As quoted in a letter from Lyndall F. Urwick to Edmund Young, February 17, 1971, supplied by Mr. Young.

44. "A Secret Is Shared," *Manufacturing Engineering,* February 1988, p. 15.

45. Henri Fayol, *Administration Industrielle et Générale,* Constance Storrs, trans. (London: Sir Isaac Pitman & Sons Ltd., 1949).

46. Fayol, *Administration,* p. 79.

47. Fayol, *Administration,* p. 84.

48. Fayol, *Administration,* p. 90.

49. Max Weber, *The Theory of Social and Economic Organizations,* A. M. Henderson and Talcott Parsons, trans. and eds. (New York: The Free Press, 1947), pp. 328–334.

50. Catheryn Seckler-Hudson, *Process of Organization and Management* (Washington, DC: Public Affairs Press, 1948).

51. Edmund Young, supplemental notes for "Management for Engineers" course taught at Fort Leonard Wood, MO, August 1988.

52. Lyndall Urwick, "The Professors and the Professionals," after dinner talk, Oxford Centre for Management Studies, Oxford, England, October 12, 1972.

53. Berkeley Rice, "The Hawthorne Effect: Persistence of a Flawed Theory," *Psychology Today,* February 1982, pp. 70–74.

54. Chester I. Barnard, *The Functions of the Executive* (Cambridge, MA: Harvard University Press, 1938).

55. Alfred P. Sloan Jr., *My Years with General Motors* (New York: Doubleday & Company, Inc., 1964).

56. Thomas J. Peters and Robert H. Waterman Jr., *In Search of Excellence: Lessons from America's Best Run Companies* (New York: Harper & Row, Publishers, Inc., 1982).

PART 2

Functions of Technology Management

CHAPTER **3**

Planning and Forecasting

PREVIEW

We begin this chapter by developing a model of the planning/decision-making process. Then we consider the importance of defining an organization's mission, its strategy, and its goals and objectives as a prelude to planning. We then introduce management by objectives. Following this, we discuss planning concepts such as premises or assumptions, the planning horizon, and the distinction between policies and procedures. We compare forecasting using sales estimates, moving average methods, and regression models, and summarize technological forecasting. Finally, we introduce a few concepts essential to managing technology, such as invention, innovation, and entrepreneurship.

NATURE OF PLANNING

Importance of Planning

The basic functions of management are commonly identified as *planning, organizing, leading* (or *directing* or *motivating*), and *controlling*. Of these, planning is said to have *primacy*—to come first. Organizing, leading, and controlling have little purpose unless they are focused on achieving desired objectives. The importance of planning in warfare was emphasized by Hsun Tzu (298-238 B.C.) in *The Art of War* (the world's oldest military treatise):

The general who wins a battle makes many calculations in his temple ere the battle is fought. The general who loses a battle makes few calculations before hand. It is by attention to this point that I can see who is likely to win or lose.[1]

Planning provides a method of identifying objectives and designing a sequence of programs and activities to achieve these objectives. Amos and Sarchet[2] define planning simply as "deciding in advance what to do, how to do it, when to do it, and who is to do it"; from that definition, planning must obviously precede doing!

The Planning/Decision-Making Process

There is a basic, logical method for solving problems that may be called, depending on the application, the planning process, the decision-making process, or the scientific method. Although there seem to be as many diagrams of this process as there are authors, a typical representation of this process follows the general logic of Figure 3-1.

A problem cannot be solved until it is recognized. When the "roof falls in" or the workers go on strike, the existence of a problem will be obvious. On the other hand, chronic (perennial) problems or opportunities often go unrecognized. American automobile manufacturers, for example, did not realize that they had a quality problem until the Japanese recognized better quality as an opportunity; the United States lost billions of dollars in markets as a result.

Once a problem is recognized, the nature of the desired solution must be defined carefully in terms consistent with the overall objectives and strategy of the organization. Assumptions about the environment (*premises*) need to be stated, and the solution found will

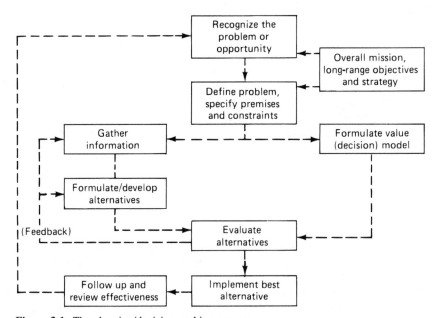

Figure 3-1. The planning/decision-making process.

be valid only if these assumptions prove true. Finally, the constraints or limitations bounding the solution must be defined.

Information bearing on alternative solutions is then gathered, and alternative solutions are formulated. This is the most creative step in problem solving, since alternatives that are not considered are lost. Simply stating an alternative is not enough—each concept must be fleshed out in enough detail that its benefits and disadvantages can be effectively evaluated. At the same time, some "value model" or measure of merit, whether quantitative or qualitative, needs to be defined against which alternatives can be evaluated. The solution that best satisfies this value model is then recommended.

Seldom is the decision process as linear as the preceding paragraph suggests. Identification of alternatives often leads to a search for more information. Evaluation often leads to modification or combination of alternatives to find a new one that combines the advantages of several. After the solution is put into effect (implemented), it is important to check back later to determine if the problem as stated was really resolved by the solution. Often there will be unexpected secondary effects that, once realized, need to be defined as a new problem, and the process begins anew. Problem solving/decision making is therefore more often an *iterative* process, involving feedback at several steps before the best resolution is found.

OBJECTIVES: THE FOUNDATION FOR PLANNING

Vision, Purpose, Mission

The book of Proverbs[3] tells us "Where there is no vision, the people perish," and we can find many examples in history to show this. In the darkest period of World War II, when Great Britain was reduced to little but determination, Winston Churchill could say:

> Let us therefore brace ourselves to our duties, and so bear ourselves that if the British Empire and its Commonwealth last for a thousand years, men still will say: *"This* was their finest hour."

In contrast, Collins and Lazier discuss as a case example the United States war in Vietnam:

> In terms of tactics and logistics, according to Harry G. Summers in the book *On Strategy*, the United States Army was successful in Vietnam. Over a million soldiers a year were transported to and from Vietnam, and they were sustained in the field better than any army in history. In tactical engagements, the army had an extraordinary success rate, with enemy forces thrown back with terrible losses in engagement after engagement. Even so, it was North Vietnam that emerged victorious. How could the United States have succeeded so well, yet failed so miserably?
>
> A wide range of authors on the subject all came to a startlingly simple conclusion: the United States didn't know precisely what it was trying to achieve, and it was therefore impossible to have an effective strategy. A 1974 survey of Army generals who had commanded in Vietnam found that almost 70 percent of them were uncertain of United States objectives.[4]

A clear vision of the basic purpose or mission for which it exists is essential to the long-term success of any enterprise. Drucker[5] cites the old German tradition of the *Unternehmer* (entrepreneur or, literally, "undertaker"), who alone or with a few top managers, were all that needed to understand the underlying purpose of a firm in earlier, simpler times. In today's complex and dynamic economy, however, the basic mission of the organization must be communicated clearly and repeatedly to the many managers and professionals whose actions determine whether these purposes are indeed achieved.

Drucker provides an example of a vision of central purpose in the statement of Theodore N. Vail of American Telephone and Telegraph Company about 1910 that "our business is service." He believes that without this vision of the social responsibility of a natural monopoly, the "Bell System" would have inevitably been nationalized in the 1930s, as was the case in other countries. Only in the late 1960s, when technology provided alternatives to Bell's monopoly, was it time to pursue a revised purpose.

Indeed, it is essential to review the basic purpose of an enterprise periodically to determine if the underlying assumptions are still valid. Drucker[6] provides a classic example of this in the story of Sears, Roebuck & Co. Julius Rosenwald provided Sears its first clear sense of mission, service to the isolated American farmer, when he purchased the firm in 1895. Rosenwald "analyzed the market, began the systematic development of merchandise sources, and invented the regular, factual mail-order catalog and the policy of 'satisfaction guaranteed or your money back.'" The Chicago mail-order plant, designed by Otto Doering in 1903 and opened in 1905, helped carry out this vision. Drucker considers it "the first modern mass-production plant . . . with an assembly line, conveyor belt, standardized, interchangeable parts—and, above all, with planned plant-wide scheduling."

By the mid-1920s farmers were no longer isolated, since automobiles and better roads allowed them to shop in the cities. Moreover, lower economic groups in the cities were now developing both the income and the desire for goods formerly sold only to the richer classes. General Robert E. Wood joined Sears at this time and controlled the company well past his retirement in 1954. His vision of "quality at a good price—the mass merchandiser for middle America" led to innovations of products redesigned for the mass market, development of suppliers to produce them, the creation of a new concept in retail stores, and the training of an army of store managers to run them.

By the early 1970s, however, Sears again needed a new vision to meet the new realities of the marketplace. New entrepreneurs recognized much earlier than Sears the concept of discount merchandising goods (often imported from the Pacific rim nations). One of these was Sam Walton, who in a few decades became America's richest person by combining in his Wal-Mart stores an understanding of the new market opportunity, initially in smaller Midwest cities, and a new emphasis on value for the customer. As a result of this lack of vision, Sears has been replaced by Wal-Mart as America's largest merchandiser. Only recently, under new CEO Arthur Martinez, is there again a new vision:

> Just over a year since he took on the task, Martinez has slashed costs by firing thousands of workers, closed a slew of stores, and dumped the company's version of a mascot—the catalog— which was losing tens of millions of dollars a year. More importantly, Martinez also has figured out how to spruce up the company's image and sell clothes to working women—and that's where the big profits are in the retail business.[7]

By 1994 Sears was again making a profit from (more but smaller) catalogs, although they now employ only 20 (down from 19,000) people in catalog operations: they lend the Sears name and the purchase data base of their 24 million credit card customers in joint ventures with carefully selected specialty merchandisers.[8]

Collins and Lazier[9] identify three components of a good "Vision Framework":

1. **Core Values and Beliefs:** a system of guiding principles and tenets: a philosophy of business and life.
2. **Purpose:** the fundamental reason for the organization's existence.
3. **Mission:** a bold, compelling, audacious goal with a clear finish line and a specific time frame. Once completed, a new mission is set.

Strategies

Military organizations have long emphasized development of *strategies* or "grand plans" to attain longer-range objectives, supported by *tactics,* the shorter-range means to carry out these strategies. A successful enterprise needs to develop effective strategies for achieving their ultimate goals, and *strategic planning* is the organized process for selecting these strategies.

Strategic planning. One approach that corporations use in strategic planning is to identify the businesses they are in and the ones they want to be in the future, and to define a strategy for getting from the first set to the second. One tool for doing this is use of the "business portfolio matrix" developed by the Boston Consulting Group, which divides the businesses a company is in into four types based on market share and market growth (Table 3-1).

TABLE 3–1 Business Portfolio Matrix

Market	Market Share	
	Small	Large
Fast-growth	Question marks	Stars
Slow-growth	Dogs	Cash cows

Obviously, if you have a large share of a fast-growing market (the *star*), it will pay you to invest more money. Where you have only a small share of a slow-growing market, you need to eliminate these *dogs.* When you have only a small share of a fast-growing market, you *may* want to make a major effort to increase your market share of these *question marks.* Finally, when you have a large share of a market that has little future, you will want to "milk" the profits from these *cash cows* and invest them in stars and, possibly, question marks.

Strategic management of technology. Any enterprise that requires staying current with technology to assure future profitability (and this includes the organizations employing most engineers) must make the strategic planning for and management of technology an integral part of their business strategy. According to Erickson et al.,[10] "the management of technology encompasses the management of research, product and process development, and manufacturing engineering." They divide technologies into three broad classes:

- **Base technologies.** These are technologies that a firm must master to be an effective competitor in its chosen product-market mix. . . . The trick for R&D management is to invest enough—but only enough—effort to maintain competence in these areas. . . .

- **Key technologies.** These technologies provide competitive advantage. They may permit the producer to embed differentiating features or functions in the product or to attain greater efficiencies in the production process . . . [and] they should be given the highest priority. . . .
- **Pacing technologies.** These technologies could become tomorrow's key technologies. . . . The critical issue in technology management is balancing support of key technologies to sustain current competitive position and support of pacing technologies to create future vitality.

Goals and Objectives

The basic vision or purpose or mission of an organization must next be interpreted in terms of goals or objectives. Although some authors distinguish between goals and objectives, I will treat them as being synonymous.

Objectives are the aims or end points toward which activity is directed. Objectives should be established before further planning is begun, since you cannot plan effectively until you know what you are trying to achieve. One must distinguish between two types of goals that often co-exist: the *official objectives* that management says it is pursuing in its public statements, and the *operative goals* that it actually is pursuing. Managers of rural electric cooperatives in the United States, when asked by the author for their underlying objectives, often replied "to provide the best possible service at the lowest possible cost." This is hardly an operative goal, since "best service" implies a high level of staffing of maintenance and repair crews, and "lowest cost" implies a lesser level. Only by examining the level actually being maintained could one deduce the operative goal.

Peter Drucker[11] believes that objectives need to be established in all areas on which the organization's survival depends. He distinguishes eight such "key result areas:"

1. **Market share.** Market share is the ratio of dollar sales of an enterprise in a particular market to the total sales of all competitive products and services in that market. Firms with a small market share usually are less profitable because they have fewer sales over which to spread the fixed costs of operation, and managers often decline to enter or remain in a specific market unless they can either achieve a satisfactory market share or can define a smaller "market segment" in which they can be a leader.
2. **Innovation.** Most successful companies, especially in the areas of technology where most engineers will work, are continually searching for new products and services. 3M, for example, requires of its 40-odd divisions that at least 25% of sales be of products introduced in the last five years. Nonetheless, some successful companies deliberately choose to be followers and to provide low-cost, high-volume products without the high expense of being first.
3. **Productivity.** Productivity measures an organization's ability to produce more goods and services per unit of input (labor, materials, and investment). In recent years *quality* has been added as a related and essential area for setting objectives. The two are not inconsistent, since higher quality usually leads to lower scrap and rework losses, fewer returns, and greater customer satisfaction, increasing productivity and profitability.

4. **Physical and financial resources.** An enterprise needs to establish goals for the resources (plant, equipment, inventory, and capital) it needs to perform effectively.

5. **Manager performance and development.** Since good management is the key to enterprise success, effective firms plan carefully to assure that managers will be available in the years ahead in the quality and quantity needed for the organization to prosper. Supporting goals are then developed in areas such as recruitment, training, and evaluation.

6. **Worker performance and attitude.** Peters and Waterman[12] found that respect for the individual employee was a common thread running through America's most successful businesses. Personnel are "crew members" at McDonald's, "hosts" at Disneyland, "ambassadors" at Six Flags, and "associates" at J.C. Penney Stores. An unfortunate outcome of Frederick Taylor's scientific management revolution was, as we saw in Chapter 2, the division of work into deciding how to do it (by management) and doing as you are told (by the workers). Today's more educated work force has much to offer the company that knows how to motivate it and challenge it effectively.

7. **Profitability.** The profitability of an enterprise is essential to its continuation, and the desired level should be set explicitly as an objective against which to measure enterprise success.

8. **Social responsibility.** Every enterprise has responsibilities as a "corporate citizen" that extend beyond the legal and economic requirements. These include responsibilities to customers, employees, suppliers, community, and society as a whole. The organization that does not at least take responsibility for its effect on the environment deserves to be penalized by society.

Management by Objectives

In this same 1954 work,[13] Drucker formulated the concept of *management by objectives* (MBO). Since then, MBO has been widely adopted to translate broad organizational goals and objectives like those discussed above into specific individual objectives. MBO can (and usually should) be employed between superior and subordinate at every level. The steps in MBO are generally as follows.

First, both superior and subordinate should have an understanding of the goals and objectives of the overall organization and those of the superior's group.

Superior and subordinate then meet to establish objectives for the subordinate's attention over the next six months or year that are consistent with group objectives. These objectives should require some effort to attain, yet not be beyond reach. They should be quantifiable if feasible (e.g., reduce scrap by 20%); if not, they should be verifiable (e.g., write a new quality assurance plan), so that it is possible to determine at the end of the period whether or not the objective has been achieved. The relative amount of input from superior and subordinate in negotiating these objectives may vary, but the result should be mutual agreement. In agreeing to an objective proposed by the superior, the subordinate may identify specific resources or authority that need to be supplied by the superior to make it possible, and this is to the advantage of both. Objectives should not be confined to tasks for the sole benefit of the superior, but should also include developmental objectives designed to strengthen the subordinate's capabilities.

The subordinate then proceeds over the ensuing period (typically, six months or a year) to carry out his or her job with an emphasis on achieving these objectives. Naturally, if problems occur or priorities change, superior and subordinate can meet at any time and may modify the objectives, but they should not be changed without such agreement.

At the end of the period, superior and subordinate meet again to evaluate the subordinate's success in meeting assigned goals. This should be a constructive process, not an excuse for placing blame. This review session should end by mutually establishing a new set of objectives for the following period, of which some may be extensions of earlier objectives, some may be new objectives, and some earlier objectives may be deemphasized.

Advantages claimed for MBO include greater commitment and satisfaction on the part of subordinates, enforced planning and prioritizing of future activities on the part of both, and a more rational method of performance evaluation based on contribution to organizational objectives.

Disadvantages include the time and paperwork involved, misuse when superiors simply assign (rather than negotiate) objectives, and the gamesmanship of subordinates who try to negotiate easy goals. There is also a tendency for subordinates to focus on the relatively few, verifiable, MBO objectives negotiated to the detriment of the many other objectives, both qualitative and quantitative, that a professional must also keep in balance.

While annual salary reviews should not be scheduled at the same time as the periodic MBO evaluation, there needs to be some relationship between the ability to meet objectives and the reward system to make MBO effective. Finally, MBO will not be a success unless it has the initiating and continuing support of higher management.

SOME PLANNING CONCEPTS

Responsibility for Planning

Planning is a continuing responsibility of every manager. The higher managers rise, the more time they must spend in planning and the further into the future they must try to foresee. Most large organizations have staff offices for planning. The planning staff can coordinate the overall planning effort, gather and analyze information on the economy, markets, and competition, and perform other assigned tasks. The ultimate responsibility for planning, however, must rest with top and middle management. Line management must establish the objectives for the enterprise, devise the overall strategy, provide planning guidelines, and periodically review and redirect the planning effort. To have purpose, plans must lead to action, and managers are unlikely to carry out with any enthusiasm plans they have not "bought into" by being part of the planning process.

Planning Premises

An essential for effective planning is establishment of the premises, or assumptions, on which planning is to be based. Weihrich and Koontz[14] define planning premises as "the anticipated environment in which plans are expected to operate. They include assumptions or forecasts of the future and known conditions that will affect the operation of plans." Examples of planning premises include assumptions about future economic conditions, government decisions (regulation, tax law, and trade policy, for example), the nature of competition, and future markets.

In managing technology it is essential to establish planning premises about the future of technology and competition. Betz[15] cites Monsanto as an example. In the early 1980s it became clear to their executives that the basic chemicals they depended on were produced with mature, well-known technologies that could be replicated anywhere, and that countries with cheap sources of raw materials (principally petroleum) would have a relative advantage. Monsanto therefore divested itself of these traditional products and made major investments in biochemical research and in acquisition of electronic companies.

Where there are uncertainties about critical premises, prudent managers develop *contingency plans* that can be implemented if indicators show a change in the environmental conditions from those on which mainstream planning is based. Modest changes in current plans may be needed to add flexibility so that a switch to a contingency plan can be made quickly if needed.

Planning Horizon

The "planning horizon" asks how far into the future one should plan. This varies greatly, depending on the nature of the business and the plan. The vendor of pins and pennants outside a baseball stadium need not plan beyond the current season. In contrast, after Julius Rosenwald bought out Sears in 1895 he built a continuing business of mail-order service to the American farmer that included the Chicago mail-order plant described previously. His planning period needed to look far enough ahead to encompass a return on this long-term investment. Koontz and Weihrich summarize this difference into the "*commitment principle:* Logical planning encompasses a future period of time necessary to fulfill, through a series of actions, the commitments involved in decisions made today."[16] High-technology products may have short effective lives, and therefore short planning horizons:

> For example, in the 1960s and 1970s, IBM introduced new mainframe computers in 1964 (Model 360), in 1970 (Model 370), in 1977 (Model 303X), in 1979 (Models 38 and 4300 and 308X), and in 1983 (Model 4381), and each of these model generations was about a 100 to 200 percent improvement over the previous models in terms of improved performance and price.[17]

The decision of a utility company to build a nuclear power plant, on the other hand, must consider at least 10 years time to obtain necessary approvals and build the plant, and several decades operation to recover the investment. Many utilities who made such decisions based on 1970 projections of energy-use growth had no way to foresee the energy conservation following the OPEC oil crisis of 1973 or the increasing public attack on nuclear reactors, and they had to cancel partially built reactors at costs of billions of dollars. The planning horizon can vary from days to years depending on the level of the manager.

Systems of Plans

Usually not just one plan is involved, but a system of them. In 1916 Henri Fayol divided his

"Plan of Action in a Large Mining and Metallurgical Firm" into "yearly forecasts" and "ten-yearly forecasts," the latter redone every five years. Current practice is not much different, involving *strategic plans* of from 3 to 15 years futurity and *operating plans,* usually one year in duration (but sometimes as much as three years in duration). Complex programs will require not just a single plan but a system of plans, each describing a related activity. For example, a complex project might require most or all of the plans listed in Table 3-2.

Policies and Procedures

Policies are guides for decision making that permit implementation of upper management objectives, with room for interpretation and *discretion* by subordinates. *Rules,* in contrast, do not permit discretion. Policies have a hierarchy of levels, just as plans do. For example, the president of an engineering firm might establish a policy that approval of "small" design changes should be delegated to an "appropriate" level to reduce the number of matters demanding higher management attention. The chief engineer of a project might then implement this by establishing a supporting policy that engineering design supervisors could approve design changes costing up to $5000 involving technical specialties in their area of responsibility (which in turn involves judgment on the part of the supervisors). To be effective, policies should be clear, flexible, and communicated throughout the organization; should involve participation in their development; and should be reviewed regularly.

A *procedure,* on the other hand, is a prescribed sequence of activities to accomplish a desired purpose. Procedures tell you "if you want to do this, do it this way." For example, while the decision to approve (release) a design drawing requires technical judgment, the established *procedure* for doing so must be followed to assure that appropriate individuals have had a chance to approve the drawing and that its official existence is communicated to those who need to be informed of it. "Standard operating procedures" (SOPs) are examples of procedures used at the operating (working) level.

FORECASTING

Henri Fayol identified the first management function as *prevoyance,* a French word meaning to "foresee" and "prepare for" action. An essential preliminary to effective planning is therefore foreseeing or *forecasting* what the future will be like. The engineer manager must be concerned with both future markets and future technology, and must therefore understand both sales and technological forecasting.

TABLE 3–2   A System of Plans for a Complex Project

Project statement of work	Production plan
Work breakdown structure	Tooling plan
Project schedule	Make-or-buy plan
Project budget	Quality assurance plan
Specifications	Facilities plan
Management plan	Training plan
Configuration management plan	Logistics support plan
Security plan	Reliability plan
System test plan	Transportation plan

Sales Estimates

The most important premise or assumption in planning and decision making is the level of future sales (or, for nonprofit activities, of future operations). Almost everything for which we plan is based on this assumption: the production level (which determines how many people we must hire and train or, if production declines, lay off); the need for new facilities and equipment; the size of the sales force and advertising budget; and new funding for purchases and for investment in inventory and accounts receivable. Let us consider the common ways this is done.

Jury of executive opinion. This is the simplest method, in that the executives of the organization (typically, the vice presidents of the various divisions) each provide an estimate (educated guess) of future volume, and the president provides a considered average of these estimates. This method is inexpensive and quick and may be entirely acceptable if the future conforms to the assumptions the executives have used in estimating.

Sales force composite. In this commonly used method, members of the sales force estimate sales in their own territory. Regional sales managers adjust these estimates for their opinion of the optimism or pessimism of individual salespeople, and the general sales manager "massages" the figures to account for new products or factors of which individual salesmen are unaware. Since the field sales force is closest to the customer, this method has much to recommend it. However, if there is any suggestion that the estimate a salesperson provides will next appear as a minimum goal they must achieve, the sales force may find it in their own best interest to "play games" with the figures.

Users' expectation. When a company sells most of its product to a few customers, the simplest method is to ask the customers to project their needs for the future period. The customers depend for their own success on reliable sources of supply, and so communication is in the best interest of both parties. For consumer goods, though, not only is such information expensive to obtain, but consumers often do not know what they will purchase in the future.

Choice of method. Companies with effective planning will combine a variety of methods to arrive at the best sales forecast. Qualitative estimates from the sales force and customer surveys may be compared with more quantitative estimates obtained from moving average or regression models (discussed in the following sections). Finally, the chief executive, with the assistance of other top officers, will establish a sales forecast to be used in future planning.

Moving Average Forecasts

Simple moving average. Where the values of a parameter show no clear trend with time, a forecast F_{n+1} for the next period can be taken as the simple average of some number n of the most recent actual values A_t:

$$F_{n+1} = \frac{1}{n} \sum_{t=1}^{n} A_t$$

For example, if sales for years 1994, 1993, 1992, and 1991 ($n = 4$) were 1600, 1200, 1300, and 1100, respectively, sales for 1995 would be forecast as

$$F_{1995} = \frac{1600 + 1200 + 1300 + 1100}{4} = 1300$$

Weighted moving average. The preceding method has the disadvantage that an earlier value (1990, for example) has no influence at all, but a value n years in the past (1991) is weighted as heavily as the most recent value (1994). We can improve on our model by assigning a set of weights w_t that total unity (1.0) to the previous n values:

$$F_{n+1} = \sum_{t=1}^{n} w_t A_t, \text{ where } \sum_{t=1}^{n} w_t = 1.0$$

Using weights of 0.4, 0.3, 0.2, and 0.1 for the most recent ($n =$) 4 past years in our example above yields

$$\begin{aligned} F_{1995} &= 0.4A_{1994} + 0.3A_{1993} + 0.2A_{1992} + 0.1A_{1991} \\ &= 0.4(1600) + 0.3(1200) + 0.2(1300) + 0.1(1100) \\ &= 1370 \end{aligned}$$

Exponential smoothing. The weighted moving average techniques have the disadvantage that you (or your computer) must record and remember n previous values and n weights for each parameter being forecast, which can be burdensome if n is large. The simple exponential smoothing method continuously reduces the weight of a value as it becomes older, yet minimizes the data that must be retained in memory. In this technique the forecast value for the next time period F_{n+1} is taken as the sum of (1) the forecasted value F_n for the current period, plus (2) some fraction α of the difference between the actual (A_n) and forecasted (F_n) values for the current period:

$$\begin{aligned} F_{n+1} &= F_n + \alpha(A_n - F_n) \\ &= \alpha A_n + (1 - \alpha)F_n \end{aligned}$$

Expanding the right-hand equation using similar expressions for F_n, F_{n-1}, and so on, gives us

$$\begin{aligned} F_{n+1} &= \alpha A_n + (1 - \alpha)[\alpha A_{n-1} + (1 - \alpha)F_{n-1}] \\ &= \alpha A_n + \alpha(1 - \alpha)A_{n-1} + \alpha(1 - \alpha)^2 A_{n-2} + \alpha(1 - \alpha)^3 A_{n-3} + \cdots \end{aligned}$$

We can see from this last equation that the weight put on past values continues to decrease but never becomes zero. To start the forecast sequence, we must set the first forecast equal to the actual value of the preceding year. For the data used in the two preceding examples, the forecasts are as shown in Table 3-3 for two selected values of α:

For example, if $\alpha = 0.3$,

$$\begin{aligned} F_{1993} &= 0.3A_{1992} + 0.7F_{1992} \\ &= 0.3(1300) + 0.7(1100) = 1160 \\ F_{1994} &= 0.3A_{1993} + 0.7F_{1993} \\ &= 0.3(1200) + 0.7(1160) = 1172 \end{aligned}$$

TABLE 3–3 Exponential Smoothing Calculation

Year (t)	Forecast F_t		Actual Value A_t
	$\alpha = 0.3$	$\alpha = 0.6$	
1991			1100
1992	1100	1100	1300
1993	1160	1220	1200
1994	1172	1208	1600
1995	1300	1443	

Regression Models

Regression models are a major class of *explanatory forecasting models,* which attempt to develop logical relationships that not only provide useful forecasts, but also identify the causes and factors leading to the forecast value. Regression models assume that a *linear relationship* exists between a variable designated the *dependent (unknown) variable* and one or more other *independent (known) variables.*

Simple regression model. The simple regression model assumes that the independent variable I depends on a single dependent variable D. Figure 3-2 gives an example in which the parameter we have been forecasting in the moving average calculations is taken as D and time, for convenience expressed as (year − 1991), is taken as I.

The regression problem is to identify a line
$$D = a + bI$$
such that the sum of the squares of the deviations between actual and estimated values (the vertical line segments in the figure) is minimized. The two constants in this least squares equation are found from

$$b = \frac{n\Sigma(I_i D_i) - \Sigma I_i \Sigma D_i}{n\Sigma(I_i^2) - (\Sigma I_i)^2}$$

$$a = \Sigma \frac{D_i}{n} - b \Sigma \frac{I_i}{n} = \bar{D} - b\bar{I}$$

where $\bar{D}$ and $\bar{I}$ are the mean values of D and I respectively, and Σ indicates a summation from $i = 1$ to n.

TABLE 3–4 Data for Regression Calculation

	I	D	DI	I^2
	0	1100	0	0
	1	1300	1300	1
	2	1200	2400	4
	3	1600	4800	9
Σ	6	5200	8500	14
Mean	1.5	1300		

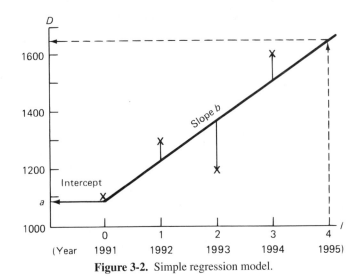

Figure 3-2. Simple regression model.

We can calculate the trend line in Figure 3-2 using the data in Table 3-4:

$$b = \frac{4(8500) - 6(5200)}{4(14) - (6)^2} = 140$$

$$a = \frac{5200}{4} - 140\left(\frac{6}{4}\right) = 1300 - 140(1.5) = 1090$$

and we can forecast a value for 1995:

$$D_{1995} = 1090 + (1995 - 1991)(140) = 1090 + 560 = 1650$$

Regressions can be used to calculate the best fit to a straight line on a normal graph (as above) or on semilog or log-log paper. The resulting forecasts can be seriously in error if the assumptions (premises) on which they are based prove to be in error. Electric utility companies came to believe over a period of several decades that the demand for electricity would increase about 7% per year (which would give a straight line on a semilog plot). Although this may not sound like a large growth rate, it means that demand should increase by a factor of about $(1.07)^{10} = 2.0$ (i.e., double) every 10 years. Since large power plants (especially nuclear ones) take at least that long to plan and construct, when the oil crisis struck in 1973 and consumers drastically reduced their use of now-expensive power through conservation, utilities were left with billions of dollars of capacity under construction that would not be useful for a long time.

Augustine gives a convincing but facetious example of the problems with such extrapolation in his delightful *Augustine's Laws*. For example, he plots the logarithm of the cost of military aircraft against the year in which they were first operational, from the Wright Flyer (about $3000 in 1912) to the F-15 (about $20,000,000 in 1975), and concludes:

In the year 2054, the entire [U.S.] defense budget will purchase just one tactical aircraft. The aircraft will have to be shared by the Air Force and the Navy 3½ days each per week except for leap year, when it will be made available to the Marines for the extra day.[18]

Multiple regression. In multiple regression, the dependent variable D is assumed to be a function of more than one independent variable I_j, such as

$$D = c_0 + c_1 I_1 + \frac{c_2}{I_2} + c_3 I_3^2 + \cdots$$

The dependent variable can be assumed to be proportional directly or inversely, proportional to a power or a root, or proportional in some other way to the independent variables, as is suggested in the equation above. Past values of dependent and independent variables are then used in *regression analysis* to reduce the independent variables to the most important ones and to find the values for the constants c_i that give the best fit. For example, a manufacturer of replacement automobile tires might find that the demand for tires varied with the cost of gasoline, the current unemployment rate, sales of automobiles two years before, and the weight of those automobiles.

Dannenbring and Starr[19] provide one (of many) convenient sources of further information on multiple regression and other explanatory forecasting models.

Technological Forecasting

Engineers usually are involved in planning environments where technology is changing, and it is essential that planning be based on the best estimate of the technology that will be available in the future. Shannon[20] bases his belief in the feasibility of technological forecasting on three premises: (1) technological events and capabilities grow in a very organized manner; (2) technology responds to needs, opportunities, and the provision of resources; and (3) new technology can be anticipated by understanding the process of innovation. Roman reports a definition of Marvin Cetron that a technological forecast is a prediction, based on confidence, that certain technical developments can occur within a specified time period with a given level of resource allocation.[21] Two types of technological forecasting should be considered: *normative* and *exploratory.*

In *normative technological forecasting,* some desired future goal is selected, and a process is developed, working backward from the future to the present, designed to achieve this goal. For example, the U.S. government might decide that it is essential to have power available from nuclear fusion in commercial quantities in the year 2020, and then work backward to establish a schedule for a full-scale demonstration plant, a smaller pilot plant, the research tasks that must precede them, and finally develop the overall budget and schedule required to reach the normative goal. President Kennedy's 1961 decision to land a man on the surface of the moon before the end of that decade certainly required normative technological forecasting.

An *exploratory technological forecast* begins with the present state of technology and extrapolates into the future assuming some expected rate of technical progress. A common forecasting method is use of a panel of experts in the technical field involved. Panels are sometimes consulted using the "Delphi process" (named after the Oracle at Delphi in ancient Greek mythology). Each expert is asked independently when some future technical breakthrough (such as the ability to select the sex of unborn children, for example) will occur, if ever. Averages of the estimates are then reported back to panel members, who are then given a chance to modify their estimate or explain why they hold their belief. A final value is adopted after several such iterations.

For example, Japan's Science and Technology Agency polled some 2800 of the nation's leading specialists in research institutes and universities to identify R&D goals for the next 30 years. The five most important (and the year they were expected to be achieved) were:[22]

1. Commercialization of technologies to eliminate air pollutants such as the oxides of nitrogen (2003).
2. Invention of a computer to operate faster than 10 teraflops (10 trillion calculations per second) (2004).
3. Discovery of the major development mechanism of cancer (2010).
4. Commercialization of effective methods to prevent the spread of cancer (2007).
5. Diffusion of global-scale environmental preservation technologies (2011).

One useful model for technological forecasting is the technology S-curve, shown in Figure 3-3. The performance gained from a new technology tends to start slowly, then rise almost exponentially as many scientists and engineers begin applying themselves to product improvement. Ultimately, as the technology becomes mature, performance gains become more and more difficult to attain, and performance approaches some natural limit. Gains then require the use of entirely new technology.

For example, the efficiency of the incandescent lamp (measured in lumens per watt) was improved from Edison's original carbon filament by the development of today's tungsten filament in 1910 by General Electric Research Laboratories, but the continuing improvement in lighting efficiency has come from the invention and improvement of the fluorescent lamp and later the mercury and sodium vapor lamps. Similarly, improvement of the cost per bit of memory storage (the inverse of performance) over time has shifted to a new and more efficient curve each time a denser integrated circuit with more memory capacity was developed. The technology forecaster might have predicted in advance that this process would continue, making new applications requiring extensive computer memory storage more and more practical. This has proven true as 128K, 256K, and then 1-, 4-, and 16-megabyte chips have been developed. Nonetheless, there is some practical physical limit to the improvement of this technology; computer chip engineers have estimated this might be a billion bits per chip!

STRATEGIES FOR MANAGING TECHNOLOGY

Invention and Innovation

American folklore idealizes its inventors—people who have come up with an idea for a novel product or process. But that is not enough. Betz gives as an example the development of the Xerox copier.[23] Chester Carlson, who had experience as a carbon chemist, a printer,

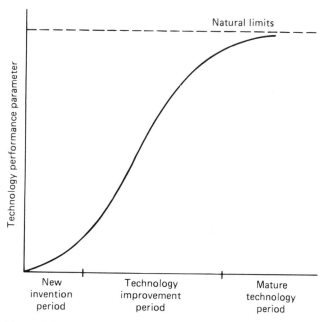

Figure 3-3. Technology S-curve.
(From Frederick Betz, *Managing Technology: Competing Through New Ventures, Innovation, and Corporate Research,* Prentice-Hall, Inc., Englewood Cliffs, NJ, 1987, p. 62; reprinted by permission of Prentice-Hall, Inc., copyright 1987.)

and a patent lawyer, sought a better way to copy legal documents. He conceived the idea of projecting an image of the work to be copied onto paper coated with ink, to hold the ink electrostatically where dark spots (letters) were projected, and then to bake these letters into the paper. He succeeded in obtaining a crude image in 1935, but the invention was neither efficient nor economical. Nonetheless, it was enough to apply for a patent, which was issued in 1942.

Carlson went from company to company looking for support for the process; he was turned down again and again. Finally, a group at Battelle Memorial Institute, a nonprofit R&D organization, agreed in 1945 to try to develop the process into a commercially practical one in return for a share of the royalties, and Joseph Wilson, president of the small Haloid Corporation, took the risk of producing the first copiers based on Carlson's patents and Battelle's developments. That company grew and grew as a result, and changed its name to Xerox after its principal product, the office copier.

The contributions of Battelle and Haloid constitute *technological innovation,* "the introduction into the marketplace of new products, processes, and services based on new technology."[24] Without innovation, inventions create little benefit. Reduction of ideas to successful products and processes is a difficult task—we will see in Chapter 9 that many different ideas must be considered and developed enough for careful evaluation to produce one successful, profitable new product.

Producing the first successful product is often not enough; innovation must continue to keep the product line competitive. In July 1981, Adam Osborne became the first to market the personal computer as a complete package (computer, disk drive, monitor, printer, and software), and it was an instant success. However, his monitor was only 5 inches (diagonally) in size and would hold only 52 characters across (versus 80 on a typed page). Competitors (especially Kaypro) moved quickly to correct this deficiency, and by September 1983 the Osborne Company was in bankruptcy. The lesson is clear—product innovation and improvement must be a continuing part of technology strategy.

Myron Tribus believes the competitiveness of an enterprise depends on two distinct thrusts, each having two characteristics: (1) doing new and better things (invention and innovation, just discussed), and (2) doing things in new and better ways (quality and productivity). He elaborates:

> Of the four characteristics, invention, innovation, quality, and productivity, the last three require the collaboration of many people. Invention, on the other hand, is usually the product of one or at the most two or three minds. Management to enhance invention, therefore, is somewhat different from management to enhance innovation, quality, and productivity. Invention is not subject to scheduling.[25]

Entrepreneurship

It takes a special kind of person to lead the innovation task successfully—the *entrepreneur,* defined by Webster's as "a person who organizes and manages a business undertaking, assuming the risk for the sake of profit."[26] Betz extends this:

> The entrepreneur is a kind of business hero; and like all heroes, they have qualities to be admired: initiative, daring, courage, commitment. These values are especially admired in turbulent business conditions, when initiative is required for survival.[27]

While the initial concept of an entrepreneur is of a person who creates a new business for personal profit, established corporations need continuing entrepreneurial activity to create the new products and new businesses that will assure future growth of the organization; the term *intrapreneurship* has been coined to describe this activity. The challenge in managing technology is to provide a climate where intrapreneurs are encouraged to take risks, are given needed resources and time, and are permitted early failures, while shifting to closer control of resources and costs as products become mature.

Managing Technological Change

Top management in technological enterprise must constantly be aware of the technologies underlying their business and the potential for change. Business history is replete with stories of companies that failed to recognize in time new technology that would replace their key products. Most of the companies that led in production of vacuum tube–based electronics are no longer important today in electronics; other companies led the transistor revolution and replaced them; in turn, a new group of firms have come to the fore with products based on large-scale integrated circuits. A few companies (General Electric is an example) have maintained leadership or have at least been vigorous early followers as technology changes, and they have prospered because they have succeeded in integrating technology strategy into business strategy.

DISCUSSION QUESTIONS

3-1. Why is planning said to have "primacy" among the managerial functions?

3-2. Develop your own model of the steps in the planning process.

3-3. Select a company (other than Sears) whose apparent mission or purpose has changed over its history and describe the change.

3-4. Pick a company that has acquired and divested businesses and estimate the reason for these moves. Were they consistent with the logic of Table 3-1?

3-5. Select a company or industry for which the strategic management of technology is important. Describe some of the *base, key,* and *pacing* technologies that are important for their strategic management of technology.

3-6. Pick a company with which you are familiar and estimate from its actions what the objectives of its management appear to be in each of Drucker's eight "key result areas."

3-7. Briefly outline the concept of management by objectives (MBO) and the steps involved in implementing this technique in organizations.

3-8. For what types of employees or positions do you think management by objectives (MBO) should prove particularly effective? Ineffective?

3-9. For a given product and company (such as automobiles from Ford) list a set of premises (assumptions) regarding such matters as the economy, competition, materials, labor, customer demand, and others that should govern their planning over the next five years.

3-10. What length of planning horizon would you recommend for planning (**a**) the forest resources of a large paper company; (**b**) construction of a new automobile plant; (**c**) creation of a new housing development of 15 homes?

3-11. Create an extended list of the plans and decisions for a large manufacturing company that will depend on the sales forecast for the next year.

3-12. Sales of a particular product (in thousands of dollars) for the years 1991 through 1994 have been 48, 64, 67, and 83, respectively.
(**a**) What sales would you predict for 1995 using a simple four-year moving average?
(**b**) What sales would you predict for 1995 using a weighted moving average with weights of 0.50 for the immediate preceding year and 0.3, 0.15, and 0.05 for the three years before that?

3-13. Using exponential smoothing with a weight α of 0.6 on actual values: (**a**) If sales are $45,000 and $50,000 for 1994 and 1995, what would you forecast for 1996? (**b**) Given this forecast and actual 1996 sales of $53,000, what would you then forecast for 1997?

3-14. In question 3-12, taking actual 1991 sales of $48,000 as the forecast for 1992, what sales would you forecast for 1993, 1994, and 1995 using exponential smoothing and a weight α on actual values of (**a**) 0.4; (**b**) 0.8?

3-15. In question 3-12, what sales would you forecast for 1995 using the simple regression (least squares) method?

3-16. Cite another example (in addition to Xerox and Osborne's computer) showing the difficulty of transforming a technical idea into a successful product (innovation).

3-17. Describe an entrepreneur you either know or have read about. What personal characteristics led to or limited his or her success?

NOTES

1. Edmund Young, "Management Thoughts for Today from the Ancient Chinese," *Management Bulletin,* October 1980, p. 29.
2. John M. Amos and Bernard R. Sarchet, *Management for Engineers* (Englewood Cliffs, NJ: Prentice-Hall, Inc., 1981), p. 51.
3. *King James Bible,* Proverbs 12:18.
4. James C. Collins and William C. Lazier, *Beyond Entrepreneurship: Turning Your Business into an Enduring Great Company* (Englewood Cliffs, NJ: Prentice-Hall, 1992), Chapter 2: "Vision," reprinted in *Engineering Management Review,* Spring 1993, pp. 61–75.
5. Peter F. Drucker, *Management: Tasks, Responsibilities, Practices* (New York: Harper & Row, Publishers, Inc., 1974), pp. 75–77.
6. Drucker, *Management,* Chapter 5.
7. John McCormick, "The Savior at Sears," *Newsweek,* November 1, 1993, p. 42.
8. "Strategies for the New Mail Order," *Business Week,* December 19, 1994, pp. 82 ff.
9. Collins and Lazier, *Beyond Entrepreneurship,* pp. 66–70.
10. Tamara J. Erickson, John F. Magee, Philip A. Roussel, and Kamal N. Saad, "Managing Technology as a Business Strategy," *Engineering Management Review,* Spring 1991, pp. 34–38.
11. Peter F. Drucker, *The Practice of Management.* ©1954 (© renewed 1982) by Peter F. Drucker; reprinted by permission of HarperCollins Publishers, Inc.
12. Thomas J. Peters and Robert H. Waterman Jr., *In Search of Excellence: Lessons from America's Best Run Companies* (New York: Harper & Row, Publishers, Inc., 1982).
13. Drucker, *Practice of Management.*
14. Heinz Weihrich and Harold Koontz, *Management: A Global Perspective,* 10th ed. (New York: McGraw-Hill Book Company, 1993), p. 185.
15. Frederick Betz, *Managing Technology: Competing Through New Ventures, Innovation, and Corporate Research* (Englewood Cliffs, NJ: Prentice-Hall, Inc., 1987), pp. 132-134.
16. Weihrich and Koontz, *Management,* p. 135.
17. Susan Fraker, "High-Speed Management for the High-Tech Age," *Fortune,* March 5, 1984, pp. 62–68, in Betz, *Managing Technology,* pp. 158–159.
18. Norman R. Augustine, *Augustine's Laws and Major System Development Programs,* revised and enlarged (Washington, DC: American Institute of Aeronautics and Astronautics, 1983), p. 55.
19. David G. Dannenbring and Martin K. Starr, *Management Science: An Introduction* (New York: McGraw-Hill Book Company, 1981), Chapter 19.
20. Robert E. Shannon, *Engineering Management* (New York: John Wiley & Sons, Inc., 1980), p. 43.
21. Daniel D. Roman, "Technological Forecasting in the Decision Process," *Academy of Management Journal,* 13:2, June 1970, pp. 127–138.
22. Hajime Nagahama, "Technopolicy," *Look Japan,* May 1993, reported in *The Futurist,* September–October 1993, p. 8.
23. Betz, *Managing Technology,* pp. 5–6.
24. Betz, *Managing Technology,* p. 6.
25. Myron Tribus, "Applying Quality Management Principles in R&D," *Engineering Management Journal,* 2:3, September 1990, p. 29.
26. *Webster's New World Dictionary of the American Language,* 2nd college ed. (Englewood Cliffs, NJ: Prentice Hall, Inc., 1986).
27. Betz, *Managing Technology,* p. 13.

CHAPTER ● 4

Decision Making

PREVIEW

In this chapter we first relate decision making to planning, then consider the types of decisions and the concept of "satisficing." A discussion of the origins of management science leads into one on modeling and on the five-step process of management science in which models are used to solve real-world problems. We discuss three types of decision making, with examples of problem-solving methods useful with each type: decision making under certainty, using linear programming; decision making under risk, using expected value, decision trees, queuing theory, and simulation; and decision making under uncertainty, using game theory. We continue with brief discussions of integrated data bases, management information and decision support systems, and expert systems, and close with a comment on the need for effective implementation of decisions.

NATURE OF DECISION MAKING

Relation to Planning

Managerial decision making is the process of making a conscious choice between two or more rational alternatives in order to select the one that will produce the most desirable consequences (benefits) relative to unwanted consequences (costs). If there is only one alternative, there is nothing to decide. The overall planning/decision-making process has already been described at the beginning of Chapter 3, and there we discussed the key first steps of setting objectives and establishing premises (assumptions). In this chapter we consider the process of developing and evaluating alternatives and selecting from among them the best alternative, and we review briefly some of the tools of management science available to help us in this evaluation and selection.

If planning is truly "deciding in advance what to do, how to do it, when to do it, and who is to do it" (as proposed by Amos and Sarchet[1]), then decision making is an essential part of planning. Decision making is also required in designing and staffing an organization, developing methods of motivating subordinates, and identifying corrective actions in the control process. However, it is conventionally studied as part of the planning function, and we will do so here.

Occasions for Decision

Chester Barnard wrote his classic book *The Functions of the Executive* from his experience as president of the New Jersey Bell Telephone Company and of the Rockefeller Foundation, and in it he pursued the nature of managerial decision making at some length. He concluded that

> the occasions for decision originate in three distinct fields: (a) from authoritative communications from superiors; (b) from cases referred for decision by subordinates; and (c) from cases originating in the initiative of the executive concerned.[2]

Barnard points out that occasions for decisions stemming from the "requirements of superior authority . . . cannot be avoided," although portions of it may be delegated further to subordinates. Appellate cases (referred to the executive by subordinates) should not always be decided by the executive. Barnard explains[3] that "the test of executive action is to make these decisions when they are important, or when they cannot be delegated reasonably, and to decline the others."

Barnard concludes that "occasions of decision arising from the initiative of the executive are the most important test of the executive." These are occasions where no one has asked for a decision, and the executive usually cannot be criticized for not making one. The effective executive takes the initiative to think through the problems and opportunities facing the organization, conceives programs to make the necessary changes, and implements them. Only in this way does the executive fulfill the obligation to *make a difference* because he or she is in that chair rather than someone else.

Types of Decisions

Routine and nonroutine decisions. Pringle et al. classify decisions on a continuum ranging from routine to nonroutine, depending on the extent to which they are *structured*. They describe *routine* decisions as focusing on well-structured situations that

> recur frequently, involve standard decision procedures, and entail a minimum of uncertainty. Common examples include payroll processing, reordering standard inventory items, paying suppliers, and so on. The decision maker can usually rely on policies, rules, past precedents, standardized methods of processing, or computational techniques. Probably 90 percent of management decisions are largely routine.[4]

Indeed, routine decisions usually can be delegated to lower levels to be made within established policy limits, and increasingly they can be programmed for computer "decision" if they can be structured simply enough. *Nonroutine* decisions, on the other hand, "deal with unstructured situations of a novel, nonrecurring nature," often involving incomplete knowledge, high uncertainty, and the use of subjective judgment or even intuition, where "no alternative can be proved to be the best possible solution to the particular problem."[5] Such decisions become more and more common the higher one goes in management and the longer the future period influenced by the decision. Unfortunately, almost the entire educational process of the engineer is based on the solution of highly structured problems for which there is a single "textbook solution." Engineers often find themselves unable to rise in management unless they can develop the "tolerance for ambiguity" that is needed to tackle unstructured problems.

Objective versus bounded rationality. Simon defines a decision as being "'objectively' rational if *in fact* it is the correct behavior for maximizing given values in a given situation."[6] Such rational decisions are made by "(a) viewing the behavior alternatives prior to decision in panoramic [exhaustive] fashion, (b) considering the whole complex of consequences that would follow on each choice, and (c) with the system of values as criterion singling out one from the whole set of alternatives." Rational decision making, therefore, consists of *optimizing,* or *maximizing,* the outcome by choosing the single best alternative from among all possible ones, which is the approach suggested in the planning/decision-making model at the beginning of Chapter 3. Instead, Simon believes that

> actual behavior falls short, in at least three ways, of objective rationality . . .:
>
> 1. Rationality requires a complete knowledge and anticipation of the consequences that will follow on each choice. In fact, knowledge of consequences is always fragmentary.
> 2. Since these consequences lie in the future, imagination must supply the lack of experienced feeling in attaching value to them. But values can be only imperfectly anticipated.
> 3. Rationality requires a choice among all possible alternative behaviors. In actual behavior, only a few of these possible alternatives ever come to mind.[7]

Managers, under pressure to reach a decision, have neither the time nor other resources to consider all alternatives or all the facts about any alternative. A manager "must operate under conditions of *bounded rationality,* taking into account only those few factors of which he or she is aware, understands, and regards as relevant." Administrators must *satisfice* by accepting a course of action that is satisfactory or "good enough," and get on with the job rather than searching forever for the "one best way." Managers of engineers and scientists, in particular, must learn to insist that their subordinates go on to other problems when they reach a solution that satisfices, rather than pursuing their research or design beyond the point at which incremental benefits no longer match the costs to achieve them.

Level of certainty. Decisions may also be classified as being made under conditions of *certainty, risk,* or *uncertainty,* depending on the degree with which the future environment determining the outcome of these decisions is known. We will compare these three categories later in this chapter.

MANAGEMENT SCIENCE

Origins

Quantitative techniques have been used in business for many years in applications such as return on investment, inventory turnover, and statistical sampling theory. However, today's emphasis on the quantitative solution of complex problems in operations and management, known initially as *operations research* and more commonly today as *management science,* began at the Bawdsey Research Station in England at the beginning of World War II. As Hicks puts it:

> In August 1940, a research group was organized under the direction of P. M. S. Blackett of the University of Manchester to study the use of a new radar-controlled antiaircraft system. The research group came to be known as "Blackett's circus." The name does not seem unlikely in the light of their diverse backgrounds. The group was composed of three physiologists, two mathematical physicists, one astrophysicist, one Army officer, one surveyor, one general physicist, and two mathematicians. The formation of this group seems to be commonly accepted as the beginning of operations research.[8]

Some of the problems this group and (several that grew from it) studied were the optimum depth antisubmarine bombs should be exploded for greatest effectiveness (20 to 25 feet) and the relative merits of large versus small convoys (large convoys led to fewer total ship losses). Soon after the United States entered the war, similar activities were initiated by the U.S. Navy and the Army Air Force. With the immediacy of the military threat, these studies involved *research* on the *operations* of existing systems. After the war these techniques were applied to longer-range military problems and to problems of industrial organizations. With the development of more and more powerful electronic computers, it became possible to model large systems as a part of the design process, and the terms *systems engineering* and *management science* came into use. Management science has been defined as having the following "primary distinguishing characteristics":

1. A systems view of the problem—a viewpoint is taken that includes all of the significant interrelated variables contained in the problem.
2. The team approach—personnel with heterogeneous backgrounds and training work together on specific problems.
3. An emphasis on the use of formal mathematical models and statistical and quantitative techniques.[9]

Models and Their Analysis

A *model* is an abstraction or simplification of reality, designed to include only the essential features that determine the behavior of a real system. For example, a three-dimensional *physical* model of a chemical processing plant might include scale models of major equipment and large-diameter pipes, but it would not normally include small piping or electrical wiring. The *conceptual* model of the planning/decision-making process in Chapter 3 certainly does not illustrate all the steps and feedback loops present in a real situation; it is only indicative of the major ones.

Most of the models of management science are *mathematical* models. These can be as simple as the common equation representing the financial operations of a company:

$$net\ income = revenue - expenses - taxes$$

On the other hand, they may involve a very complex set of equations. As an example, the *Urban Dynamics* model was created by Jay Forrester to simulate the growth and decay of cities.[10] This model consisted of 154 equations representing relationships between the factors that he believed were essential: three economic classes of workers (managerial/professional, skilled, and "underemployed"), three corresponding classes of housing, three types of industry (new, mature, and declining), taxation, and land use. The values of these factors evolved through 250 simulated years to model the changing characteristics of a city. Even these 154 relationships still proved too simplistic to provide any reliable guide to urban development policies (see Babcock[11] for a discussion).

Management science uses a five-step process that begins in the real world, moves into the model world to solve the problem, then returns to the real world for implementation (the explanation below is, in itself, a conceptional model of a more complex process):

Real world

1. Formulate the problem (defining objectives, variables, and constraints).

Simulated (model) world

2. Construct a mathematical model (a simplified yet realistic representation of the system).

3. Test the model's ability to predict the present from the past, and revise until you are satisfied.

4. Derive a solution from the model.

5. Apply the model's solution to the real system, document its effectiveness, and revise further as required.

The Analyst and the Manager

To be effective, the management science analyst cannot just create models in an "ivory tower." The problem-solving team must include managers and others from the department or system being studied—to establish objectives, explain system operation, review the model as it develops from an operating perspective, and help test the model. The user who has been part of model development, has developed some understanding of it and confidence in it, and feels a sense of "ownership" of it is most likely to use it effectively.

The manager is not likely to have a detailed knowledge of management science techniques, nor the time for model development. Today's manager should, however, understand the nature of management science tools and the types of management situations in which they might be useful. Increasingly, management positions are being filled with graduates of management (or engineering management) programs that have included an introduction to the fundamentals of management science and statistics. Regrettably, all too few operations research or management science programs require the introduction to organization and behavioral theory that would help close the manager/analyst gap from the opposite direction.

There is considerable discussion today of the effect of computers and their applications (management science, decision support systems, expert systems, etc.) on managers and organizations. Certainly, workers and managers whose jobs are so routine that their decisions can be reduced to mathematical equations have reason to worry about being replaced by computers. For most managers, however, modern methods offer the chance to reduce the time one must spend on more trivial matters, freeing up time for the types of work and decisions that only people can accomplish.

TOOLS FOR DECISION MAKING

Categories of Decision Making

Decision making can be discussed conveniently in three categories: decision making under certainty, under risk, and under uncertainty. The *payoff table,* or *decision matrix,* shown in Table 4-1 will help in this discussion. Our decision will be made among some number m of alternatives, identified as $A_1, A_2, \ldots, A_m$. There may be more than one future "state of nature" N (the model allows for n different futures). These future states of nature may not be equally likely, but each state N_j will have some (known or unknown) probability of occurrence p_j. Since the future must take on one of the n values of N_j, the sum of the n values of p_j must be 1.0.

The *outcome* (or payoff, or benefit gained) will depend on both the alternative chosen and the future state of nature that occurs. For example, if you choose alternative A_i and state of nature N_j takes place (as it will with probability p_j), the payoff will be outcome O_{ij}. A full payoff table will contain m times n possible outcomes.

Let us consider what this model implies, and the analytical tools we might choose to use under each of our three classes of decision making.

Decision Making Under Certainty

Decision making under *certainty* implies that we are certain of the future state of nature (or assume that we are). (In our model, this means that the probability p_1 of future N_1 is 1.0, and all other futures have zero probability.) The solution, naturally, is to choose the alternative A_i that gives us the most favorable outcome O_{i1}. Although this may seem like a trivial exercise, there are many problems that are so complex that sophisticated mathematical techniques are needed to find the best solution.

TABLE 4-1 Payoff Table

Alternative	State of Nature/Probability					
	N_1 p_1	N_2 p_2	...	N_j p_j	...	N_n p_n
A_1	O_{11}	O_{12}	...	O_{1j}	...	O_{1n}
A_2	O_{21}	O_{22}	...	O_{2j}	...	O_{2n}
...	...	...	...	...	...	...
A_i	O_{i1}	O_{i2}	...	O_{ij}	...	O_{in}
...	...	...	...	...	...	...
A_m	O_{m1}	O_{m2}	...	O_{mj}	...	O_{mn}

Linear programming. One common technique for decision making under certainty is called *linear programming*. In this method, a desired benefit (such as profit) can be expressed as a mathematical function (the value model or *objective function*) of several variables. The solution is the set of values for the independent variables that serves to maximize the benefit (or, in many problems, to minimize the cost) subject to certain limits (*constraints*).

Simple example. For example, consider a factory producing two products, product X and product Y. The problem is this: If you can realize $10 profit per unit of product X and $14 per unit of product Y, what is the production level of x units of product X and y units of product Y that maximizes the profit $P?$ That is, you seek to

$$\text{maximize } P = 10x + 14y$$

As illustrated in Figure 4-1, you can get a profit of

- $350 by selling 35 units of X or 25 units of Y
- $700 by selling 70 units of X or 50 units of Y
- $620 by selling 62 units of X or 44.3 units of Y, or (as in the first two cases as well) any combination of X and Y on the *isoprofit line* connecting these two points.

Your production, and therefore your profit, is subject to resource limitations, or *constraints*. Assume in this example that you employ five workers: three machinists and two assemblers, and that each works only 40 hours a week. Products X and/or Y can be produced by these workers subject to the following constraints:

- Product X requires three hours of machining and one hour of assembly per unit.
- Product Y requires two hours of machining and two hours of assembly per unit.

These constraints are expressed mathematically as follows:

1. $3x + 2y \leq 120$ (hours machining time)
2. $x + 2y \leq 80$ (hours assembly time)

Since there are only two products, these limitations can be shown on a two-dimensional graph (Figure 4-2). Since all relationships are linear, the solution to our problem will fall at one of the corners. To find the solution, begin at some feasible solution (satisfying the

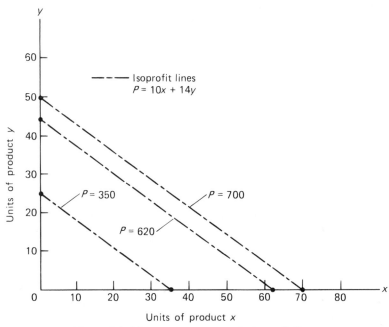

Figure 4-1. Linear program example: isoprofit lines.

given constraints) such as $(x,y) = (0,0)$, and proceed in the direction of "steepest ascent" of the profit function (in this case, by increasing production of Y at \$14 profit per unit) until some constraint is reached. Since assembly hours are limited to 80, no more than $^{80}/_{2}$, or 40, units of Y can be made, earning $40 \times \$14$, or \$560 profit. Then proceed along the steepest allowable ascent from there (along the assembly constraint line) until another constraint (machining hours) is reached. At that point $(x,y) = (20,30)$ and profit $P = (20 \times \$10) + (30 \times \$14)$, or \$620. Since there is no remaining edge along which profit increases, this is the optimum solution. The \$620 isoprofit line from Figure 4-1 has been repeated in Figure 4-2 to illustrate that the maximum profit solution has been reached. In this particular case the optimum solution is at the point the constraints intersect (where we take all constraints as equalities), but this is not always the case. For example, if the unit profit for Y increased from \$14 to \$21, you would maximize profit $(P = 10x + 21y)$ by making 40 units of Y and none of X.

Computer solution. About fifty years ago George Danzig of Stanford University developed the *simplex method,* which expresses the foregoing technique in a mathematical algorithm that permits computer solution of linear programming problems with many variables (dimensions), not just the two (assembly and machining) of this example. Even so, practical problems exist that are too large for even the fastest mainframe computers. For example, an AT&T model of current and future telephone demand among the 20-odd countries rimming the Pacific Ocean has 42,000 variables, and it takes 4 to 7 hours for a single run; one model of the domestic long-distance network has 800,000 variables and would take

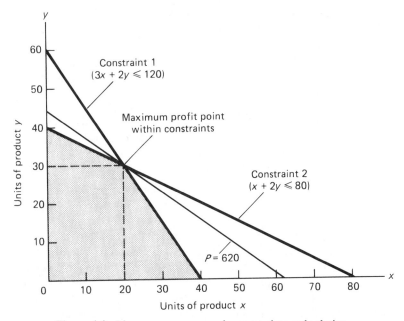

Figure 4-2. Linear program example: constraints and solution.

weeks for one solution. Fortunately, Narendra Karmarkar at AT&T's Bell Laboratories has developed[12] a shortcut method based on "an arcane discipline known as projective geometry" that is 50 to 100 times as fast, finding a solution to the larger problem above in less than an hour. The new method "showed AT&T how to squeeze an extra 9% to 10% of capacity out of its $15 billion system"—an impressive justification for developing this management science methodology!

Another classic linear programming application is the oil refinery problem, where profit is maximized over a set of available crude oils, process equipment limitations, products with different unit profits, and other constraints. Other applications include assignment of employees with differing aptitudes to the jobs that need to be done to maximize the overall use of skills; selecting the quantities of items to be shipped from a number of warehouses to a variety of customers while minimizing transportation cost; and many more. In each case there is one best answer, and the challenge is to express the problem properly so that it fits a known method of solution.

Decision Making Under Risk

Nature of risk. In decision making under *risk* one assumes that there exist a number of possible future states of nature N_j, as we saw in Table 4-1. Each N_j has a known (or assumed) probability p_j of occurring, and there may not be one future state that results in the best outcome for all alternatives A_i. Examples of future states and their probabilities are:

- Alternative weather (N_1 = rain; N_2 = good weather) will affect the profitability of alternative construction schedules; here, the probabilities p_1 of rain and p_2 of good weather can be estimated from historical data.

- Alternative economic futures (boom or bust) determine the relative profitability of conservative versus high-risk investment strategy; here, the assumed probabilities of different economic futures might be based on the judgment of a panel of economists.

Expected value. Given the future states of nature and their probabilities, the solution in decision making under risk is the alternative A_i that provides the highest *expected value E_i*, which is defined as the sum of the products of each outcome O_{ij} times the probability p_j that the associated state of nature N_j occurs:

$$E_i = \sum_{j=1}^{n} (p_j O_{ij})$$

Simple example. For example, consider the simple payoff table of Table 4-2, with only two alternative decisions and two possible states of nature. Alternative A_1 has a constant cost of $200, and A_2 a cost of $100,000 if future N_2 takes place (and none otherwise). At first glance alternative A_1 looks like the clear winner, but consider the situation when the probability (p_1) of the first state of nature is 0.999 and the probability (p_2) of the second state is only 0.001. The expected value of choosing alternative A_2 is only

$$E(A_2) = 0.999(\$0) - 0.001(\$100,000) = \$-100$$

Note that this outcome of $-100 is not possible: the outcome if alternative A_2 is chosen will be a loss of either $0 or $100,000, not $100. However, if you have many decisions of this type over time and you choose alternatives that maximize expected value each time, you should achieve the best overall result. Since we should prefer expected value E_2 of $-100 to E_1 of $-200, we should choose A_2, other things being equal.

But first, let us use these figures in a specific application. Assume that you own a $100,000 house and are offered fire insurance on it for $200 a year. This is twice the "expected value" of your fire loss (as it has to be to pay insurance company overhead and agent costs). However, if you are like most people, you will probably buy the insurance because, quite reasonably, your attitude toward risk is such that you are not willing to accept loss of your house! The insurance company has a different perspective, since they have many houses to insure and can profit from maximizing expected value in the long run, as long as they don't insure too many properties in the path of the same hurricane or earthquake.

Another example. Consider that you own rights to a plot of land under which there may or may not be oil. You are considering three alternatives: doing nothing ("don't drill"),

TABLE 4–2   Example of Decision Making Under Risk

	N_1 $p_1 = 0.999$	N_2 $p_2 = 0.001$
A_1	$ –200	$ –200
A_2	0	–100,000

TABLE 4–3 Well Drilling Example—Decision Making Under Risk

Alternative	State of Nature/Probability			Expected Value
	N_1: Dry Hole $p_1 = 0.6$	N_2: Small Well $p_2 = 0.3$	N_3: Big Well $p_3 = 0.1$	
A_1: Don't drill	$ 0	$ 0	$ 0	$ 0
A_2: Drill alone	−500,000	300,000	9,300,000	720,000
A_3: Farm out	0	125,000	1,250,000	162,500

drilling at your own expense of $500,000, and "farming out" the opportunity to someone who will drill the well and give you part of the profit if the well is successful. You see three possible states of nature: a dry hole, a mildly interesting small well, and a very profitable gusher. You estimate the probabilities of the three states of nature p_j and the nine outcomes O_{ij} as shown in Table 4-3.

The first thing you can do is eliminate alternative A_1, since alternative A_3 is at least as attractive for all states of nature and is more attractive for at least one state of nature. A_3 is therefore said to *dominate* A_1.

Next, you can calculate the expected values for the surviving alternatives A_2 and A_3:

$$E_2 = 0.6(-500,000) + 0.3(300,000) + 0.1(9,300,000)$$
$$= \$720,000$$
$$E_3 = 0.6(0) + 0.3(125,000) + 0.1(1,250,000)$$
$$= \$162,500$$

and you choose alternative A_2 if (and only if) you are willing and able to risk losing $500,000.

Decision trees provide another technique used in finding expected value. They begin with a single *decision node* (normally represented by a square or rectangle), from which a number of decision alternatives radiate. Each alternative ends in a *chance node,* normally represented by a circle. From each chance node radiate several possible futures, each with a probability of occurring and an outcome value. The expected value for each alternative is the sum of the products of outcomes and related probabilities, just as calculated previously. Figure 4-3 illustrates use of a decision tree in our simple insurance example.

The conclusion reached is identical mathematically to that obtained from Table 4-2. Decision trees provide a very visible solution procedure, especially when a sequence of de-

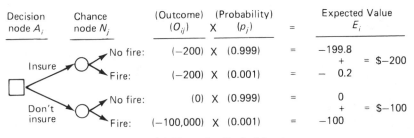

Figure 4-3 Example of a decision tree.

TABLE 4–4   Typical Waiting-Line Situations

Organization	Activity	Arrivals	Servers
Airport	Landing	Airplanes	Runway
College	Registration	Students	Registrars
Court system	Trials	Cases	Judges
Hospital	Medical service	Patients	Rooms/doctors
Personnel office	Job interviews	Applicants	Interviewers
Supermarket	Checkout	Customers	Checkout clerks
Toll bridge	Taking tolls	Vehicles	Toll takers
Toolroom	Tool issue	Machinists	Toolroom clerks

cisions, chance nodes, new decisions, and new chance nodes exist. For example, if you are deciding whether to expand production capacity in December 1996, a decision a year later, in December 1997, as to what to do then will depend both on the first decision and on the sales enjoyed as an outcome during 1997. The possible December 1997 decisions lead to (a larger number of) chance nodes for 1998. The technique used starts with the later year, 1998 (the farthest branches). Examining the outcomes of all the possible 1998 chance nodes, you find the optimum second decision and its expected value, in 1997, for each 1997 chance node, that is, for each possible combination of first decision in December 1996 and resulting outcome for 1997. Then you use those values as part of the calculation of expected values for each first-level decision alternative in December 1996.

Queuing (waiting-line) theory.   Most organizations have situations where a class of people or objects arrive at a facility of some type for service. The times between arrivals (and often the time required for serving each arrival) are not constant, but they can usually be approximated by a probability distribution. The first work in this field was by the Danish engineer A. K. Erlang, who studied the effect of fluctuating demand for telephone calls on the need for automatic dialing equipment. Table 4-4 lists some other common examples of waiting lines.

The essence of the typical queuing problem is identifying the optimum number of servers needed to reduce overall cost. In the toolroom problem, machinists appear at random times at the window of an enclosed toolroom to sign out expensive tools as they are needed for a job, and attendants find the tools, sign them out, and later receive them back. The production facility is paying for the time of both toolroom attendants and the (normally more expensive) machinists, and therefore it wishes to provide the number of servers that will minimize overall cost. In most of the other cases in the table, the serving facility is not paying directly for the time lost in queues, but it wishes to avoid disgruntled customers or clients who might choose to go elsewhere for service. Mathematical expressions for mean queue length and delay as a function of mean arrival and service rates have been developed for a number of probability distributions (in particular exponential and Poisson) of arrival and of service times.

Simulation.   There are many situations where the real-world system being studied is too complex to express in simple equations that can be solved in a reasonable time. In many oth-

TABLE 4–5 Data for Risk as Variance Example

Project X		Project Y	
Probability	Cash Flow	Probability	Cash Flow
0.10	$3000	0.10	$2000
0.20	3500	0.25	3000
0.40	4000	0.30	4000
0.20	4500	0.25	5000
0.10	5000	0.10	6000

ers, the variability involved cannot be expressed in simple probability distributions. A common approach in such cases is to construct a mathematical model that simulates the operation of the real system by mathematically describing the behavior of individual parts and the interactions between the parts. Stochastic (variable) activities in the model (such as time between arrivals, expected revenue, or tensile strength) can be represented by probability distributions or frequency tables.

A computer program can then be developed that will cause the model to perform one cycle of operation. Wherever a probabilistic value is called for in the calculation, the program will call for a random number and use it to identify the value to insert in the calculation. (For example, if daily sales of an item were two units 10% of the time, one 30% of the time, and none 60% of the time, the program could call for a single random digit and insert the sales quantity "2" in the calculation if the random digit were a 0, "1" if the digit were 1, 2, or 3, and "0" if it were 4 through 9.) Although programs for this purpose can be written in common languages such as FORTRAN or BASIC, special-purpose simulation languages such as GPSS, SIMSCRIPT, or SLAMII are much more efficient and powerful for this purpose.

The outcome of a single run of a program with many probabilistic values is not significant in itself, but with today's fast computers the problem can be economically rerun 100 or 1000 times to develop a probability distribution of the final outcome. Conditions simulated in the model can then be changed and the modified model exercised again until a satisfactory result is obtained. The policy expressed in the most successful version of the model can then be tested in the real world; its success there will depend largely on how well the critical factors in the real world have been captured in the model.

Risk as variance. Another common meaning of risk is variability of outcome, measured by the variance or (more often) its square root, the standard deviation. Consider two investment projects X and Y,[13] having the discrete probability distribution of expected cash flows in each of the next several years shown in Table 4-5.

Expected cash flows are calculated the same as expected value:

$$E(X) = 0.10(3000) + 0.20(3500) + 0.40(4000) + 0.20(4500) + 0.10(5000)$$
$$= \$4000$$
$$E(Y) = 0.10(2000) + 0.25(3000) + 0.30(4000) + 0.25(5000) + 0.10(6000)$$
$$= \$4000$$

Although both projects have the same mean (expected) cash flows, the expected values of the variances (squares of the deviations from the mean) differ as follows (see also Figure 4-4):

$$V_X = 0.10(3000 - 4000)^2 + 0.20(3500 - 4000)^2 + \cdots + 0.10(5000 - 4000)^2$$
$$= 300,000$$
$$V_Y = 0.10(2000 - 4000)^2 + 0.25(3000 - 4000)^2 + \cdots + 0.10(6000 - 4000)^2$$
$$= 1,300,000$$

The standard deviations are the square roots of these values:

$$\sigma_X = \$548, \; \sigma_Y = \$1140$$

Since project Y has the greater variability (whether measured in variance or in standard deviation) it must be considered to offer greater *risk* than does project X.

Decision Making Under Uncertainty

Uncertainty occurs when there exist several (i.e., more than one) future states of nature N_j, but the probabilities p_j of each of these states occurring are not known. A different kind of logic is used here, based on attitudes toward risk.

Methods used. The optimistic decision maker may choose the alternative that offers the highest possible outcome (the "maximax" solution); the pessimist may choose the alternative whose worst outcome is "least bad" (the "maximin" solution); a third decision maker may choose a position somewhere between optimism and pessimism; a fourth may simply assume that all states of nature are equally likely (the so-called "principle of insufficient reason"), set all p_j values equal to $1.0/n$, and maximize expected value based on that assumption; a fifth decision maker may choose the alternative that has the smallest difference between the best and worst outcomes (the "minimax regret" solution).

For example, consider the well-drilling problem (Table 4-3) if the probabilities p_j for the three future states of nature N_j cannot be estimated. In Table 4-6 the "Maximum" column lists the best possible outcome for alternatives A_2 and A_3; the optimist will seek to "maximax" by choosing A_2 as the best outcome in that column. The pessimist will look at the

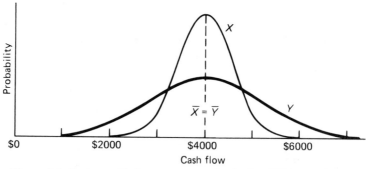

Figure 4-4. Projects with the same expected value but different variances.

TABLE 4–6 Decision Making Under Uncertainty Example

Alternative	Maximum	Minimum	Hurwicz ($\alpha = 0.2$)	Equally Likely
A_2	$9,300,000*	$–500,000	$1,460,000*	$3,033,333*
A_3	1,250,000	0*	250,000	458,333

*Preferred solution.

"Minimum" column, which lists the worst possible outcome for each alternative, and he or she will "maximin" by choosing A_3 as having the best [algebraic] worst case. (In this example both maxima came from future state N_3 and both minima from future state N_1, but this sort of coincidence does not usually occur.)

A decision maker who is neither a total optimist nor a total pessimist may be asked to express a "coefficient of optimism" as a fractional value α between 0 and 1 and then to

maximize [α(best outcome) + (1 – α)(worst outcome)]

The outcome using this "Hurwicz" approach and a coefficient of optimism of 0.2 is shown in the third column of Table 4-6; A_2 is again the winner.

If decision makers believe that the future states are "equally likely," they will seek the higher expected value and choose A_2 on that basis:

$$E_2 = \frac{-500,000 + 300,000 + 9,300,000}{3} = \$3,033,333$$

$$E_3 = \frac{0 + 125,000 + 1,250,000}{3} = \$458,333$$

(If, on the other hand, they believe that some futures are more likely than others, they should be invited to express their best estimates as p_j values and solve the problem as a decision under risk!)

The final approach to decision making under uncertainty involves creating a second matrix, not of outcomes, but of regret. Regret is quantified to show how much better the outcome might have been if you had known what the future was going to be. If there is a "small well" under your land and you did not drill for it, you would regret the $300,000 you might have earned. On the other hand, if you farmed out the drilling your regret would be only $175,000 ($300,000 less the $125,000 profit sharing you received). Table 4-7 provides this regret matrix and lists in the right-hand column the maximum regret possible for each alternative. The decision maker who wishes to minimize the maximum regret (minimax regret) will therefore choose A_2.

TABLE 4–7 Well Drilling Example—Regret Matrix

Alternative	State of Nature			Maximum Regret
	N_1: Dry Hole	N_2: Small Well	N_3: Big Well	
A_1: Don't drill	$ 0	$300,000	$9,300,000	$9,300,000
A_2: Drill alone	500,000	0	0	500,000
A_3: Farm out	0	175,000	8,050,000	8,050,000

Game theory. A related approach is *game theory,* where the future states of nature and their probabilities are replaced by the decisions of a competitor. Begley and Grant explain:

> In essence, game theory provides the model of a contest. The contest can be a war or an election, an auction or a children's game, as long as it requires strategy, bargaining, threat and reward.[14]

They cited as an example the conflict between the [former] Soviet Union and the United States (Table 4-8). The best overall solution (second best for each party) is disarmament of both parties, but this requires trust (unlikely under 1982 conditions) or foolproof verification. As a result, both parties settled in the past for the arm–arm option, only the third best for each party, resulting in the arms race and the absurd proliferation of nuclear weapons.

In other situations, game theory leads to selecting a mixture of two or more strategies, alternated randomly with some specified probability. Again, Begley and Grant provide a simple example:

> In the children's game called Odds and Evens, for instance, two players flash one or two fingers. If the total is 2 or 4, Even wins; if [it is] 3, Odd wins. A little analysis shows that the winning ploy is to randomly mix up the number of fingers flashed. For no matter what Odd does, Even can expect to come out the winner about half the time and vice versa. If Even attempts anything trickier, such as alternating 1s and 2s, he can be beaten if Odd catches on to the strategy and alternates 2s and 1s.[15]

COMPUTER-BASED INFORMATION SYSTEMS

Integrated Data Bases

Until recent years, each part of an organization maintained separate files and developed separate information forms for its specific purposes, often requiring the same information to be entered again and again. Not only is this expensive, but when the same information is recorded separately in several places it becomes difficult to keep current and reliable. The computer revolution has made it possible to enter information only once in a shared data base—where it can be updated in a single act, yet still be available for all to use.

The American Society for Engineering Management, for example, began in 1980 with a mailing list keypunched on computer cards (later, directly entered into a mainframe

TABLE 4–8 Political Application of Game Theory

U.S. Strategy	Soviet Strategy	
	Arm	Disarm
Arm	Third-best for U.S. Third-best for U.S.S.R.	Best for U.S. Worst for U.S.S.R.
Disarm	Worst for U.S. Best for U.S.S.R.	Second-best for U.S. Second-best for U.S.S.R.

Source: Sharon Begley with David Grant, "Games Scholars Play," *Newsweek,* September 6, 1982, p. 72.

computer memory), but with all other information copied as required into handwritten or typed files. In late 1987 it switched to a PC/XT personal computer with a central data base containing about 35 items on each member (name, title, number, home and office addresses and telephones, engineering degrees and registration, offices held, dues status, and others). Simple commands cause this desktop wonder to spew forth mailing labels, members in a particular local section or joining in a particular year, tabulations of registered professional engineers with the master's level as their highest degree, or almost any permutation of the data entered into the base.

Hospitals once prepared separate forms repeating patients' names, addresses, illness, doctor, and other common items many different times. Today most of this is entered once at admission into a central computer base, supplemented with notes from the nursing floor, laboratory, and other locations, and these data are processed in medical records and used in billing without repeating data entry.

The CAD/CAM revolution in design and manufacture provides a much more sophisticated example. Designs are now created on the computer, and this same record is used by others to analyze strength, heat transfer, and other design conditions; then it is transformed into instructions to manufacture the item on numerically controlled machines, and to test the item for conformance to design. A small class of graduate students invited to provide additional examples of the use of a common data base (discussion question 4-13) cited:

- A pharmaceutical company with a data base on each batch of product, including raw material lots used, production date, equipment and personnel, test results, and shipping destination used for quality analysis, financial analysis, and (if needed) product recall.
- The University of Missouri–Rolla (like most large schools) has a standard student data base used (and contributed to) by the registrar, financial aid office, cashier, placement office, academic departments, and finally, the alumni office.
- Union Pacific uses a common data base to keep track of load location (for tracing), trip distance (for billing), and car location (for maintenance).
- The laser scanners at supermarket checkout stations increase checking efficiency, eliminate individual price tags, speed price changes, update inventories, and are used to evaluate personnel and stocking and display decisions.
- Wal-Mart stores order merchandise from warehouses by "wanding" the bar code of a desired item and entering quantity into a terminal. At the warehouse this action is used to automatically update the data base, call for repurchase, and print an order request with bar-coded shipping labels. Once these carton labels are applied, computerized conveyor systems automatically read the store number and start each carton on its way to the proper shipping door for deliveries to that store.

Management Information/Decision Support Systems

Traditionally, top managers have relied primarily on oral and visual sources of information: scheduled committee meetings, telephone calls, business luncheons, and strolls through the workplace, supplemented by the often condensed and delayed information in written reports and periodicals. Quite recently, the existence of computer networks, centralized data bases,

TABLE 4–9   Effect of Management Level on Decisions

Management Level	Number of Decisions	Cost of Making Poor Decisions	Information Needs
Top	Least	Highest	Strategic
Middle	Intermediate	Intermediate	Implementation
First-line	Most	Lowest	Operational

and user-friendly software has provided a new source of prompt, accurate data to the manager. A recent survey showed that 93% of senior executives used a personal computer, 60% of them for planning and decision support.

Contemporary authors distinguish two classes of application of computer-based management systems:

> *Management Information Systems* (MIS) focus on generating better solutions for structured problems, as well as improving efficiency in dealing with structured tasks. On the other hand a *Decision Support System* (DSS) is interactive and provides the user with easy access to decision models and data in order to support semistructured and unstructured decision-making tasks. It improves effectiveness in making decisions where a manager's judgment is still essential.[16]

As one rises from front-line supervisor through middle management to top management, the nature of decisions and the information needed to make them changes (see Table 4-9). The higher the management level the fewer decisions may be in number, but the greater is the cost of error. A carefully constructed master data base should be capable of providing the detailed current data needed for operational decisions as well as the longer-range strategic data for top management decisions.

Expert Systems

As part of the fast-developing field of *artificial intelligence* (AI), a new type of computer model is being developed with the purpose of making available to average or neophyte practitioners in many fields the skill and know-how of experts in the field. These *expert systems* are created by reviewing step by step with the experts the reasoning methods they use in a particular application, and reducing these to an *inference engine* that, combined with a *knowledge base* of facts and rules and a *user interface,* may be consulted by someone newer to the field who wants guidance. Special computer languages have been developed for this purpose, including M1 and LISP (which is often employed on a special computer called the "LISP machine"). Expert systems (and the entire field of AI) are rapidly being developed, and their overall impact on engineering management can only be speculated upon at this point.

IMPLEMENTATION

Decisions, no matter how well conceived, are of little value until they are put to use—that is, until they are implemented. Koestenbaum puts it well:

Leadership is to know that decisions are merely the start, not the end. Next comes the higher-level decision to sustain and to implement the original decision, and that requires courage.

Courage is the willingness to submerge oneself in the loneliness, the anxiety, and the guilt of a decisionmaker. Courage is the decision, and a decision it is, to have faith in the crisis of the soul that comes with every significant decision. The faith is that on the other end one finds in oneself character and the exhilaration of having become a strong, centered, and grounded human being.[17]

DISCUSSION QUESTIONS

4-1. Give some examples of each of the three "occasions for decision" cited by Chester Barnard. Explain in your own words why Barnard thought the third category was most important.

4-2. (**a**) Explain the difference between "optimizing" and "satisficing" in making decisions, and (**b**) distinguish between routine and nonroutine decisions.

4-3. Use a concrete example showing the five-step process by which management science uses a simulation model to solve real-world problems.

4-4. You operate a small wooden toy company making two products: alphabet blocks and wooden trucks. Your profit is $30 per box of blocks and $40 per box of trucks. Producing a box of blocks requires one hour of woodworking and two hours of painting; producing a box of trucks takes three hours of woodworking but only one hour of painting. You employ three woodworkers and two painters, each working 40 hours a week. How many boxes of blocks (B) and trucks (T) should you make each week to maximize profit? Solve graphically as a linear program and confirm analytically.

4-5. A commercial orchard grows, picks, and packs apples and pears. A peck (quarter bushel) of apples takes 4 minutes to pick and 5 minutes to pack; a peck of pears takes 5 minutes to pick and 4 minutes to pack. Only one picker and one packer are available. How many pecks each of apples and pears should be picked and packed every hour (60 minutes) if the profit is $3/peck for apples and $2/peck for pears? Solve graphically as a linear program and confirm analytically.

4-6. Solve the drilling problem (Table 4-3) using a decision tree.

4-7. You must decide whether to buy new machinery to produce product X or to modify existing machinery. You believe the probability of a prosperous economy next year is 0.6 and of a recession is 0.4. Prepare a decision tree and use it to recommend the best course of action. The applicable payoff table of profits (+) and losses (-) is

	N_1 (prosperity)	N_2 (recession)
A_1 (buy new)	$+950,000	$–200,000
A_2 (modify)	+700,000	+300,000

4-8. If you have no idea of the economic probabilities p_j in question 4-7, what would be your decision based on uncertainty using (**a**) maximax, (**b**) maximin, (**c**) equally likely, and (**d**) minimax regret assumptions?

4-9. You are considering three investment alternatives for some spare cash: Old Reliable Corporation stock (A_1), Fly-By-Nite Air Cargo Company stock (A_2), and a federally insured savings certificate (A_3). You expect the economy will either "boom" (N_1) or "bust" (N_2), and you estimate that a boom is more likely ($p_1 = 0.6$) than a bust ($p_2 = 0.4$). Outcomes for the three alternatives are expected to be (1) $2000 in boom or $500 in bust for ORC; (2) $6000 in boom but $-5000 (loss) in bust for FBN; and (3) $1200 for the certificate in either case. Set up a payoff table (decision matrix) for this problem and show which alternative maximizes expected value.

4-10. If you have no idea of the economic probabilities p_j in question 4-9, what would be your decision based on uncertainty using (**a**) maximax, (**b**) maximin, (**c**) equally likely, and (**d**) minimax regret assumptions?

4-11. Your company has proposed to produce a component for an automobile plant, but it will not have a decision from that plant for six months. You estimate the possible future states and their probabilities as: receive full contract (N_1, with probability $p_1 = 0.3$); receive partial contract (N_2, $p_2 = 0.2$); and lose award (no contract) (N_3, $p_3 = 0.5$). Any tooling you use on the contract must be ordered now. If your alternatives and their outcomes (in thousands of dollars) are as shown in the following table, what should be your decision?

	N_1	N_2	N_3
A_1 (full tooling)	+800	+400	−400
A_2 (minimum tooling)	+500	+150	−100
A_3 (no tooling)	−400	−100	0

4-12. From another reference provide the problem statement and the solution for a typical queuing (waiting-line) problem.

4-13. Describe an example from an organization you know or have read about where a common data base is used for a number of different purposes.

NOTES

1. John M. Amos and Bernard R. Sarchet, *Management for Engineers* (Englewood Cliffs, NJ: Prentice-Hall, Inc., 1981), p. 51.
2. Chester I. Barnard, *The Functions of the Executive* (Cambridge, MA: Harvard University Press, 1938), p. 190.
3. Barnard, *Functions,* p. 191.
4. Charles D. Pringle, Daniel F. Jennings, and Justin G. Longnecker, *Managing Organizations: Functions and Behaviors* (Columbus, OH: Merrill Publishing Company, 1988), p. 131.
5. Pringle et al., *Managing Organizations,* p. 131.
6. Herbert A. Simon, *Administrative Behavior: A Study of Decision-Making Processes in Administrative Organization,* 3rd ed. (New York: Macmillan Publishing Company, 1976), p. 80.
7. Simon, *Administrative Behavior,* p. 81.
8. Philip E. Hicks, *Introduction to Industrial Engineering and Management Science* (New York: McGraw-Hill Book Company, 1977), p. 42.
9. Pringle et al., *Managing Organizations,* p. 154.
10. Jay W. Forrester, *Urban Dynamics* (Cambridge, MA: The MIT Press, 1969).
11. Daniel L. Babcock, "Analysis and Improvement of a Dynamic Urban Model," unpublished Ph.D. dissertation, University of California, Los Angeles, 1970.
12. "The Startling Discovery Bell Labs Kept in the Shadows," *Business Week,* September 21, 1987, pp. 69, 72, 76.

13. Example taken from supplemental class notes used by Professor Edmund Young in teaching from the manuscript of this text, September 1988.

14. Sharon Begley with David Grant, "Games Scholars Play," *Newsweek,* September 6, 1982, p. 72.

15. Begley and Grant, "Games Scholars Play," p. 72.

16. Gus W. Grammas, Greg Lewin, and Suzanne P. DuMont Bays, "Decision Support, Feedback, and Control," in John E. Ullmann, ed., *Handbook of Engineering Management* (New York: John Wiley & Sons, Inc., 1986), Chapter 11.

17. Peter Koestenbaum, *The Heart of Business: Ethics, Power, and Philosophy* (Dallas, TX: Saybrook, 1987), p. 352.

CHAPTER 5

Organizing

PREVIEW

We begin this chapter by distinguishing between the legal forms of organization: proprietorship, partnership, and corporation. Next we discuss the organizing process, and compare the various logics of subdivision, or *departmentation*. We discuss factors determining effective span of control, and the nature of line, staff, and service relationships. We describe the effect of technology on organization structure, and finally, introduce a few of the more modern organizational forms.

NATURE OF ORGANIZING

Legal Forms of Organization

As background to the study of organization, let us first compare the legal forms in which a business can be organized: the sole proprietorship, the partnership, the corporation, and the cooperative. The *sole proprietorship* is a business owned and operated by one person. It is simple to organize and to shut down, has few legal restrictions, and the owner is free to make all decisions. Profit from it is taxed only once—on the Schedule C (Profit or Loss from Business or Profession) attachment to the owner's individual income tax form. However, the owner faces unlimited liability for the debts of the business, he or she may find it difficult to raise capital to fund growth of the business, and the duration of the business is limited to the life of the proprietor.

The *partnership* is an "association of two or more partners to carry on as co-owners of a business for profit" (Uniform Partnership Act). The partnership is almost as easy to organize as a proprietorship and has relatively few legal restrictions. Partnerships permit pooling the managerial skills and judgments and the financial strengths of several people who have a direct financial interest in the enterprise, but suffer the disadvantages of divided decision-making authority and potential damage to the business when partners disagree. Although a partnership files a tax return to allocate partnership profit (or loss), it does not pay taxes—the partners do on their individual tax forms, whether they actually receive the profit or leave it in the enterprise to grow further. Normally, partners have unlimited liability for partnership debts. In a *limited partnership* there must be at least one *general partner* with unlimited liability, but the rest may be *limited partners,* who are financially liable only to the extent of their investment in the venture.

Corporations are legal entities owned by shareholders, who in general have no liability beyond loss of the value of their stock. Corporations have perpetual life (as long as they submit an annual report to the state in which they are chartered), and they find it easier to raise money, transfer ownership, and change management. It is more difficult and expensive to organize a corporation, but the main disadvantage is that corporate income is taxed twice: once as corporation income tax the year the profit is made, and again as personal income tax when the after-tax profit is distributed as dividends. Also, corporations are subject to many state and federal controls not affecting other forms of business. (Under certain conditions, corporations with no more than 35 shareholders, all U.S. residents, may elect to be treated as "Subchapter S" corporations and avoid double taxation.)

Cooperatives are a special type of organization owned by users or customers, to whom earnings are usually distributed tax free in proportion to patronage. For example, about 1000 rural electric cooperatives distribute electricity over much of America's nonmetropolitan land area; each customer of this service buys a share initially for a few dollars, and he or she can cast one vote to elect the board members who manage the cooperative.

While sole proprietorships are the most common form of business organization in sheer numbers, most large organizations are corporations.

Organizing Defined

Weihrich and Koontz[1] believe that people "will work together most effectively if they know the parts they are to play in any team operation and how their roles relate to one another. . . . Designing and maintaining these systems of roles is basically the managerial function of organizing." They continue:

> For an *organizational role* to exist and be meaningful to people, it must incorporate (1) verifiable objectives, which . . . are a major part of planning; (2) a clear idea of the major duties or activities involved; and (3) an understood area of discretion or authority, so that the person filling the role knows what he or she can do to accomplish goals. In addition, to make a role work

out effectively, provision should be made for supplying needed information and other tools necessary for performance in that role.

It is in this sense that we think of *organizing* as (1) the identification and classification of required activities, (2) the grouping of activities necessary to attain objectives, (3) the assignment of each grouping to a manager with the authority (delegation) necessary to supervise it, and (4) the provision for coordination horizontally (on the same or similar organizational level) and vertically (for example, corporate headquarters, division, and department) in the organization structure.[2]

Organizing by Key Activities

Effective organizing must first consider the basic mission and long-range objectives established for the organization and the strategy conceived to accomplish them. Peter Drucker[3] recommends first identifying the *key activities,* which he terms the "load-bearing parts of the structure." He poses three questions to help identify the key activities:

1. In what area is excellence required to obtain the company's objectives?
2. In what areas would lack of performance endanger the results, if not the survival, of the enterprise?
3. What are the *values* that are truly important to us in this company?

Once the key activities have been established, Drucker suggests "two additional pieces of work: an analysis of decisions and an analysis of relations."[4] In *decision analysis* one must first identify what decisions are needed to attain effectiveness in key activities. Then the nature of these decisions is established in terms of their *futurity* (the period in the future to which they commit the company), the *impact* they have on other functions, their *frequency* (recurrent decisions can be made at lower levels once policies for them have been thought through), and the extent to which they involve ethical, social, and political considerations. *Relations analysis,* on the other hand, asks with whom the person in charge of an activity will have to work, and it seeks to assure "that the crucial relations, that is, the relationship on which depend its success and the effectiveness of its contribution, should be easy, accessible, and central to the unit."

In the 1990s, more and more organizations are being restructured into teams that include the specialists needed to carry through a project or solve a problem, and that are delegated the authority (*empowered*) to make the necessary decisions. In the modern concept of *concurrent engineering* (discussed in Chapter 10), teams of design engineers, marketing people, and production specialists work together to launch new products earlier that better meet customer needs.

TRADITIONAL ORGANIZATION THEORY

Patterns of Departmentation

Organizations are divided into smaller units (a process called departmentation or departmentalization) using a number of different logics. Figure 5-1, which illustrates some of the more common methods, will be used to help us "grow" a company in the container business.

Let us assume that you have a large collection of compact disks (CDs) and that you also enjoy woodworking. You begin to make some attractive wooden cabinets for your CDs in your basement or garage. They are admired by your friends and neighbors, who buy some, and then you find several local stores who want to carry them. You now are an *entrepreneur* and have a business. As demand increases, you need some help in the shop, and you hire several local people (Tom, Dick, and Mary) who will, naturally, take direction from you (Figure 5-1a).

As you grow, you find yourself away from the plant (now moved to a local industrial park) for extended periods selling your product and arranging financing. You appoint the most experienced worker as foreman, and later as production manager. You hire salespeople to help sell your product and, as they increase in number, appoint one as sales manager. A local CPA agrees to work half time as your finance manager, and an engineering student moonlights as your designer. You have now established a pattern of *functional departmentation,* which is the first logic of subdivision for most new organizations, and which is present at some level in almost any organization. Functional subdivision need not be confined to a single level (as in Figure 5-1b). Marketing is often divided into sales, advertising, and market research. Production may be broken into component production, assembly, and finishing.

As your business grows, you may also become interested in producing clear plastic storage boxes for computer diskettes. You soon discover that production methods for plastic boxes are very different from those for wooden cabinets, and you organize separate production shops under separate supervisors to produce the two kinds. Next, you discover that your diskette boxes appeal to a different market, and you need a different group of salespeople. Then you find that the sales force dealing with diskette boxes needs much closer contact with your plastic box production foreman than they do with salespeople selling CD cabinets to the consumer, but that the chain of command through the general sales manager, then you, and finally the overall production manager makes decision making slow and difficult. You may now be ready to reorganize by *product* as in Figure 5-2a.

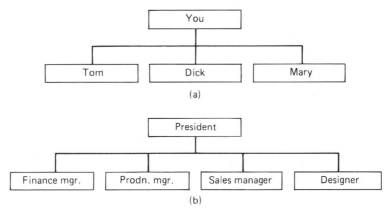

Figure 5-1. Methods of departmentation: (a) primitive organization; (b) functional departmentation.

The separate CD Cabinet and Diskette Box Divisions will begin with their own manufacturing and marketing functions, and later you may add accounting and personnel functions to each division. Obtaining bank loans, selling stock, and other financial activities are best handled centrally, you will need consistent personnel policies in both divisions, and you need some top-level advice on new markets and new technical advances; thus, you will need to organize a *staff* at the corporate level in addition to your product divisions.

In the days of the pony express, long-distance communication was slow and unreliable. If an enterprise on the (American) east coast wished to set up a west coast division, they would have to give the regional manager broad authority, perhaps even creating *geographic* or *territorial* divisions as in Figure 5-2b. Today, however, managers communicate by telephone and fax machine across the world almost as easily as with the next building, and they can jet anywhere in the continent in a day for more protracted personal meetings. As a result, communication per se is no longer the most important logic for top-level organization. However, you may find that accommodating regional differences is the key to ef-

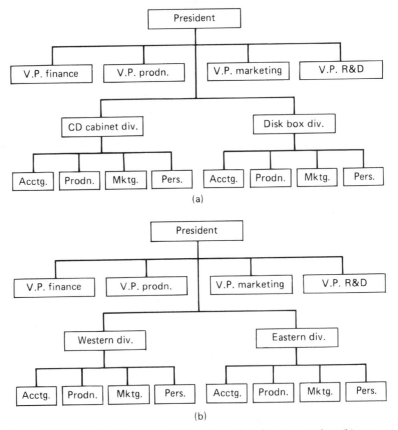

Figure 5-2. Methods of departmentation: (a) product departmentation; (b) geographic departmentation. Acctg., accounting; Prodn., production; Mktg., marketing; Pers., personnel.

fective management, and you may then create regional (geographic) divisions. A company that builds housing developments may find, for example, that regional differences in housing styles, construction codes, marketing media, and methods of mortgage financing are very important, and it may set up separate geographic divisions each responsible for construction and marketing in their own region. Geographic subdivision is more common at lower levels; sales forces are commonly divided by region, for example, for more efficient and more personal customer contact.

Sometimes the *type of customer* is a more important consideration than location, and departmentation by customer is indicated. For example, creating weapons systems for the U.S. Department of Defense often requires state of the art technology and a special understanding of military procurement and product use. On the other hand, products that will be used by other industries in producing their own goods need to be cost-effective and rugged; consumer goods need to be attractive in appearance and price. Many firms successful in producing military goods have tried marketing to the consumer with poor success, because it requires a different mental outlook and frequently a separate organization devoted to that market. Even when the product is the same or similar (washing machines, for example), separate sales forces may be desirable for regional sales to consumers through local distributors and for large-volume national sales to major store chains or to the federal government for military housing use.

Another departmentation logic is by *process* or *equipment* (especially where equipment is so expensive, immobile, or critical that it must be operated centrally). Computing used to be done centrally when it was confined to a single (very expensive) mainframe computer in each company; today, the economics of personal computers has spread usage to many departments (and often every desk), with a staff computer group providing support and service. Where manufacturing or service is carried on around the clock, operating personnel may be grouped by *shift* or *time.* Subdivision by sheer *numbers,* as in the biblical example below, is indicated only when a large number of people must perform very similar and routine tasks, and this is becoming increasingly less common.

As one might expect, enterprises may combine several or all of these methods in designing their organization. In Figure 5-3 we see functional subdivision at the top level, prod-

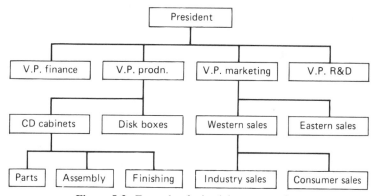

Figure 5-3. Example of mixed departmentation.

uct and process subdivision in manufacturing, and geographic and customer departmentation in marketing.

Span of Control

As soon as a new organization grows to significant size, subordinate managers must be appointed to help the top manager manage. This need was recognized as soon as large groups of people began working toward a common purpose, and it was clearly expressed in early biblical writings:

> And it came to pass on the morrow, that Moses sat to judge the people: and the people stood by Moses from the morning unto the evening. . . . And Moses' father-in-law said unto him "The thing thou doest is not good. Thou wilt surely wilt away, both thou and this people that is with thee: for this thing is too heavy for thee; thou art not able to perform it thyself alone." . . . So Moses hearkened to the voice of his father-in-law, and did all that he had said. And Moses chose able men out of all Israel, and made them heads over the people, rulers of thousands, rulers of hundreds, rulers of fifties, and rulers of tens. And they judged the people at all seasons: the hard causes they brought unto Moses, but every small matter they judged themselves.[5]

Significance. The question is not *whether* intermediate managers are needed, but how many. This depends on the number of people reporting directly to each manager, referred to as the *span of management* or *span of control*. For example, if a simple hierarchical organization with only 64 workers (nonmanagers) and a chief executive officer has a span of

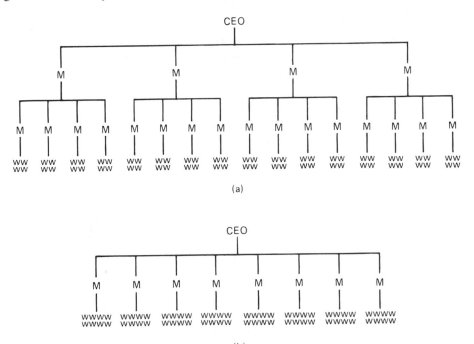

Figure 5-4. Control spans of (a) four and (b) eight compared. M, manager; w, worker.

only two subordinates per manager at every level, it will need 62 managers at five intermediate [middle] levels of management between worker and CEO; with a span of four (Figure 5-4a), it will have 20 managers at two levels; with a span of eight (Figure 5-4b), it will
have only a single level of eight managers.

Many armies are organized on a span of control of about four: four squads per pla-
toon, four platoons per company, four companies per battalion, and so on. Wren[6] reports that
the biblical span of 10 (rulers of thousands, hundreds, and tens) was adopted independently
by the Egyptians, by the Roman legions (with their *centurions* commanding 100 soldiers),
by the Tatars (Tartars) of Mongolia, and by the Incas of what is now Peru and Chile, peo-
ples who had nothing in common other than 10 fingers to "count off." The Roman Catholic
Church, on the other hand, is said to have 2000 persons reporting to the pope (assisted by
the papal office).

Narrow spans of control (tall organizations) are not only expensive because of the cost
of having so many managers, but the multiple levels can increase communication and decision time and stifle initiative because of the temptation of a manager with few subordinates to micromanage (interfering in decisions that should be made at lower levels). Excessively wide spans, on the other hand, can leave managers with inadequate time to
supervise the activities they are responsible for and leave subordinates with inadequate access to their busy supervisor.

Factors determining effective spans. What, then, does determine a desirable span?
Graicunas, a Lithuanian engineer and management consultant, believed that this depended
on the number of *relationships* that existed between manager and subordinates individually
and in various combinations, and among the subordinates themselves. He calculated the
number of relationships for a manager with *n* subordinates as

$$n[2^{(n-1)} + n - 1]$$

so that every subordinate added *more than doubled* the number of relationships the manager
had to be concerned with and, Graicunas assumed, the difficulty of the job. However, many
of Graicunas's relationships are not significant in a particular application, and effective span
of control depends on many factors other than the simple number of subordinates.

Studies of effective spans have identified the following conditions as affecting the
number of people a manager can effectively supervise:

- *Subordinate training.* The more completely subordinates are trained for their jobs,
  the less demands they place on supervisors.
- *Nature of jobs supervised.* The simpler the tasks supervised, the greater the simi-
  larity between the jobs supervised, and the less subordinates work at dispersed locations, the easier it is to supervise more people. On the other hand, when subordinates need frequent contact with people in other parts of the organization to do
  their job effectively (as may planners and coordinators), supporting these relationships can increase the supervisor's burden.
- *Rate of change of activities and personnel.* Events move more rapidly in some
  types of organization than in others. An army must be staffed for the rate of decision making required in a combat situation (and for the rapid turnover of com-

manders). On the other hand, changes of policy and procedure in the Roman Catholic Church take place only after many years (or even centuries), and priests are well educated and have relatively few different assignments in a lifetime of service; thus, there are very few levels—typically parish pastor, bishop, and sometimes an archbishop—between the individual believer and the papal office.

- *Clarity of instruction and delegation.* The more clearly the work to be done can be described, and the more completely the supervisor delegates to the subordinate the resources needed to accomplish this well-defined job, the less subsequent supervision should be required (and the more people the supervisor should be able to handle).
- *Staff assistance.* Usually, administrative activity is not confined to the manager, but involves some (or all) of the time of one or more other people. While most managers have "access to" clerical and secretarial support, higher-level managers usually have "administrative assistants" or "assistants to" the manager of considerable capability who relieve the manager of much office (and sometimes personal) routine, expanding the time they have available for work only they can do. Even the first-line supervisor often delegates some of the short-term leadership: the engineering design supervisor may have 20 engineers, but some will be more experienced "lead engineers" who are responsible for the day-to-day activities of younger engineers and technicians in completing a common task. In the military, senior managers commonly have deputies who are fully qualified to act in their absence, almost doubling the effective leadership potential of the office.

Effective management spans do vary with level within organizations. First-line supervisors, who are concerned with their direct subordinates but not with lower levels, usually have larger spans than do middle managers. Spans of CEOs may vary substantially, depending on the managerial style of the incumbent. Finally, the skill and experience of the manager does, of course, have an effect on the number of people that he or she can supervise.

Current trends in spans. As we will see at the end of this chapter (under "Impact of the Information Revolution"), the availability of computerized information systems throughout the modern organization is drastically decreasing the number of management levels and therefore increasing the effective span of control. Indeed, Byrne reports[7] spans "have risen to as high as 30 to 1 in divisions of companies such as Ameritech, the Baby Bell," and that a ratio lower than 10 to 1 "is a warning of arterial sclerosis."

Line and Staff Relationships

Traditionally, the "line functions" in an organization were those that accomplished the main mission or objectives of the organization, and these were thought to include production, sales, and finance in the typical manufacturing organization. "Staff functions," on the other hand, were those that helped the line accomplish these objectives by providing some sort of advice or service. A useful distinction may be made between *personal staff,* such as the "assistant to" who does troubleshooting or special assignments for a single manager, and *specialized staff,* who serve the entire organization in an area of special competence. Examples of specialized staff organizations include personnel, procurement, legal counsel, and market research. In today's more complex knowledge-based organizations, the activities of "staff specialists" may be as essential to the ultimate success of the organization as "line workers," and these distinctions have become blurred.

Much more fruitful is examining the type of *relationship* involved in a particular transaction:

Line relationships are superior–subordinate relationships and can be traced in a "chain of command" from the organization president through a succession of levels of managers to the lowest worker.

Staff relationships are *advisory* in nature. Four types of staff relationships, arranged in order of increasing levels of influence, are (1) providing advice only on request, (2) recommending where the staff office deems appropriate, (3) "consulting authority," in which line managers must consult (but need not obey) staff in their area, and (4) "concurring authority," in which the staff specialist has a veto authority over the line manager.

Functional (specialized) authority is a special type of staff authority over others who are not their line subordinates. It is as binding as line authority, but does not carry the right to discipline for violation. Usually, it controls "how to" accomplish some action falling in the area of responsibility of the staff office, and it is delegated to staff because of the need for uniformity or special expertise. Examples include specification of budget formats by the financial officer and of criteria for documenting research findings or for reducing product liability by the legal counsel.

Service relationships are "facilitative activities" that are centralized for economy of scale, uniformity, or special capability, but are only supportive of the main mission. Examples include custodial, security, and medical services.

A manager may, at different times, exhibit all of these relationships. For example, a personnel manager will exert line authority over direct subordinates in his or her office, provide staff advice to the chief executive on the need for instituting an affirmative action program, exercise functional authority by defining how job descriptions must be filled out, and provide a service to the entire organization by maintaining employee records.

Friction between line and staff personnel occurs for many reasons. Staff specialists, hired because of their college training in a needed new discipline, may have little understanding of the problems and realities of the line organization. Line managers, on the other hand, are often older and have longer tenure in the organization but may be less educated and have little understanding of the expertise of the staff specialist and the need the organization has for it. Each side needs to listen to the other with courtesy and mutual respect for the good of the total organization. Military officers tend to have assignments alternating between command (line) and staff responsibilities, and they are often sent to "Command and Staff School" or some equivalent in midcareer; as a result, they have a better chance at understanding both sides of this relationship.

Corporate restructuring in the 1990s is reducing the size of specialist staff organizations at the corporate and divisional levels. Instead, individual specialists become members of working teams that, as a group, are empowered to get the work of the organization accomplished with much less need for approvals "up the chain of command." As a result, specialists can integrate their knowledge into work as it is being done, avoiding much of the friction, misunderstanding, and wasted or repeated effort of the past.

TECHNOLOGY AND MODERN ORGANIZATION STRUCTURES

The Woodward and Aston Studies

The nature of manufacturing processes and the size of the organization have in the past exerted considerable influence on organizational design. In the 1950s, *Joan Woodward* and her associates studied the operations of about 100 manufacturing firms in the South Essex region of England, gathering data on manufacturing methods, organization, communication, and performance. She reported[8] some significant differences when she organized these firms into categories of increasing complexity of manufacturing process technology. She classified 80 of these firms into three broad classes (the rest employing combinations of these or remaining unclassified):

1. *Unit:* production of units to customer's orders, prototypes, large equipment in stages, or small batches to customer's orders (sometimes known as job-shop operation)—24 firms. Production runs in this group are too small to justify specialized manufacturing equipment, procedures, or tooling, and production is normally carried out by skilled craftsmen using general-purpose equipment and their past experience.
2. *Mass:* production of large batches, often on an assembly line, and mass production—31 firms. Long production runs justify special production methods, specially designed equipment, and elaborate methods of scheduling and programming. Jobs, on the other hand, tend to be standardized, repetitive, and to use less skilled workers, whose efforts are regulated by the speed of the assembly line. Automobiles and household appliances often are made by these methods.
3. *Process:* continuous process production systems such as those used in the petroleum and chemical industries—25 firms. These normally involve high capital investment per worker and are highly automated. Skilled workers are needed to monitor and maintain these complex production systems.

Woodward found two characteristics that increased continually as manufacturing complexity increased from unit to mass to process technologies: (1) the number of levels of management, and (2) the span of control *for chief executives*. As shown in Table 5-1, however, in many ways the unit and process technologies at the low and the high ends of the technology scale were similar to each other and different from the large batch and mass production technologies in the middle of the scale. For example, the quantity production technologies showed a much higher median number of employees reporting to each first-line supervisor, had the most highly developed line–staff organization with the largest number of staff specialists, employed highly developed production control systems, used the greatest amount of formal written documentation, and had the least pleasant organizational climate. They also favored *mechanistic* management systems, which are like classic bureaucracies: centralized, formalized, with standardized jobs.

TABLE 5–1 Organization Characteristics Versus Production Technology

Production Technology	Unit	Mass	Process
Number of firms observed	24	31	25
Levels of management (mode)	3	4	6
Span of control—chief executives (median)	4	7	10
Span of control—first-line supervisors (median)	23	49	13
Typical management system	Organic	Mechanistic	Organic
Development of staff activities	Limited	High	Limited
Predicting, scheduling, and control systems	Limited	Extensive	Integral
Communications	Verbal	Written	Verbal
Pleasantness and openness of organizational climate	Greater	Less	Greater

Source: Adapted from Joan Woodward, *Industrial Organization: Theory and Practice*, Oxford University Press, Oxford, 1965.

Firms using unit or process production technologies, on the other hand, were more likely to favor *organic* management systems, which are more decentralized with less rigidly defined jobs, less attention to rank, and a great deal of lateral (as opposed to vertical) communication. They also enjoyed a more pleasant and relaxed organizational climate, more reliance on verbal communication, and fewer people reporting to first-line supervisors. Control systems were less necessary in unit manufacturing and integral with (built into) the continuous processing equipment.

Woodward's observations were on small to medium-sized companies (only 13 of the 100 had more than 1000 employees, the largest being under 9000) located in southeastern England, and these were studied in the 1950s. Whereas some later studies have supported Woodward's findings, others have produced conflicting results.

Perhaps the most prominent of the studies dissenting from Woodward's findings was by a group of scholars at the *University of Aston,* Birmingham, England. They investigated a group of 46 firms in the vicinity of Birmingham with from 240 to more than 25,000 employees. In studying the operations technology of these firms they used a 10-step *production technology* scale quite similar to Woodward's. They did agree with Woodward that mass production firms had larger spans of control for first-line supervisors, with more staff specialists for control and greater distinctions between line and staff than did unit or process firms. However, they found that the size of the firm (in number of employees) correlated better with other parameters than did the type of technology. Thus the "larger" firms (25,000 employees) were more likely to have high levels of specialization, standardization, formalization, and centralization, regardless of the type of technology. Further work indicates that *both* the technological complexity of Woodward and the organizational size dimension of the Aston group must be considered in effective organization design.

Project and Other Temporary Organizations

Max Weber's characterization of *bureaucracy* was outlined in Chapter 2, and this formal, rather rigid organizational style still has value for carrying out highly structured, routine functions over long periods of time. However, conventional bureaucracy can carry with it pressures to conform that stultify innovation and growth, as Whyte showed in his classic book *The Organization Man.*[9] Far more flexible structures are needed for many of the tasks of today and tomorrow. This led Toffler to predict the "breakdown of bureaucracy" in his 1970 best-seller *Future Shock:*

> We are, in fact, witnessing the arrival of a new organizational system that will increasingly challenge, and ultimately supplant bureaucracy. This is the organization of the future. I call it "Adhocracy." . . .
>
> The result is that man's organizational relationships today tend to change at a faster pace than ever before. . . . The high rate of turnover is most dramatically symbolized by the rapid rise of what executives call "project" or "task-force" management. Here teams are assembled to solve specific short-term problems. . . . What we see here is nothing less than the creation of a disposable division—the organizational equivalent of paper dresses or throw-away tissues.[10]

The proposal teams gathered together by aerospace companies to respond to a major military request for proposal (RFP) provide an excellent example of the "disposable organization"; as many as a thousand people, often from several cooperating companies, may come together for one to three months for this specific purpose and then disband back to their original organizations. Project management organizations and their operations are of especial importance to engineers, and two chapters (14 and 15) are dedicated to project management. Particular attention is placed in Chapter 15 on *matrix management* organizations, which are frequently used in project management.

There are other modern examples of temporary or "team" organizational structures; Cleland and Kerzner[11] provide descriptions of production teams, worker-management teams, product-design teams, quality circles, crisis-management teams, and task forces. We discuss use of product-design teams for design review and configuration management in Chapter 10, and the function of quality circles in Chapter 12.

Impact of the Information Revolution

Modern computer and telecommunications technologies are rapidly changing our organizations in ways that we do not yet fully understand. Not since the introduction of the electric motor into industry has there been an innovation so universal in its scope. Lund and Hansen[12] believe that the time horizons between design and production are collapsing because the design data base, once created, is available for design analysis and evaluation, creating prototypes, control of ultimate production, and even planning and control of quality inspection. Quicker start-ups and product changes reduce the optimal size of production runs, reducing the resources tied up in in-process and finished-goods inventory. Product life cycles will be shortened in many industries. The successful firms will be those evidencing the flexibility, adaptability, and quick response that computer-based technologies can provide.

Lund and Hansen also "see a diminishing of the size and importance of centralized corporate headquarters"[13] as operating decisions are pushed to lower levels (and simpler ones are automated). Computer-literate executives will be able to draw figures from central data bases as needed to analyze a given situation, reducing the need for the intermediaries that gather and analyze data today. Tom Peters predicts an even greater impact—the "complete destruction of hierarchy as we have known it . . . the biggest change in organization in thousands of years"[14] because of the access all employees will now have to all the company's information.

As computer-based automation replaces conventional processes, it will sharply reduce the number of workers (and their foremen) needed per unit of output. Factory workers will be *monitoring* the production process rather than forming part of it, and they will need at least the following skills:

1. Visualization (ability to manipulate mental patterns)
2. Conceptual thinking (or abstract reasoning)
3. Understanding of process phenomena (machine fundamentals and machine/material interactions)
4. Statistical inference (appreciation of trends, limits, and the meaning of data)
5. Oral and visual communication
6. Attentiveness
7. Individual responsibility[15]

If this view is correct, there will be little future in industry for the uneducated employee and mechanistic leadership favored by Frederick Winslow Taylor and, at least until recently, General Motors, in the factory of the future.

Peter Drucker provided, in 1988, an excellent forecast of today's emerging organizational styles:

> The typical large business 20 years hence will have fewer than half the levels of management of its counterpart today, and no more than a third the managers. In its structure, and in its management problems and concerns, it will bear little resemblance to the typical management company, circa 1950, which our textbooks still consider the norm. Instead it is far more likely to resemble organizations that neither the practicing manager nor the management scholar pays much attention to today: the hospital, the university, the symphony orchestra. For like them, the typical business will be knowledge-based, an organization composed largely of specialists who direct and discipline their own performance through an organized feedback from colleagues, customers, and headquarters. For this reason, it will be what I call an information-based organization.
>
> . . . [In the new information-based organization] it becomes clear that both the number of management levels and the number of managers can be sharply cut. The reason is straightforward: it turns out that whole layers of management neither make decisions nor lead. Instead, their main, if not their only, function is to serve as "relays"—human boosters for the faint, unfocused signals that pass for communication in the traditional pre-information organization.[16]

Drucker sees four special problems for management as particularly critical in the new information-based organization:

1. Developing rewards, recognition, and career opportunities for specialists [since opportunities for promotion into the management hierarchy will drastically decrease].
2. Creating unified vision in an organization of specialists.

3. Devising the management structure for an organization of task forces.
4. Ensuring the supply, preparation, and testing of top management people [since the progression of middle management levels that provided this training in the past have diminished].[17]

By late 1994, this revolution in American organizations was well under way. A special *Business Week* report on "Rethinking Work"[18] discusses some of the salient aspects:

1. Virtual disappearance of job security, replaced by shared responsibility: "employers have an obligation to provide opportunity for self-improvement; employees have to take charge of their own careers."
2. Increasing demand for well-paid professional and technical workers; decreasing demand for operators, laborers, craftsmen, clerical staff, and farm workers.
3. Reduced real wages (purchasing power down from 1973 to 1993 by 23% for high school dropouts, 15% for high school graduates, 8% for college graduates, and 5% for those with two years graduate work), increasing the need for the two-income family.
4. Continuing "downsizing" of staff, with the surviving personnel working longer hours under higher stress.
5. Increases in part-time, contract, and self-employed workers who are paid only when needed without the fringe benefits that often add 40% to payroll cost.

As a result of these drastic changes, American industry in the mid-1990s was again internationally competitive, offering continuing opportunity for those who have the skills and training needed. However, each individual must take personal responsibility for their own career, to assure they continue to acquire the new knowledge and skills they will need. We discuss some of the ways the engineering professional can do this in Chapter 16.

DISCUSSION QUESTIONS

5-1. You have begun a small but growing business. What advantages and disadvantages should you consider before changing it from a sole proprietorship to a corporation?

5-2. Select four different businesses and identify some of their "key activities" by posing Drucker's three questions.

5-3. Under what conditions might each of the following logics of departmentation be desirable: functional, geographic, customer, product, and process?

5-4. Chart an organization with which you are familiar that has at least three organizational levels, and identify the number of people reporting to each manager at each level. How do the spans of control at the lowest level compare with those at higher level(s)?

5-5. From question 5-4 select a particular manager and his or her group. Analyze the difficulty of the manager's job based on the criteria (subordinate training, nature of jobs supervised, etc.) provided in the text. In your opinion, does the manager's actual span of control reflect the difficulty of that position?

5-6. Distinguish between functional (specialized) staff authority and traditional line authority.

5-7. According to the Woodward and Aston studies, what conditions lead to a formalized, standardized organizational environment?

5-8. Describe from your experience or reading a temporary organization or task force formed to accomplish some specific purpose. How was it formed, organized, and ultimately disbanded?

5-9. What changes in organization structure might you expect as a result of the information revolution?

5-10. If the development of the information-based organization continues to have the effect on management predicted by Drucker, what will be the impact on career expectations of engineers and other specialist professionals?

5-11. Choose an enterprise with which you are familiar that has undergone significant recent reorganization. Compare the new and old organizations with regard to (**a**) size and influence of specialized staff, (**b**) management levels, (**c**) typical spans of control, and (**d**) responsibility delegated to nonmanagerial professionals. What other changes occurred in the reorganization?

5-12. Discuss the strategy you propose to use in your personal career to assure you will remain in demand in a changing, competitive world.

NOTES

1. Heinz Weihrich and Harold Koontz, *Management: A Global Perspective,* 10th ed. (New York: McGraw-Hill Book Company, 1993), p. 244.

2. Weihrich and Koontz, *Management,* p. 244.

3. Peter F. Drucker, *Management: Tasks, Responsibilities, Practices* (New York: Harper & Row, Publishers, Inc., 1974), pp. 530–531.

4. Drucker, *Management,* pp. 542–549.

5. *King James Bible,* Exodus, Chapter 18.

6. Daniel A. Wren, *The Evolution of Management Thought,* 3rd ed. (New York: John Wiley & Sons, Inc., 1987), pp. 15–21.

7. John A. Byrne, "Belt-Tightening the Smart Way," *Business Week,* Enterprise/1993 issue, p. 35.

8. Joan Woodward, *Industrial Organization: Theory and Practice* (Oxford: Oxford University Press, 1965).

9. William H. Whyte Jr., *The Organization Man* (Garden City, NY: Doubleday & Company, Inc., 1956).

10. Alvin Toffler, *Future Shock* (New York: Random House, Inc., 1970; paperback, New York: Bantam Books, 1971), pp. 125, 132–133 (Bantam).

11. David I. Cleland and Harold Kerzner, *Engineering Team Management* (New York: Van Nostrand Reinhold Company, Inc., 1986), pp. 6–17.

12. Robert T. Lund and John A. Hansen, *Keeping America at Work: Strategies for Employing the New Technologies* (New York: John Wiley & Sons, Inc., copyright © 1986), pp. 64–72.

13. Lund and Hansen, *Keeping America at Work,* pp. 78–85.

14. Thomas J. Peters, commentary on the *Nightly Business Report* (public television program), June 8, 1988.

15. Lund and Hansen, *Keeping America at Work,* pp. 92–93.

16. Peter F. Drucker, "The Coming of the New Organization," *Harvard Business Review,* January–February 1988, pp. 45–46.

17. Drucker, "The New Organization," p. 50.

18. "Rethinking Work: The New World of Work," *Business Week,* October 17, 1994, pp. 76–87.

CHAPTER ⬤6

Some Human Aspects of Organizing

PREVIEW

We begin this chapter by considering the steps in staffing technical organizations. The first step is human resource planning, in which the type and number of people needed in the next six months to a year is established. Next we study the process of personnel selection, including discussion of effective résumés (and cover letters), the employment application, campus interviews for engineering graduates, reference checks, plant visits, and the job offer. We then consider the process of orienting and training the new employee and appraising his or her performance.

In the second major section of the chapter we consider the nature of authority, the sources of authority and power, and the importance of status and culture. Next we study the system of assignment, delegation, and accountability. We consider the importance of delegation and the difficulty engineers often have in practicing it, and look at the subject of decentralized management. The chapter closes with discussion of committees and meetings: reasons for using them, problems they present, and methods of making them effective.

STAFFING TECHNICAL ORGANIZATIONS

The management function of staffing involves finding, attracting, and keeping personnel of the quality and quantity needed to meet the organization's goals. Staffing is included in some management textbooks as part of the organization function and in others as a separate function, but the same steps are required. Effective staffing requires first identifying the nature and number of people needed, planning how to get them, selecting the best applicants, orienting and training them, evaluating their performance, and providing adequate compensation.

Human Resource Planning

Hiring technical professionals. Hiring a laborer when jobs are scarce may involve just a call to the nearest union hall, but hiring quantities of engineers and other professionals, whether new college graduates or experienced professionals with specific skills, requires planning ahead from six months to more than a year. Planning for the overall personnel (or human resource) needs of a large high-technology firm can therefore be quite complex. Following is the process used in one division of a large aerospace firm to come up with the required quantity and quality of technical personnel.

1. Document the number of technical personnel of each classification presently on hand.
2. Estimate the number of professionals of each type needed in the near future (six months to a year) to meet firm contracts and likely potential business.
3. Estimate the expected attrition in the current staff including (a) resignations as a function of the national demand for scientists and engineers and the relationship between your salary scale and that of your competition; (b) transfers out to other divisions and promotion to higher positions; and (c) retirements, deaths, and leaves of absence.
4. Establish the need for increased personnel as
   increase (4) = need (2) − personnel on hand (1) + attrition (3)
   Subdivide this increase (4) into (5) new college hires, (6) experienced professionals, (7) technician support, and (8) other sources.
5. Each 100 new college hires may require making 200 offers, as a result of 400 candidates visiting your plant or division, stemming from 600 campus interviews. The campus interviews, in turn, might require scheduling trips to 20 campuses to interview 10 students in each of three interview days. (The factors quoted here will vary with the economy, industry, and employer.)
6. Develop a hiring plan to acquire experienced personnel using national and local hiring, employment agencies and "headhunters," career centers, and employee referrals.
7. Develop a plan to acquire needed technicians from two- and four-year technical institutes, B.A. and B.S. graduates in physics and math, discharged military technicians, advertisements, state and commercial employment services, and employee referral.
8. Needs that cannot be met by sources (5), (6), and (7), especially those of too short a duration to justify permanent hiring, can be met by scheduling overtime, hiring contract (temporary) engineers, borrowing engineers from other company divisions, and contracting work to other company divisions or to other companies.[1]

Hiring managers. A similar plan must be developed for staffing management positions.

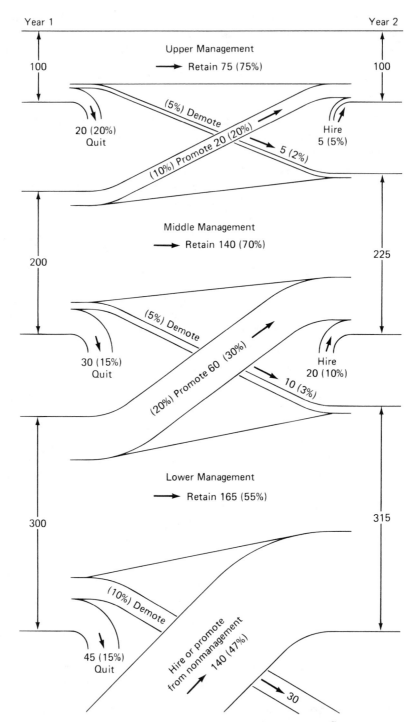

Figure 6-1. Illustration of typical annual management flow.

Figure 6-1 illustrates what the managerial staffing needs might be from one year to the next for an organization employing about 300 first-line managers, 200 middle managers, and 100 upper-level managers. Most middle and upper management positions are shown being filled by promotion, although a few hires at these levels will always be needed where the organization does not already have someone with the right skills. A healthy organization will have a large annual requirement for new first-line supervisors, many of whom will be promoted within the company from employees experienced in a specialty, but often with little experience in management.

Job requisition/description. A manager wishing to fill a professional position normally must fill out a form known variously as a job description or job requisition, which then is approved by higher management and given to the personnel department as guidance in their search for candidates that might be considered for the position. Table 6-1 illustrates a typical job requisition.

Selection

Selecting those applicants who will be offered jobs from among the many contacted in the search described above is essentially a filtering process. Résumés and/or applications are reviewed, potential candidates are screened in campus or telephone interviews, references are checked, and applicants who pass through these screens are invited to the company for interviews (and sometimes testing) before job offers are made.

Résumé and cover letter. For most engineering professionals the first impression is normally made by the résumé, which is submitted with a cover letter in response to an advertisement or as an initial inquiry. The *cover letter* should be addressed to the appropriate individual by name, not "Personnel Director" or "To Whom it May Concern" (call the company if you do not have a name). Normally, it begins by identifying the position or type of work you are applying for and, if appropriate, where you heard of the opening. A second paragraph can state why that company and position interest you, and describe *concisely* (a sentence or two) the education, experience, and other abilities that have prepared you for the position you seek. A closing paragraph can refer to the attached résumé, thank the recipient for his/her consideration of it, and (if appropriate) indicate you will call within a specified

TABLE 6–1 Example of Job Description/Requisition

Job Requisition

Title of Position: Research Engineer
Educational Requirements: B.S. in chemical engineering or equivalent
Experience: At least two years in chemical processing, with pilot-plant operation and process development experience preferred.
Description of Duties:
    1. Supervise pilot-plant operations for producing new organic intermediates.
    2. Identify and recommend process improvements, including conversion of existing batch methods to a continuous process.
    3. Work with production engineering to design manufacturing plant.
Will Report To: Manager of Chemical Process Research
Salary Range: $45,000 to $50,000 per year.

time to inquire about a possible interview. The cover letter must be impeccable in appearance, grammar, and spelling. A quality cover letter should encourage the recipient to give your résumé fair consideration; with a poor one, your résumé may not be read.

The *résumé* itself includes all or most of the following:

1. Name, address, and telephone number(s).
2. Current job position and/or status (such as "graduating senior").
3. Current and longer-term employment objectives.
4. Summary of education (formal degrees and continuing education).
5. Employment experience, with the most recent employment first, emphasizing accomplishment (the longer you are out of college, the more likely this is to precede education in a résumé).
6. Publications, significant presentations, and patents.
7. Significant honors and awards.
8. Professional affiliations.
9. References (who have agreed to serve) or, in your initial inquiry, a statement here or in the cover letter that references are available on request.

Writing an effective résumé is an important skill that many engineers do not master easily. An effective résumé normally should not exceed two pages (résumés applying for academic positions are an exception, since publications and presentations are listed there in detail). The résumé should be well organized, concise, faultless in grammar and spelling, and attractively printed on quality paper. It should also emphasize (without being dishonest) those parts of your education and experience most applicable to the position applied for, and so an individual may need several versions of a résumé. Fortunately, with today's word processing software and laser printers, this is becoming easier. Most bookstores will carry books on résumé writing and other aspects of job hunting; Lewis[2] is a good example.

Employment application. If the résumé leads to further interest from a potential employer, the applicant will normally have to fill out (neatly, of course) much of the same information on an employment application, arranged in a standard form familiar to interviewers from that organization. The application will normally not ask for race, religion, citizenship or national origin, height, weight, age or date of birth, marital status, or age of children or require a photograph because of antidiscrimination laws, but the applicant may choose to include some of these in his or her résumé.

Campus interview. The newly graduating engineer typically makes the first contact with potential employers in the campus placement interview. Indeed, about half of all campus interviews are with engineering students, even though they make up only from 4 to 10% of all students.[3] Interview outcomes are a complex dynamic of the attributes of the applicant, of the interviewer, and of the situation (the physical setting and the economic demand for engineers, for example). The interviewer needs to learn enough about the applicant to recommend for or against an invitation for a plant visit, and the applicant needs to learn about the employment opportunities and other advantages (and limitations) of working for the employer. The applicant is well advised not only to read the potential employer's placement brochure in advance, but also to learn more about the company with the help of the university reference library and discussion with classmates and professors who may know something about the organization. Some students are uncomfortable in early interviews and do not sell themselves well; many colleges provide the chance for mock interviews, often using television tape, to help develop this skill. Engineers need to learn to conduct interviews as well, since they may find themselves interviewing candidates at their plant or back on campus after a few years experience.

Reference checks. Before inviting an applicant for a site visit, a prospective employer commonly checks the references given in a résumé or application, or requests them if they have not already been provided. References for the new graduate include professors and supervisors from part-time jobs; for the experienced engineer they will be primarily past and (if your employer knows of your search) current supervisors and co-workers. References may be checked by mail or telephone. Mail reference (background) checks are cheaper but slower, and normally they involve completion of a form.

The applicant must decide in advance whether to waive the right to see the completed reference forms; the employer will naturally give more credence to references that have been written in confidence. References often will respond with more candor or reveal more by their inflection and hesitation in telephone discussions, but they may be hard to reach. Of course, the best reference is eyeball-to-eyeball discussion with a prior supervisor—this is feasible for intraplant transfers but otherwise is used only by the federal government where the proposed position involves security clearance. An increasing problem with references is the fear of liability if a bad reference is given. Some employers have a policy of confirming only that a former employee worked under a given job title during a given calendar period, although these same employers may try to get the maximum information in reference checks on people they are interviewing.

Site (plant) visits. When a company has a strong interest in an engineer or other professional, he or she will normally be invited for a visit to a chosen company location at company expense. The applicant should be interviewed by three or four people on the staff, including one or more supervisors with open position(s) for whom the candidate is being considered, and at least part of the visit should involve a tour of the area in which the candidate might work. The candidate's reaction to the work observed and the type of questions asked give insight into his or her interest and suitability for the position; at the same time the candidate can gain insight into the work being done and judge from work observed and answers to his/her questions whether that might be the right assignment. At some point the personnel office will provide information on company benefit programs and answer questions on general company policies. Tuition support of graduate courses, for example, is a policy that many new engineering graduates inquire about.

Applicants for nonprofessional positions such as hourly production or clerical jobs are more likely to visit the company employment office as a first step (or following a newspaper ad or telephone inquiry). Here, the personnel office provides the major screening. The applicant may be required to take tests, which may be of general intelligence, aptitude, ability (such as a typing test), personality, or interest. Any tests used should first be *validated* to show that test outcomes are related to successful job performance, since there are legal and ethical problems in using tests that might tend to exclude specific groups of applicants without carefully documented reasons. Engineering students have already been thoroughly tested in establishing their academic record, and preemployment tests are uncommon.

At the end of an interview or plant visit it is perfectly proper for the applicant to inquire "when do you expect to make a hiring decision" or "If I haven't heard from you by [date], may I call you?" A prompt letter thanking the interviewer for courtesies extended and expressing continuing interest in the company is generally appropriate.

Starting salary. If an employer is interested in an applicant, sooner or later he or she will ask "What salary do you expect?" Often this will occur toward the end of a site visit, and the applicant should be prepared. A candidate who replies "Whatever is your going rate," will probably be offered the bottom of the range. Since future salary adjustments in most companies are typically small percentage adjustments to current salary, inequities in starting salary can be adjusted only slowly. Kennedy suggests:

> The answer should be "I understand that market in this area for entry-level [your specialty] engineering jobs is X to Y dollars. I'm told I should expect to be paid market." This is both accurate and polite. The irony of the whole salary negotiation game is that if you hold out for market, the company thinks you're better and brighter than someone who actually hands them a bargain.[4]

College placement offices should have copies of the College Placement Council's latest *CPC Salary Survey,* published each September to summarize initial salary offers reported over the previous academic year. For the 1993–94 recruiting year, 42.3% of all reported offers to bachelor's degree candidates were to those with technical majors. Starting offers vary with engineering discipline: in the 1993–94 academic year, the CPC reported average offers to B.S. degree candidates of about $30,000 per year to civil and architectural engineers and to engineering technology majors, from $33,300 to $35,000 to industrial, computer, electrical/electronic, and mechanical engineers, and about $39,000 to chemical and petroleum engineers. Offers to women were up to 4% higher than those men received in the major engineering fields, up to 6% less in a few fields (architectural, mining, and geological engineering), and 3% less in engineering technology.[5]

Experienced engineers will measure their expectations based on the years since their bachelor's degree, graduate degrees if any, the quality of their experience, local cost of living, and other factors. The Engineering Workforce (formerly Manpower) Commission (EWC) of the American Association of Engineering Societies (AAES) surveys employers annually. Figure 6-2 summarizes salaries earned by 100,278 full-time engineering graduates, including those in managerial and nontechnical positions, at 236 U.S. private sector employers as of February 1, 1994. The table shows salaries (excluding bonuses) exceeded by 90%, 75%, 50%, 25%, and 10% of engineers versus years since their bachelor's degree. The EWC report further divides data by industrial sector, degree level, metropolitan area, and supervisory responsibility (some or none).

Another annual survey often summarized in engineering journals is that of approximately 55,000 members of the National Society of Professional Engineers (NSPE), who are disproportionately engineers (especially civil engineers) in private practice or consulting or-

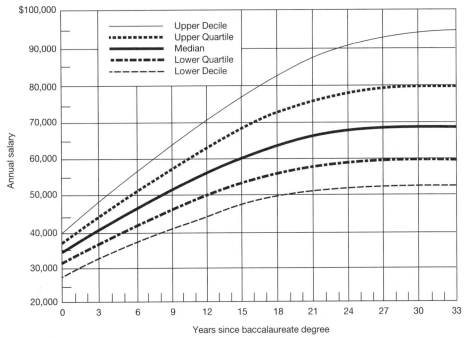

Figure 6-2. Salary curves for graduate engineers, all degree levels, employed by 236 private sector organizations.
(From *Engineers' Salaries: Special Industry Report 1994,* Engineering Workforce Commission, American Association of Engineering Societies, Inc., Washington, DC, 1994, p. 17. Reproduced by permission.)

ganizations. The 19% of NSPE members responding to the (January 1,) 1994 survey reported median income of about $2000 less than that of Figure 6-2 for the first four years, comparable in midcareer, and more (an average of $77,000 instead of $69,000) for engineers with more than 27 years of experience.[6] Salary varies by the region of the country and by metropolitan area: the previous (1993) NSPE survey reported median income of about $75,000 in suburban New York City and San Francisco versus $53,500 in Dayton and Cleveland, OH. The cost of living varies, however: that in Washington, DC, was 135% and in Anchorage, AK, and San Diego, CA, 129% of the average U.S. city, and that in Cookesville, TN, only 84%.

Job offer. The employment offer is a standard format letter offering a specific position and identifying salary, reporting date, position and title, the person the candidate will report to, and often provisions for moving expenses. The candidate should acknowledge the offer immediately. A candidate with other potential offers in process may ask for a reasonable delay (and then call the other company(ies) and say "I have an interesting offer from X Corporation and they are pressing me—when might I hear of your interest?") A candidate who already has a better offer from someone else can reply "I've been offered $Y by Z company—I'd rather work for you, but this is a factor I'll have to consider in my answer." Striking a balance between demanding too much and selling oneself too cheaply requires the candidate to have a clear understanding of his or her true worth in the current job market.

Orientation and Training

When a new employee reports to work, the employing organization needs to help the new-comer become part of the organization by introducing him or her to the policies and values of the organization as a whole and the specific requirements of the person's new department and job. The personnel department normally has the responsibility to tell the newcomer about fringe benefits such as medical insurance, vacations, tuition reimbursement, pensions, and the like. This can be accomplished with a short one-on-one discussion on the first day as the new employee processes through personnel or a more formal presentation periodically for all new employees; in either case, most organizations of any size will provide every em-ployee a current edition of an employees' handbook describing benefit programs.

Inculcating the values of the organization, such as attitudes toward quality, safety, and customers, is a more difficult task of involving establishing attitudes. While presentations can be made to new employees by management emphasizing these values, they must be ev-ident in their practice by members of the organization to be given credence. Some large or-ganizations will spend from three months to a year in rotating the new employee through a variety of departments and jobs to orient the individual to the organization before placing him or her in the first "permanent" assignment. Occasionally, a fast-growing organization will have a formal orientation program set up in which functional managers will briefly de-scribe the nature and function of their departments. More often the new employee will be assigned directly to a department and supervisor.

In any event, the immediate supervisor of the new employee bears the major respon-sibility for introducing him or her to the new group and the specific job assignment. Su-pervisors tend to be busy with current problems the new employee cannot help with until "brought up to speed," and so there is a tendency to hand the new hire a 6-inch stack of re-ports to read for familiarization and get back to the immediate problem. After several days of such isolation the new hire begins to wonder why he or she is there! The more astute su-pervisor realizes that there will always be current problems and spends some time getting the new employee started and thinking through some initial assignments that will assist in the orientation process. Often, other employees in the group will be asked to assist by tak-ing the new employee along on visits to other departments, introducing the new hire, and in the process providing insight into current activities of the immediate group and its rela-tionships with the larger organization. Frequently, a specific senior member of the group will be assigned primary responsibility for orienting the new employee.

In a more comprehensive sense, orientation and training can be considered to include the total *socialization* of the new employee to the environment and culture of his or her new organization. Pringle et al. describe this well:

> The socialization process, culminating with the employee's transformation from an "outsider" to an organizational "insider," may require anywhere from a month to a year, depending on the particular organization and the individual. Socialization encompasses such formal and informal activities as learning the job and developing appropriate skills, forming new interpersonal re-

lationships, and accepting the organization's culture and norms. From the organization's perspective, effective socialization results in order and consistency in behavior.[7]

Appraising Performance

There are several reasons for requiring formal appraisal of an employee's performance. In a 1984 member survey by the American Management Association, 86% of those responding reported using performance appraisal in determining compensation [pay and bonuses], 65% for counseling, 64% to assist training and development, 45% for promotion, 43% for staff planning, and 30% for retention/discharge decisions.[8] A written record of performance in some consistent form is especially important in large organizations where personnel are frequently transferred, such as the military service, and in bureaucratic organizations such as civil service to justify terminating (firing) poor performers and rewarding exceptional ones.

Perhaps the oldest and most common technique for performance appraisal is the conventional *rating scale,* in which an employee is given a rating by checking one of five or more level-of-performance boxes for each of a series of attributes. For example, clerical and other hourly workers in the University of Missouri system are rated in five steps from "Outstanding" to "Inadequate" in each of (1) knowledge of the work, (2) quality of the work, (3) quantity of the work, (4) attendance and punctuality, (5) carrying out instructions, and (6) an overall appraisal. Sometimes each box in the matrix of attributes and ratings has a word description to help the rater, as shown in the apocryphal example in Table 6-2.

The conventional rating system is easy to develop and easy to grade, but it offers a number of problems. Some raters suffer from a "halo effect," in which they assign the same rating to every category; some from a "recency effect," in which they base their rating only on the most recent part of the rating period. Raters differ in their interpretation of "outstanding" and the other categories, and some are more lenient than others in their ratings. Moreover, raters find that they are competing with their peers in trying to justify promotion or other benefits for their employees, and thus they soon recognize the competitive need to inflate ratings. In the U.S. Army at one time, over 90% of all officers were rated in the top "outstanding" category, and the designation of being merely "above average" threatened a military career. One approach in this computer age is to include the average of all rating values issued by that rater as a standard of comparison, but that is complicated and seldom used.

Table 6-3, in which ratings of 80 employees are compared, provides some alternative methods. The "pure rating" again places no limit on the fraction of employees who can be rated "superior." In the "forced ranking," or "ladder," approach only one person can be placed on each step of the ladder, and the rater is forced to discriminate between employees but has no way to indicate employees considered equal or significant gaps in ability between two people in the sequence. The "modified ranking" satisfies these last two objections. In the percentile, or "forced distribution," approach, 40% of employees must be placed in the third category (average), 20% each in the second and fourth, and only 10% in the first and fifth categories.

Most forms used for appraisal of professionals in large organizations involve a combination of methods. In a 1972 survey of practices of Fortune 500 companies, Hatlan[9] found that 42% included a "weighted checklist" (conventional rating scale), and 42% some form of ranking or forced distribution. About 25% included a self-appraisal, 23% included

TABLE 6–2 Rating Scale for Cartoon Heroes

Performance Factors	Far Exceeds Requirements	Exceeds Requirements	Meets Requirements	Needs Improvement	Does Not Meet Requirements
Quality	Leaps tall buildings with a single bound	Must take a running start to leap over tall buildings	Can leap only over short buildings	Crashes into buildings when attempting to leap over them	Cannot recognize buildings at all
Timeliness	Is faster than a speeding bullet	Is as fast as a speeding bullet	Not quite as fast as a speeding bullet	Would you believe a slow bullet	Wounds self with bullet when attempting to shoot
Initiative	Is stronger than a locomotive	Is stronger than a bull elephant	Is stronger than a bull	Takes bull by the horns	Shoots the bull
Ability	Walks on water consistently	Walks on water in emergencies	Washes with water	Drinks water	Has water on the knee
Communications	Talks with God	Talks with angels	Talks to himself	Argues with himself	Loses those arguments

appraisal by several other managers who were able to observe the individual, and 14% some form of peer evaluation. About 15% emphasized "critical incidents," which are specific examples of good or poor performance, but only 8% emphasized a free-form essay on subordinate performance (which tests the expository skill of the rater as much as anything). Hatlan found, however, that the overwhelming majority (97%) of appraisal systems surveyed incorporated some form of management by objectives (MBO), which we discussed in Chapter 3.

Ranking or forced distribution methods have some logic to them, but in a culture that prefers to believe (with the Garrison Keiler's *Prairie Home Companion* radio show) that "all our children are above average," it has serious drawbacks. Many supervisors like to believe that their employees are all superior. Engineers who are doing effective work are hardly motivated by an "average" classification and may be encouraged to look elsewhere if placed in category 4 out of 5. Moen describes actions of companies such as General Motors and American Cyanamid Company in eliminating use of forced distribution systems, and he paraphrases the appeal of quality guru Deming for more motivating appraisal systems:

> W. Edwards Deming is asking for a transformation of the American style of management through the application of his 14 points. The most serious disease that stands in the way of this

TABLE 6–3 Examples of Different Performance Rating Systems

Pure Rating		Forced Ranking	Modified Ranking		Percentile	
Superior Performance	Exceeds Expectations				Top (10%)	2nd (20%)
Derek	Ampère	1. Monroe	100. —		Ampère	James
Monroe	da Vinci	2. Derek	99. Monroe		Derek	.
	Einstein	3. Ladd	98. —		Einstein	.
	Galileo	4. Garbo	97. Derek		Garbo	.
	Garbo	5. Einstein	96. —		Hemingway	.
	.	6. Ampère	95. Ladd, Garbo		Ladd	.
	.	7. Truman	94. Ampère,		Monroe	.
	.	8. Hemingway	Truman,		Taylor	.
	Truman	9. James	Einstein		Truman	.
	Volta	.	93. Hemingway,			
	Zukendorf	.	James			
		.	.			
		.	.			
Needs Improvement	Unsatisfactory	.	.		4th (20%)	Bottom (5%)
Abraham	Balderston	.	.		.	Balderston
.	Mozart	77. Newton	5. Glockenspiel,		.	Glockenspiel
.	.	78. Glockenspiel	Newton		.	Mozart
.		79. Mozart	4. —		.	Newton
.		80. Balderston	3. Mozart		.	
.			2. —		.	
Zen			1. Balderston		.	

Source: Jack Balderston, Philip Birnbaum, Richard Goodman, and Michael Stahl, *Modern Management Techniques in Engineering and R&D*, Van Nostrand Reinhold Company, Inc., New York, 1984, Table 14–2, p. 280.

transformation is the current practice of the performance appraisal (PA) system. Deming says the effects of this disease are devastating North America. He suggests that the [old] systems of rewards nourish the win-lose philosophy and that they destroy people. Companies must adopt a win-win philosophy of cooperation, participation, and leadership directed at continuous improvement of quality.[10]

The primary emphasis in appraisal today, therefore, is on the contribution made toward achieving organizational objectives, which is the reason that personnel are employed to begin with. And with the increased emphasis on teamwork, there is greater emphasis on rewarding team members for team (or even total organization) performance rather than just individual performance.

AUTHORITY AND POWER

Other important human considerations in organizations, once they have been properly staffed, include the nature of authority and power and their effective delegation. These are considered in this section and the next.

Nature of Authority

Formal authority. The traditional view of authority is "legitimate power," the *right,* based on one's position in an organization, to direct the work activities of subordinates. In the United States, formal authority over employees of corporations is thought to stem from society as a whole, through the guarantee of private property in the Constitution of the United States. Individuals invest their assets in corporate stock and elect a board of directors, delegating to them the right to manage their invested assets. The board, in turn, elects the executive officers of the corporation and they appoint subordinate managers, delegating authority to appoint lower-level managers. In this way the direction received by the lowest-level employee from his or her supervisor or foreman can be traced to the ownership authority of stockholders. Similarly, authority over government workers stems from national or state constitutional authority conferred on the legislative and executive branches of government, who in turn delegate authority and direction to the leaders of military and government agencies.

Acceptance theory of authority. Chester Barnard, on the other hand, believed that authority originates when subordinates choose to accept the directives of superiors. According to Barnard:

> If a directive communication is accepted by one to whom it is addressed, its authority for him is confirmed or established. It is admitted as the basis for action. Disobedience of such a communication is a denial of authority for him. Therefore, under this definition the decision as to whether an order has authority or not lies with the persons to whom it is addressed, and does not reside in "persons of authority" or those who issue orders.[11]

Despite this, we know that the overwhelming majority of requests or directives from superiors are, indeed, complied with. When a person enters employment with an organization, he or she is tacitly agreeing to accept any directives toward which the employee feels no strong objection. Barnard modifies his "acceptance theory" by postulating a "zone of indifference," which he explains as follows:

> If all the orders for action reasonably practical be arranged in the order of their acceptability to the person affected, it may be conceived that there are a number which are completely unacceptable, that is, which certainly will not be obeyed; there is another group somewhat more or less on the neutral line, that is, either barely acceptable or barely unacceptable; and a third group unquestionably acceptable. This last group lies within the "zone of indifference." The person affected will accept orders lying within this zone and is relatively indifferent as to what the order is so far as the question of authority is concerned.[12]

Sources of Power

French and Raven[13] have divided the sources of power and influence into five types:

1. *Legitimate or position power (authority),* stemming from one's appointment or election as leader.
2. *Reward power,* the power to reward others for cooperation.
3. *Coercive or punishment power,* stemming from fear of punishment.
4. *Expert power,* stemming from a person's capability and reputation.
5. *Referent power,* based on an attraction to or identification with another individual (or the program or cause that person is leading) that makes the follower want to behave or believe as the other does; it is similar to what is commonly called *charisma,* a special personal gift for inspiring others that is easier to give examples of than to define.

Thamhain[14] bases his "System I" style of engineering program management on the first three of these five "bases of influence" (legitimate, reward, and coercive power), which derive *primarily* from one's formal position, and which are normally sufficient to obtain adequate (if not enthusiastic) response in traditional bureaucratic structures. In many jobs that engineers hold in modern "high-tech" organizations, and especially in project management, the formal authority granted is not enough to persuade others to get the job done. In this case the combination of expert and referent powers Thamhain calls "System II" style, which stem primarily from one's personal capabilities and reputation, are necessary for effective leadership. Even when System I power is ample, the addition of System II influence makes the manager even more effective.

Pringle et al.[15] list some additional sources of power that have been suggested by others: (6) power through access to important individuals, (7) power obtained through ingratiation or praise, (8) manipulative power, (9) power of persistence or assertiveness, and (10) power gained through forming coalitions. Engineers may feel that they should automatically be granted enough power to get the job done and may find the "office politics" involved in acquiring power distasteful. Humphrey takes a more pragmatic approach:

> While power is the ability to cause action, politics is the art of obtaining power. Power and politics are important management concerns because they form the basis for all dealings between managers.[16]

Status and Culture

Status refers to one's standing within a group or society in general, and it may lead to deference or special privileges. Two types may be distinguished. *Functional status* derives from one's type of work or profession; it explains the deference shown the physician in a hospital or (sometimes) the professor in a nonacademic setting. A scientist may try to maintain his or her status, for example, by hanging a lab coat in a conspicuous spot and demanding a blackboard on the wall (and certainly a computer in his or her office!). The other is *scalar status,* due to one's level in the organization. In some companies an engineer may begin in the "bullpen" (a sea of desks in a common room), share a head-high enclosure with one other person as a senior engineer, get a private office with a desk *and* table as a supervisor, and as a higher manager have both an adjoining conference room and a secretary. At some point the engineer may no longer be required to "clock in" his or her time card and, at a higher level, may finally get a key to the mythical executive washroom.

In some organizations such trappings are deliberately avoided to lessen the *social distance* between different levels of the organization and to promote close cooperation between all members of the "team." The president may close the executive lunchroom, eliminate time clocks entirely, and walk in from the edge of the parking lot. One may describe the collectivity of such practices and habits as the *corporate culture;* corporate executives should try to foster in their organization the culture that will be most effective in achieving the goals of the organization.

DELEGATION

Assignment, Delegation, and Accountability

Three interrelated concepts of importance are the *assignment of duties, delegation of authority,* and *exaction of accountability,* as shown in Figure 6-3. Managers use their authority to *assign duties* to subordinates, making them *responsible for* carrying out the specified activities. This assignment proceeds in stages from top management down. A company president may assign responsibility for all technical matters to the vice president for research and engineering; the vice president may assign responsibility for all project matters to a chief project engineer, who in turn assigns the duty of carrying out a specific project to engineer X.

Once a subordinate has been assigned tasks to perform, it is important to provide him or her with the resources needed to carry out the assignment. This is called *delegation of authority* and can include authority over people who will be needed to carry out the assign-

Figure 6-3. Assignment, delegation, and accountability.

ment as well as financial authority to acquire the equipment, perform the travel, or make other commitments of resources needed. Like assignment of duties, delegation of authority proceeds in stages from top management down. It is an essential management precept that "authority should be commensurate with responsibility," so that a subordinate has enough authority to carry out assignments effectively. Unfortunately, in many cases (especially in managing projects) the engineer is not given enough authority, and he or she must rely on personal influence, persuasion, or the threat (veiled or not) of appeal to higher authority.

When the manager has assigned duties to a subordinate and delegated authority to carry them out, he or she is still not through. The manager must exact (insist on or require) *accountability* from the subordinate by making the subordinate *responsible to* the manager for carrying out the duties and reporting progress periodically. The manager has now made the subordinate "responsible for" the task and "responsible to" report progress, but the manager is still accountable (responsible) to the next higher level of executive to assure that the task is effectively carried out—hence the saying "you can't delegate responsibility."

Reasons for Delegation

Delegation relieves the manager of work the subordinate is capable of doing, substituting the need to assure that the work is actually done. The subordinate, on the other hand, is given a chance to develop his or her skills by being delegated more and more responsible problems. While some subordinates prefer the security associated with very detailed supervision, those with the most future potential will respond favorably to the delegation of increasing responsibility and initiative. Further, delegation tends to locate decision making closer to the work being performed, and this often results in more practical and prompt decisions.

Barriers to Delegation for Engineers

The engineer has been trained in a rigorous discipline and has been held responsible for every calculation and every decimal place through four or more years of college and subsequent years of engineering practice. When an engineer becomes a manager, however, he or she must now be responsible for the work of other people, and this can be especially threatening to the engineer. The engineer-manager has the responsibility to train new subordinates carefully (often with the help of his or her more experienced subordinates) and to assign jobs within the capability of the subordinate.

Just as a mother needs to cut the "apron strings" that limit the growth of a child's capability, the manager needs to give subordinates increasing room to grow in capability, which comes only through practice in carrying out increasingly difficult assignments. This requires the manager to let subordinates do their own work, even though the manager might do it quicker or in some way better (but the subordinate has more time, and as long as the subordinate's decision is appropriate, it need not be the same one the manager would make). The manager must realize that subordinates will make errors (just as managers do), and learn to trust subordinates as they gain skill, yet institute a set of broad controls to assure that those decisions that are truly critical are properly reviewed.

Insecure managers load themselves with their subordinates' problems through inadequate delegation. Oncken and Wass in their classic "Management Time: Who's Got the

Monkey"[17] give examples of how this can happen: Subordinate meets manager in the corridor with "Boss, we've got a problem." The manager responds with "I'm in a rush—I'll get back to you" or "Send me a memo." In the first case the "monkey" (responsibility for the next action on the problem) has just jumped from the back of the subordinate to that of the manager; in the second case it will come riding in that afternoon on the requested memo.

Oncken and Wass recommend, instead, calling the subordinate in for an interview whose stated purpose "is to take a monkey (problem), place it on the desk between them, and figure out together how the next move might conceivably be the subordinate's." They propose that the manager establish the following ground rules for such interviews:

> At no time while I am helping you with this or any other problem will your problem become my problem. The instant your problem becomes mine, you will no longer have a problem. I cannot help a man who hasn't got a problem.
>
> When this meeting is over, the problem will leave this office exactly the way it came in—on your back. You may ask my help at any appointed time, and we will make a joint determination of what the next move will be and which of us will make it.[18]

Oncken and Wass recognize five degrees of initiative: (1) *wait* until told (lowest initiative); (2) *ask* what to do; (3) *recommend,* then take resulting action; (4) *act,* but advise at once; and (5) *act on own,* then routinely report (highest initiative). Managers need to eliminate levels 1 and 2 as early as possible, requiring the "completed staff work" of level 3 (bringing a suggested solution with each problem), and progressing to levels 4 and 5 in most problems as soon as the experience of the subordinate justifies this.

Decentralization

As organizations become larger, it no longer is effective (or even feasible) to make all decisions at the top. Alfred Sloan Jr. recognized this when he introduced *decentralized management* to General Motors about 1920.[19] This concept, which permitted the tremendous early growth achieved by GM, essentially involves the widespread use of delegation throughout the organization. Lower-level decisions can usually be made more rapidly and can often be better than higher-level decisions because they are made closer to the problem.

In times of growth, when opportunities abound if seized promptly, decentralized management can be very effective. This is especially true where the *profit center* concept can be implemented, and the lower-level manager can be given responsibility for the major factors (usually both production and sales) that determine the profit contribution from a particular product, be held accountable for results, and be rewarded for success. Recent reshaping of corporate structures has resulted in elimination of several levels of middle management and concurrent increases in the number of people reporting directly to each surviving manager. Today's managers must learn, just to survive in the modern organization, to delegate more and to coach rather than command subordinates.

The hazard inherent in decentralization is loss of control at the top, and Sloan's contribution was the effective balance of decentralized management with *centralized control* of key decisions (often the allocation of major financial resources). If top management does not retain this control, decisions made at lower levels can bankrupt the total company. Especially in times of recession and financial losses, where expenses must be cut and hard decisions on reducing operations and personnel must be made, effective top management may have to institute some *recentralization,* taking back some decision-making authority that was earlier delegated in order to avert disaster.

COMMITTEES AND MEETINGS

The committee is probably the most maligned, yet one of the most common, forms of organization structure. Milton Berle is credited with the description "the committee is a group of men who keep minutes and waste hours." A camel is derisively described as "a horse designed by a committee," but there are transportation problems a camel can solve that a horse cannot!

A committee is created when two or more people are officially designated to meet to pursue some specific purpose. Committees may be found in every type of organization: large and small, public and private, profit-making, governmental, and volunteer. Some (standing) committees may have indefinite life and may be required in an organization's bylaws; others (ad hoc) may be appointed for a specific purpose and be discharged when the purpose is met.

Reasons for Using Committees

Committees provide some definite advantages over actions by single individuals. Some of the more important reasons for using committees follow.

Policy making and administration. The highest level in most organizations is a policy-making committee, which may be called the board of directors, city council, or by some other name. Such a group typically meets monthly or quarterly. Between such meetings, operating decisions are often made by a subset of this group called an *executive committee* or by a general management committee consisting of the major executive officers.

Representation. Organizations have many committees composed of representatives selected from each organizational unit affected by a particular class of problems. Universities abound in such committees, from the academic senate and graduate council to the tenure committee, publications committee, and many more, and they are present in all organizations of any size that make any pretense of participative management. Committee members are supposed to reflect the opinions and needs of the units that sent them in group deliberation. In engineering design, for example, the *configuration control board* needs to know the impact on cost and schedule of a proposed system design change from all affected areas. Representatives from production, training, documentation, scheduling, and subcontracting as well as affected design engineering groups may be part of that committee to assure that the effect on their functional responsibility is considered before a change is made.

Sharing knowledge and expertise. Engineers meet many situations where no one person has the knowledge necessary to solve a complex problem or carry out a complex function. The *engineering design review,* for example, requires the participation of reliability, quality, safety, and manufacturing engineers and other specialists in addition to the original designers to assure that a complex new system design is ready for production.

Securing cooperation in execution. Committees consisting of the leaders of affected groups or their appointed representatives can identify any problems created by a proposed change in operation. In the ensuing discussion their viewpoint is fully aired, and when the change takes place, they should at least feel that they had their "day in court." Japanese companies are famous for the (in Western eyes) interminable meetings used to achieve consensus; once consensus is achieved, however, implementation may be very rapid and trouble-free. The American model, on the other hand, values decisive executive decision making, but this speed is often at the expense of a lack of cooperation or even opposition when the executive tries to impose this solution on managers who had no part in the decision.

Pooling of authority. Sometimes no one person has been delegated the necessary authority to solve a problem. A committee of people who collectively have authority over the resources needed to solve a problem can do so by joint decision.

Training of participants. Developing personnel for greater management responsibility should be a conscious activity of all managers. Service on a committee consisting of representatives of various organizations is excellent education for potential managers: they are given the chance to consider viewpoints they are not exposed to in their daily work, and they gain experience in expressing and defending ideas and solving interdisciplinary problems. Some organizations will deliberately create "junior management committees," staff them with a rotating membership of promising young employees, and assign them real problems that higher management does not have time to consider fully. More often, no new committees are created, but managers will treat committee assignments as opportunities to give needed experience to employees who would benefit from this wider viewpoint.

Problems with Committees

There are real dangers in the use of committees. First, perhaps, is the tendency to produce compromise solutions to which no one on the committee has any serious objection (although often solutions about which no one can be very enthusiastic) in order to expedite the committee decision and "get back to work." Thus committee decisions are often inferior to those of the most competent group member. Another problem is the inability to hold any one person responsible (accountable) for the committee decision. Both problems are largely resolved when the committee is considered an advisory group and the convening authority (who often chairs the committee) makes the final decision and is accountable for it.

Another serious problem with committees is the delay often experienced in resolution of a problem by a committee. Where quick action is needed, organizations built on committees may lose opportunities to more agile (frequently smaller) organizations.

Making Committees Effective

Committee purpose and chair.[20] The first step in making a committee effective is careful definition of the purpose the committee is to serve, its authority, and its expected life span. Just as important is selection of the proper person as committee chair—someone who has the skill and clout to run an effective meeting (see "Conduct of the Meeting," below). These actions are the responsibility of the *convening authority,* who establishes the need for the committee and normally "kicks off" the first meeting but does not necessarily serve as chair or even as a member of the committee.

Committee size and membership. The typical committee has relatively few members. In a survey of 1658 committees, Tillman[21] reported an average membership of eight, but respondents averaged 4.6 members when asked to identify an ideal committee size. From small-group research, "there does appear to be logical and empirical support for groups of five members as a suitable size, if the necessary skills are possessed by the five members. . . . Where problems are complex [as they are in many engineering situations], relatively larger groups have been shown to produce better quality decisions."[22] Committee members should be selected carefully—they should be expected to contribute to but not dominate committee activity, and they should collectively represent the skills needed for problem solution and/or the organizations whose resources are needed to solve the problem faced and whose cooperation in implementing the solution is essential.

Preparation for the meeting. Preparation is largely the responsibility of the committee chairman. He or she should publish the date and time of the meeting well in advance so that those attending can put it on their calendars. An *agenda* for the meeting should be in members' hands a few days to a week in advance, and it should include any background material that needs to be read in advance and identification of specific information members need to bring. Finally, the chairman should assure that the meeting room selected is conducive to effective interaction, and that desired supplies (pads, pencils, late handouts, coffee, or soda) will be on hand.

Conduct of the meeting. Benne and Sheats[23] believe that every group leader has two types of functions: *task functions,* to get the work done; and *group relations functions,* to maintain constructive relations among the members to facilitate attaining group objectives. They list as task functions: (1) initiating (tasks, goals, ideas, or procedures); (2) information or opinion seeking; (3) information or opinion giving; (4) clarifying (the status of discussion); (5) summarizing; and (6) testing for consensus. As group relations functions they list: (7) encouraging (the contributions of others); (8) expressing group feelings; (9) harmonizing (to reconcile disagreement); (10) modifying (one's own position in light of group progress); (11) gatekeeping (to keep communication channels open); and (12) evaluating (group function, production, commitment, and results). Groups are more effective where members assist the chairman in carrying out these functions.

Meeting follow-up. The chairman should assure that concise, readable *minutes* of the meeting are supplied to all attendees and other interested persons, preferably by the day after the meeting. Minutes should identify the *action items* for which each attendee promised to be responsible, their deadline dates, and the time and place of the next meeting if appropriate. The chairman of an ad hoc committee should be sensitive to the time at which the committee has met its objectives or outlived its usefulness, and he or she should recommend its dissolution to the convening authority when this occurs.

DISCUSSION QUESTIONS

6-1. Outline the steps a large high-technology organization goes through to identify its plan for personnel acquisition for the next year. Identify the uncertainties that apply to each step.

6-2. Prepare a résumé of your qualifications meeting the criteria described in this chapter. What (if anything) might you like to add to or delete from this résumé format?

6-3. Company representatives take a wide variety of approaches to campus interviews. If you can remember such interviews (or other interviews if you are beyond your campus days), critique the approaches used by several such interviewers and your own preparation for and responses to them.

6-4. If you have been invited to a site (plant) visit as a result of a campus (or other) interview, what preparation can you make to optimize the outcome of the visit?

6-5. Describe a performance appraisal technique or form with which you are familiar and assess its strengths and weaknesses.

6-6. Give an example in which Barnard's "acceptance theory" of authority seems to apply especially well.

6-7. What are some conditions under which a formal or System I style of leadership would be most effective, and where would you rely on System II style?

6-8. It is a management dictum that authority should be equal to responsibility. Identify situations where this is not true, and suggest how this weakness might be compensated for.

6-9. Is it reasonable that managers from backgrounds other than engineering might find delegation easier? Support your conclusion.

6-10. Identify several committees from organization(s) with which you are familiar and discuss the reasons for using a committee in each application.

6-11. Summarize the responsibilities of (**i**) the committee chair and (**ii**) a committee member (**a**) before, (**b**) during, and (**c**) after a committee meeting.

6-12. Describe the purpose, size, preparation for, conduct, and follow-up involved in a committee meeting with which you are familiar, and critique each of these.

NOTES

1. E. J. Horton, Jr., personal communication.

2. Adele Lewis, *The Best Résumés for Scientists and Engineers* (New York: John Wiley & Sons, Inc., 1988).

3. Engineering bachelor degrees as a percentage of all bachelor degrees fluctuate substantially, largely as a function of the apparent demand for engineers at the time of college entry. This ratio declined from a post–Korean War high of 9.6% in 1959 to a low of 3.9% in 1976 (reflecting a low in demand for engineers in 1970) to a high of 7.8% in 1984, and it has declined slightly from there. For year-by-year data see Fig. 1 in Betty M. Vetter, "Demographics of the Engineering Student Pipeline," *Engineering Education,* May 1988, p. 734.

4. Marilyn Moats Kennedy, "How to Talk Money in a Tough Market," *Graduating Engineer,* February 1993, pp. 51–56.

5. *CPC Salary Survey, September 1994,* College Placement Council, Bethlehem, PA, pp. 1–7.

6. "National Salary Survey," *Engineering Horizons,* Fall 1994, pp. 30–31, summarized from the 28th Income and Salary Survey, National Society of Professional Engineers, Alexandria, VA, 1994.

7. Charles D. Pringle, Daniel F. Jennings, and Justin G. Longnecker, *Managing Organizations: Functions and Behaviors* (Columbus, OH: Merrill Publishing Company, 1988), p. 250.

8. Evelyn Eichel and Harry E. Bender, *Performance Appraisal: A Study of Current Techniques* (New York: American Management Association, Research and Information Service, 1984), as summarized in Ronald D. Moen, "The Performance Appraisal System: Deming's Deadly Disease," *Quality Progress,* November 1989, p. 62.

9. James T. Hatlan, "Managerial Appraisals—A Systems View," unpublished Master's thesis, University of Missouri—Rolla, 1972.

10. Ronald D. Moen, "The Performance Appraisal System: Deming's Deadly Disease," *Quality Progress,* November 1989, p. 62.

11. Chester I. Barnard, *The Functions of the Executive* (Cambridge, MA: Harvard University Press, 1938), p. 163.

12. Barnard, *Functions,* pp. 168–169.

13. John R. P. French Jr. and Bertram Raven, "The Bases of Social Power," in D. Cartwright, ed., *Studies in Social Power* (Ann Arbor, MI: Research Center for Group Dynamics, 1959).

14. Hans J. Thamhain, *Engineering Project Management* (New York: John Wiley & Sons, Inc., 1984), pp. 218–219.

15. Pringle et al., *Managing Organizations,* p. 300.

16. Watts S. Humphrey, *Managing for Innovation: Leading Technical People* (Englewood Cliffs, NJ: Prentice-Hall, Inc., 1987), p. 157.

17. William Oncken Jr. and Donald L. Wass, "Management Time: Who's Got the Monkey?" *Harvard Business Review,* 52:6, November–December 1974, pp. 76–80.

18. Oncken and Wass, "Management Time," p. 79.

19. Alfred P. Sloan Jr., *My Years with General Motors* (New York: Doubleday & Company, Inc., 1964).

20. The author recognizes the modern, if awkward, usage of "chairperson," but considers that the older forms of "chairman" and "chair" have always referred to both men and women.

21. Rollie Tillman Jr., "Problems in Review: Committees on Trial," *Harvard Business Review,* 38 May–June 1960, pp. 6–12, 162–172.

22. A. C. Filley, "Committee Management: Guidelines from Social Science Research," *California Management Review,* 13:1, Fall 1960, p. 15.

23. K. D. Benne and P. Sheats, "Functional Roles of Group Members," *Journal of Social Issues,* 4, Spring 1948, pp. 41–49.

CHAPTER 7

Motivating and Leading Technical People

PREVIEW

In the first section of this chapter (on *motivation*) we first consider McGregor's two contrasting viewpoints (Theories X and Y) on the nature of the individual who is to be motivated and led. Then we study two types of approaches to understanding motivation: content theories and process theories. Content theories are based on human needs; we discuss three: Maslow's hierarchy of needs, Herzberg's two-factor theory, and McClelland's acquired need theory. Process theories assume behavior is determined by expected outcomes; we look at four: Adams's equity theory, Vroom's expectancy theory, the Porter–Lawler extension of them, and Skinner's behavior modification.

In the second section we consider the nature of *leadership,* beginning with traditional trait theory and its application to the engineering manager. Several related approaches emphasize two dimensions, one concerned with tasks and one with people. These approaches include The Leadership Grid®, the Michigan and Ohio State studies, and life-cycle theory. We then discuss the leadership continuum (a situational approach) and Cribbin's characterizations of "effective" and "merely successful" leadership styles.

In the third and final section we consider briefly the nature of technical professionals such as engineers, scientists, and technicians, and we discuss how they can most effectively be motivated and led.

MOTIVATION

Introduction

Some definitions. To have an effective technical organization we need to understand the nature of motivation, especially as it applies to technical professionals. Berelson and Steiner have defined *motive* as "an inner state that energizes, activates, or moves (hence 'motivation'), and that directs or channels behavior toward goals."[1] Robbins defines *motivation* in an organizational sense as "the willingness to exert high levels of effort to reach organizational goals, conditioned by the effort's ability to satisfy some individual need."[2]

Campbell et al. define motivation in terms of three measures of the resulting behavior:

1. The *direction* of an individual's behavior (measured by the choice made when several alternatives are available)
2. The *strength* of that behavior once a choice is made
3. The persistence of that behavior[3]

Shannon concludes that "there is only one way to get people to do what you would like them to do, and that is by making them *want* to do it. Motivation flows from within the individual."[4] Therefore, we need to learn why people want to do things, and how they can be persuaded (or motivated) to do those things that will enhance organizational goals.

McGregor's Theory X and Theory Y. The way we try to motivate someone depends on our assumptions about their basic nature. Douglas McGregor postulated two contrasting sets of assumptions about the average worker, calling them Theory X and Theory Y. In his Theory X, he painted a dismal picture of the nature of the average person and its implications for the task of management:

The conventional conception of management's task in harnessing human energy to organizational requirements can be stated briefly in terms of three propositions. In order to avoid the complications introduced by a label, let us call this set of propositions "Theory X":

1. Management is responsible for organizing the elements of productive enterprise—money, materials, equipment, people—in the interest of economic ends.
2. With respect to people, this is a process of directing their efforts, motivating them, controlling their actions, modifying their behavior to fit the needs of the organization.
3. Without this active intervention by management, people would be passive—even resistant—to organization needs. They must therefore be persuaded, rewarded, punished, controlled—their activities must be directed. This is management's task. . . .

Behind this conventional theory there are several additional beliefs—less explicit, but widespread:

4. The average person is by nature indolent—he works as little as possible.
5. He lacks ambition, dislikes responsibility, prefers to be led.
6. He is inherently self-centered, indifferent to organizational needs.
7. He is by nature resistant to change.
8. He is gullible, not very bright, the ready dupe of the charlatan and the demagogue.[5]

However depressing, this view appears to fit the model Frederick Taylor (Chapter 2) had of the worker as lazy, not very bright, but willing to produce more if the management elite defines exactly how the job should be done and provides incentive pay as an inducement. McGregor, however, suggests that such behavior is not necessarily inherent in human beings, and he concludes that the carrot-and-stick approach of relying on wages for motivation "does not work at all once man has reached an adequate subsistence level and is motivated primarily by higher needs" (see "Content Theories" in the next section). He continues:

> For these and many other reasons, we require a different theory of the task of managing people based on more adequate assumptions about human nature and human motivation. I am going to be so bold as to suggest the broad dimensions of such a theory. Call it "Theory Y," if you will.
>
> 1. Management is responsible for organizing the elements of productive enterprise—money, materials, equipment, people—in the interest of economic ends. [Identical to (*1*) for Theory X]
> 2. People are *not* by nature passive or resistant to organizational needs. They have become so as a result of experience in organizations.
> 3. The motivation, the potential for development, the capacity for assuming responsibility, the readiness to direct behavior toward organization goals are all present in people. Management does not have to put them there. It is the responsibility of management to make it possible for people to recognize and develop these human characteristics for themselves.
> 4. The essential task of management is to arrange organizational conditions and methods of operation so that people can achieve their own goals *best* by directing *their own* efforts toward organizational objectives.[6]

McGregor summarized by saying that "Theory X places exclusive reliance upon external control of human behavior, while Theory Y relies heavily on self-control and self-direction. It is worth noting that this difference is the difference between treating people as children and treating them as mature adults." Peter Drucker later emphasized the necessity of abandoning Theory X, especially in managing *knowledge workers* such as engineers and scientists:

> What has happened is that the general social and economic environment has changed to make Theory X tools obsolete inside most organizations. The traditional carrot and stick do not work. In developed countries, they do not work for manual workers. And nowhere do they work for knowledge workers. . . .
>
> The knowledge worker confronts a very different situation. He is the new majority. . . . This historic shift in the nature of work makes Theory Y a necessity. The knowledge worker simply does not produce under Theory X. Knowledge has to be self-directed; the knowledge worker has to take responsibility. The point was nicely driven home in the cartoon of a young man slouched in his soap company office blowing smoke rings at a big THINK sign. One older man passing the door said to another, "But how can we be sure that Smith thinks soap?" Only Smith, of course, can know whether he thinks soap, or whether that is the best way to do his job.[7]

Content versus process theories.   Theories trying to explain how people are motivated are commonly divided into two categories. *Content theories* are based on human needs and people's (often unconscious) efforts to satisfy them. *Process theories,* on the other hand, assume that behavioral choices are made more rationally, based on the expected outcomes. We examine each category in turn, with special emphasis on their application to the technical professional.

Content Theories

Maslow's hierarchy of needs. One of the earliest and most influential content theories is the concept of Abraham H. Maslow that "human needs arrange themselves in hierarchies of prepotency. That is, the appearance of one need usually rests on the prior satisfaction of another."[8] Maslow identified the following five needs, which are often portrayed in stair-step function as in Figure 7-1:

1. *Physiological needs:* air, water, food, shelter, sex
2. *Safety needs:* safe work, with security that the physiological needs will continue to be met (through job tenure and medical, unemployment, and disability insurance and retirement provisions)
3. *Love needs:* affectionate relations with friends, family, and people in general, and group acceptance
4. *Esteem needs:* self-respect or self-esteem, and the esteem of others (expressed in reputation, prestige, and recognition)
5. *Self-actualization (or self-fulfillment) needs:* the desire to become everything one is capable of becoming (to become actualized in what one is potentially)

Maslow believed that this was an approximate order of need satisfaction for most people, although there were exceptions. The "mad scientist" working alone in the corner appears to value self-esteem and self-fulfillment needs ahead of the need for love and affection. Porter et al. believe that Maslow is at least right in distinguishing between lower-order (physiological and safety) needs and the other three:

> There is strong evidence to support the view that unless the existence needs are satisfied none of the higher-order needs will come into play. There is also some evidence that unless security needs are satisfied, people will not be concerned with higher order needs. . . . There is, however, little evidence to support the view that a hierarchy exists once one moves above the security level.[9]

Figure 7-1. Maslow's hierarchy of needs.

Over most of human history the concern for survival (meeting the physiological needs) has been paramount. Industrial workers in the developed nations today, however, commonly find their physiological needs satisfied and most security needs met through fringe benefits, except where layoff is a threat. The higher-order needs can be fulfilled either at work or in society. Self-actualization, for example, can be achieved through hobbies, education and personal growth, and charitable or religious activities in the community as well as through achievement at work. The approval of friends and respect of the community can substitute in part for lack of recognition at work. The challenge to effective management is to find ways in which the higher-order needs of the individual can be satisfied in the process of achieving the objectives of the organization.

Herzberg's two-factor theory. Frederick Herzberg studied the factors affecting job attitudes and found that they could be divided into two groups: those that provided motivation when they were present, and those "hygiene" factors that led to job dissatisfaction when they did not meet expectations. In Herzberg's own words:

> The growth or *motivator* factors that are intrinsic to the job are [in order of decreasing importance]: achievement, recognition for achievement, the work itself, responsibility, and growth or advancement. The dissatisfaction-avoidance or *hygiene* factors that are extrinsic to the job include [also in order of decreasing importance]: company policy and administration, supervision, interpersonal relationships, working conditions, salary, status, and security.
>
> A composite of the factors that are involved in causing job satisfaction and job dissatisfaction, [was] drawn from samples of 1,685 employees. . . . The results indicate that motivators were the primary cause of satisfaction, and hygiene factors the primary cause of unhappiness on the job. The employees, studied in 12 different investigations, included lower-level supervisors, professional women, agricultural administrators, men about to retire from management positions, hospital maintenance personnel, manufacturing supervisors, nurses, food handlers, military officers, engineers, scientists, housekeepers, teachers, technicians, female assemblers, accountants, Finnish foremen, and Hungarian engineers.
>
> They were asked what job events had occurred in their work that had led to extreme satisfaction or extreme dissatisfaction on their part. . . . of all the factors contributing to job satisfaction, 81% were motivators. And of all the factors contributing to the employees' dissatisfaction over their work, 69% involved hygiene elements.[10]

Herzberg's hygiene factors correspond well to the lower three of Maslow's needs (physiological, safety/security, and love/relationship), and his motivators with the upper two (esteem and self-fulfillment). Herzberg considered salary as primarily a hygiene factor, and certainly it leads a person to be dissatisfied when their salary is less than they think is merited, or when they are given a smaller raise than the employee at the next desk. However, salary is "a way of keeping score," and a healthy raise can be clear recognition for one's work and in that sense motivating. Bonuses and profit sharing can be motivating as well. For example, Worthington Industries has a profit-sharing program that can amount to half an employee's total compensation. John H. McConnell, Worthington's founder, chairman, and CEO, reports:

> Our people care about quality. If customers don't accept our shipments, part of the cost comes out of each of our pockets. So people take the time to do their job right the first time. Our rejection rate is less than one percent, compared to an industry average of three to five percent.[11]

Herzberg developed the methodology of *job enrichment* to increase the content of motivators in a job. Examples of job enrichment actions include reducing the number and frequency of controls, making the worker responsible for checking his or her own work, establishing a direct relationship between the worker and the customer or user of that work (whether internal or external), and in other ways increasing authority and autonomy.

Job enrichment and the underlying two-factor theory have attracted many disciples, who have applied it in a wide variety of environments. There have been quite a few critics as well. Myers believes that people may be categorized as either *motivation seekers,* who respond well to job enrichment, or *maintenance seekers,* who "are motivated primarily by the nature of their environment and tend to avoid motivational opportunities . . . are chronically preoccupied and dissatisfied with maintenance [hygiene] factors surrounding their job . . . [and] realize little satisfaction from accomplishment and express cynicism toward the positive values of work and life in general."[12] In an extended study at Texas Instruments, Myers found that engineers, manufacturing supervisors, hourly male technicians, and especially scientists tended to be motivation seekers, whereas female assemblers tended to be maintenance seekers.

Based on analysis of job-enrichment efforts and attitude surveys involving primarily blue-collar workers, Fein reported that essentially all job-enrichment efforts are initiated by management, not by a desire of workers or their unions to make jobs more meaningful, and concluded:

> For the most part [blue-collar] workers are satisfied with the nature of their work. What they find most discomforting is their pay, their job security, and many of the work rules with which they must cope.[13]

In the 1980s many American companies, especially the automobile industry, tried to reduce the number of categories of production workers by asking workers to learn several jobs, so that they could be used more flexibly and labor cost could be reduced. In essence, this amounted to job enrichment for the benefit of corporate profit and, ultimately, the survival of the plants against foreign competition. American union workers fought this attempt bitterly.

McClelland's trio of needs. David McClelland[14] and others have proposed that there are three major motives or needs in work situations:

    1. *Need for achievement* is the drive or desire to excel, to accomplish something better than has been done in the past. People with a high need for achievement tend to be entrepreneurs, setting moderately difficult goals, taking moderate risks to achieve them, and taking personal responsibility for getting things done. Although McClelland estimates that only about 10% of the American population has a high need for achievement, he has shown that the need for achievement can be increased through proper training.[15]

    McClelland also believed that the higher the need for achievement in the total society, the greater the prosperity of the country.[16] Civilizations and nations that expect a lot from their youth have high-achieving societies. The academic achievement demanded of most children in Japan—and of college-bound children in many other countries—presents a serious challenge to American primary and secondary schools.

2. *Need for power* is the desire to control one's environment, including resources and people. Persons with a high need for power are more likely to be promoted to managerial positions and are likely to be successful managers if they master self-control and use their power for the good of the organization rather than solely for personal ends.

3. *Need for affiliation* is the need for human companionship and acceptance. People with a strong need for affiliation want reassurance and approval, are concerned about other people, and perform well as coordinators, integrators, counselors, and in sales positions.

The need for affiliation might be compared with Maslow's third level (love), the need for power with his fourth level (ego or esteem), and the need for achievement with the fifth level (self-actualization). However, McClelland's point is that different people have different needs, not just the same needs in a clear hierarchy of importance. For example, an engineer with a high need for achievement may achieve success in technical assignments in the process of satisfying this need, and he or she might be promoted into a management position as a result. If this need for achievement is combined with a low need for power, the engineer will often peak earlier in his or her career and at a lower level, since the need for achievement can be satisfied by the work itself rather than (as with need for power) requiring continuing promotions. Again, engineering jobs that put a premium on coordination and cooperation, such as today's team management organizations or the matrix organizations common in project management (Chapter 15), certainly require a blend of need for achievement and for affiliation.

Process Theories

Process theories treat human needs as just one part of the mechanism that people use in choosing their behavior, and these theories place greater emphasis on the expectation of favorable consequences or rewards. We consider four theories in this group: equity theory, expectancy theory, the Porter–Lawler extension of the first two, and behavior modification.

Equity theory. Developed by J. Stacey Adams,[17] this theory is based on the simple belief that people want to be treated fairly relative to the treatment of others. Adams describes this comparison in terms of input/outcome ratios. Inputs are a person's contribution to the organization in terms of education, experience, ability, effort, and loyalty. Outcomes are the obvious rewards of pay and promotion and the more subtle ones of recognition and social relationships. A person who feels underrewarded compared with someone else may (1) put forth less effort, (2) press for a higher salary (or a bigger office or a reserved parking place), (3) distort the perceived ratio by rationalizing, or (4) leave the situation (quit or transfer). Conversely, a person who feels overrewarded *may* be motivated to contribute more.

For example, in the fall of 1988 the University of Missouri, pleading lack of resources, offered faculty an average raise of 2%. This was met with only routine grumbling until it was discovered that the top administrators of the university had been awarded raises averaging over 10% by the governing Board of Curators "to avoid losing them to other schools." The resulting furor raised in faculty organizations was moderated only when the president of the university donated his raise to the campus library fund and the university made increases in faculty salaries its top financial priority.

Engineering faculty inevitably make another comparison—their financial rewards compared with the pay and promotions of their peers in industry—and this is a major cause of the shortage of American-born engineering faculty and doctoral students training to replace them. Humanities professors, on the other hand, fail to understand why engineering faculty should earn more (based on the industrial market argument) than they do.

Another example arises from the "two-tier" pay systems that have arisen in an attempt by companies to provide a more competitive payroll. For example, several years ago airlines with unionized flight attendants earning $36,000 a year were able to negotiate a rate for new hires rising only to half that level as long as existing union members were not affected. The position was still desirable enough to attract new candidates, who would have been satisfied indefinitely except for the obvious inequity of working with others who were earning twice the income for no more effort. In the biblical story of the talents (Matthew 25:14–30), the workers who arrived early in the field were perfectly agreeable to working through the heat of the day for the unit of money (one talent) offered until they found that workers starting in the last hour of the afternoon were being given the same pay.

Expectancy theory. Formulated in 1964 by Victor Vroom,[18] expectancy theory relates the effort a person puts forth to the expectation of achieving some desired goal. As illustrated in Figure 7-2, this involves a combination of two *expectancies:*

Effort-to-performance expectancy is a person's perception of the probability that his or her effort will lead to high performance, usually in meeting an organizationally desired goal. The ability to achieve high performance (first-order outcomes in the model) is considered a function of individual ability and the environment (tools, resources, and opportunity), in addition to the effort applied.

Performance-to-outcome expectancy, also known as *instrumentality,* is the person's perception that attaining the performance described above will lead to intrinsic and extrinsic rewards (second-order outcomes). Intrinsic rewards are intangibles such as a feeling of accomplishment or sense of achievement; extrinsic awards are tangible results such as pay or promotion. Both the effort-to-performance and the performance-to-outcome expectancy may be rated on a scale from 0.0 (no relationship) to 1.0 (certainty).

Valence measures the strength of a person's desire for these outcomes (which may be positive or negative) and is related to the individual needs we have already considered.

According to this theory, motivation can be calculated as the product of the values assigned to these three factors. For example, a student's motivation to study for a final examination may be a function of (1) the expectation that study (effort) can lead to a good grade in the final, (2) the expectation that a good grade in the final can influence the grade for the course, and (3) the value the student places on earning high grades.

Although it is difficult to make use of this model quantitatively, it offers some qualitative suggestions to the manager. Effort-to-performance expectancy may be increased by

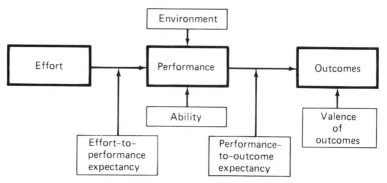

Figure 7-2. Expectancy theory of motivation.

assuring that the person understands the objectives that he or she is asked to achieve and has the training, self-confidence, and organizational support to achieve them. Performance-to-outcome expectancy can be increased by trying to discover what the person values as an outcome, and trying to provide that reward for good performance. Choate believes that U.S. industry is at risk because it does a poor job in providing this motivation:

> Dan Yankelovich, for example, reports that only 9 out of 100 American workers think that if they work harder and smarter they'll get any benefit out of it. By contrast, 93 out of 100 Japanese workers think that if they make an extra effort, they'll get extra benefits out of it. That's an important difference.[19]

The Porter–Lawler extension.[20] Lyman W. Porter and Edward E. Lawler proposed an extension of the expectancy model involving employee satisfaction. It may be compared to Figure 7-3 as follows:

- Personal effort, abilities and traits, and role perceptions (the employee's belief that certain tasks need to be done to do his or her job effectively) determine performance.
- Performance, in turn, leads to intrinsic and extrinsic rewards, as in the expectancy model.
- The perceived equity (fairness) of these rewards determines the satisfaction the employee gains from the work.
- This satisfaction colors the value placed on the rewards anticipated for future cycles of work, and therefore it influences future effort.

Behavior modification. This theory, also known as reinforcement theory or operant conditioning, has its foundations in the work of B. F. Skinner.[21] Behavior is followed by an event (*reinforcement*) that affects the probability that the behavior is repeated. Four major types of reinforcement are available to the manager:

1. *Positive reinforcement* increases the probability that desired behavior will be repeated by providing a reward (praise, recognition, raise, promotion, or other). When a sea lion jumps through a hoop it is given a fish as positive reinforcement; even with professional employees the principle is the same, but one hopes the implementation is more subtle.

2. *Negative reinforcement,* or *avoidance,* seeks to increase the probability that desired behavior will be repeated by letting the employee escape from undesired consequences.

3. *Punishment* seeks to decrease the probability that undesired behavior will be repeated by imposing penalties (undesired consequences) such as reprimands, discipline, or fines. Because punishment often leads to resentment and even poorer performance, managers try to use it as a last resort.

4. *Extinction* seeks to decrease the probability that undesired behavior will be repeated by ignoring it and withholding positive reinforcement. For example, an employee's dumb remark at a meeting might be ignored, but the next commonsense suggestion made may be followed with "good thinking." (The reverse is also true: repeated failure to recognize *desired* behavior can lead an employee to think that it is not important and to stop doing it.)

LEADERSHIP

Nature of Leadership

Leadership is the process of getting the cooperation of others in accomplishing a desired goal. Sir William Slim, commander of the British Army that defeated the Japanese in Burma in World War II, defined leadership as that "mixture of persuasion, compulsion, and example that makes men do what you want them to do." Harry Truman explained: "You know what makes leadership? It is the ability to get men to do what they don't want to do and like it." In a more subtle vein, Barney Frank said: "The great leader is the one who can show people that their self-interest is different from that which they perceived."

People become leaders by appointment or through emergence. *Formal,* or "titular," leaders are appointed branch *manager* or committee *chair* or team *captain* and have the advantage of formal authority (including the power to reward and punish), but this only gives them the *opportunity* to prove themselves effective at leadership. Although good leaders prefer to influence others through persuasion where possible, I must agree with Robert Shannon that "it is much easier to be persuaded by a person with power than by one without power."[22]

Emergent, or *informal,* leaders evolve based on their expertise or referent power as it is expressed in the process of group activity. Even as children we find certain individuals emerging as the ones whose suggestions for the games to play or the mischief to get into are followed, and throughout life we find certain people "take the lead" and are accepted as informal leaders. When the emergent leader is then appointed or elected as formal leader, he or she has a double opportunity to be effective. Recognizing this, many organizations conduct "assessment centers" to evaluate potential leaders, and these include group situations where no leader is appointed in order to see who emerges to lead the resolution of a jointly assigned problem.

Leadership traits. Early researchers into the nature of leadership tried to identify the personal characteristics, or *traits,* that made for effective leaders. For example, Peterson and Plowman list the following 18 attributes as being desirable in a leader:

- *Physical qualities* of health, vitality, and endurance;
- *Personal attributes* of personal magnetism, cooperativeness, enthusiasm, ability to inspire, persuasiveness, forcefulness, and tact;
- *Character attributes* of integrity, humanism, self-discipline, stability, and industry; and
- *Intellectual qualities* of mental capacity, ability to teach others, and a scientific approach to problems.[23]

Harris[24] had this list of 18 qualities and attributes evaluated by a group of 176 engineers, mostly electrical, mechanical, and aerospace engineers working for high-technology firms in the Dallas, TX, area. There were two phases to this research. In the first phase 130 engineers, divided into three different ranges of engineering experience, were asked to rate each of the 18 characteristics individually as they perceived their necessity for effective leadership in the engineering environment. The results appear in Table 7-1.

The attribute considered *most necessary* by less experienced engineers was *ability to inspire,* whereas engineers with intermediate (6 to 15 years) experience most valued *enthusiasm* in a leader. The attribute in a leader that apparently becomes more highly valued with experience is *integrity,* rated in eighth place by young engineers, fourth place by those with intermediate experience, and first place by engineers with more than 15 years of experience. The attribute considered *least necessary* was *health,* followed (in seventeenth place) by forcefulness, then by humanism (empathy) and vitality.

In the second phase of the research, Harris asked an additional 46 engineers who repeated the evaluation above to rate their current engineering managers on the same scale. He then calculated the difference between the mean ranking of the perceived necessity for each quality or attribute with the mean rating of current engineering managers. He found that engineering managers most exceeded the perceived need in the following categories (identified by the *t*-score of the difference of means):

- 6.95 Health
- 4.12 Endurance
- 3.79 Scientific approach to problems
- 3.69 Vitality
- 3.67 Forcefulness

On the other hand, these engineering managers were least successful in meeting expectations in the following categories:

- -9.16 Ability to inspire
- -5.36 Tact
- -4.82 Persuasiveness
- -4.17 Stability
- -2.88 Enthusiasm

Harris summarizes his research: "The results quite clearly show that engineers want and expect excellent leaders. The results also show that they are not getting what they want."

TABLE 7–1 Highest- and Lowest-Ranked Qualities and Attributes in Engineering Managers

Group I 0–5 Years Engineering Experience		Group II 6–15 Years Engineering Experience		Group III > 15 Years Engineering Experience	
		Highest-Ranked Qualities and Attributes			
1.	Ability to inspire	1.	Enthusiasm	1.	Integrity
2.	Persuasiveness	2.	Stability	2.	Ability to inspire
3.	Mental capacity	3.	Self-discipline	3.	Tact
4.	Self-discipline	4.	Ability to inspire	4.	Stability
5.	Enthusiasm	5.	Integrity	5.	Self-discipline
6.	Tact	6.	Mental capacity	6.	Persuasiveness
7.	Stability	7.	Persuasiveness	7.	Industry
8.	Integrity	8.	Cooperativeness	8.	Enthusiasm
9.	Cooperativeness	9.	Ability to teach	9.	Mental capacity
		Lowest-Ranked Qualities and Attributes			
18.	Health	18.	Health	18.	Health
17.	Forcefulness	17.	Vitality	17.	Forcefulness
16.	Personal magnetism	16.	Forcefulness	16.	Ability to teach
15.	Humanism	15.	Personal magnetism	15.	Personal magnetism
14.	Vitality	14.	Humanism	14.	Humanism
13.	Endurance	13.	Endurance	13.	Cooperativeness
12.	Industry	12.	Industry	12.	Vitality
11.	Scientific approach to problems	11.	Scientific approach to problems	11.	Scientific approach to problems
10.	Ability to teach	10.	Tact	10.	Endurance

Source: E. Douglas Harris, "Leadership Characteristics: Engineers Want More from Their Leaders," *Proceedings of the Ninth Annual Conference, American Society for Engineering Management*, Knoxville, TN, October 2–4, 1988, ASEM, 1988, pp. 209–216.

When Harris repeated this research with European engineers he got similar results, except that he found engineers in Europe were even less satisfied with their managers than were engineers in Texas.

Connolly[25] discusses studies showing that neither appointed nor informal leaders need be much above the average intelligence of the group. He shows that the development and acceptance of emergent leaders are facilitated by social skills, by technical skills in the specific tasks facing the group, and by being at the hub of a communication net.

Myers–Briggs preferences. The *Myers–Briggs Type Indicator (MBTI)*®[26] measures personal preferences on four scales, each made up of two opposite preferences:

1. *Extraversion E* (focused on the outer world of people and things) versus *Introversion I* (focused on the inner world of ideas and impressions);
2. *Intuition N* (focused on the future, with a view toward patterns and possibilities) versus *Sensing S* (focused on the present and on concrete information gained from the senses);

3. *Thinking **T*** (basing decisions on logic and on objective analysis of cause and effect) versus *Feeling **F*** (basing decisions on values and on subjective evaluation of person-centered concerns); and

4. *Judging **J*** (preferring to have things settled—a planned and organized approach to life) versus *Perceiving **P*** (preferring to keep your options open—a flexible and spontaneous approach to life).

Engineers and scientists frequently are evaluated as *ENTJ* or *INTJ;* successful engineering managers often are *ENTJ;* researchers in technical areas (and the engineering deans who are often chosen from them) *INTJ.* Only about 2% of the total population test as being in these two categories; this helps explain why engineering faculty, with preferences toward organized, logical, and theoretical presentation, often fail to "reach" those engineering students whose preferred modes of learning differ; it also helps explain why many political decisions on technical issues just "don't make sense" to the logical engineer!

The difficulty in trait or preference theories is that for every characteristic proposed, one can find no shortage of undeniably effective leaders who seem weak in that area.

People/Task Matrix Approaches

The Leadership Grid®. Robert R. Blake and Jane S. Mouton[27] developed the Managerial Grid®, an approach to analyzing the style of management (that is, collective leadership) in terms of two dimensions: *concern for people* and *concern for production* (now *concern for results*). The latest version of this approach is the Leadership Grid® shown in Figure 7-3. This approach assumes that "9,9 Team Management," in which individual objectives are achieved in the process of achieving organizational goals, is the ultimate in effective management. The grid can be used with related analyses and interventions to achieve *organizational development* by helping the management of client organizations identify their current management style and then work toward the recommended 9,9 style. This approach assumes that concern only for people (1,9) leads to a workplace that is enjoyable but not productive, that concern only for results (9,1) leads to a nonresponsive Theory X work force, that settling for "adequate" performance and morale in 5,5 Middle of the Road Management (once called Organization Man or Bureaucratic Management) leads only to mediocrity, and that low concern for both people and results is a sign of "impoverished management."

Michigan and Ohio State studies. The late Rensis Likert and associates at the University of Michigan Institute for Social Research conducted a series of studies comparing the effectiveness of *job-centered* and *employee-centered* supervision. The first type (job-centered) emphasizes the work to be performed, while the second (employee-centered) emphasizes development of effective work groups. In a related approach, researchers at Ohio State University studied the four possible combinations of high or low *initiating structure* with high or low *consideration* (see Table 7-2). Initiating structure refers to the task actions of the leader to define who does what and how, while consideration is a measure of the leader's concern for followers.

Intuitively, one would expect the "high–high" leadership style in each of these cases to be the most attractive, but this does not always prove true. Connolly states:

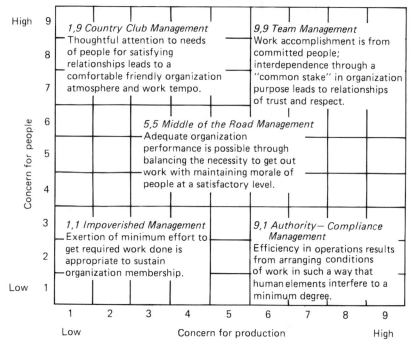

Figure 7-3. The Leadership Grid® Figure.
(From *Leadership Dilemmas—Grid Solutions,* by Robert R. Blake and Anne Adams
McCanse, Gulf Publishing Company, Houston TX, copyright 1991 by Scientific Methods
Inc., p. 29; reproduced by permission of the owners).

There are, then, fairly clear exceptions to the rule that high-high is the best management style. Structuring behavior helps only if the task is unstructured. Consideration helps satisfaction only if there is no adequate alternative source of satisfaction (such as interesting work). Providing either structure or consideration when not needed is unhelpful and may actually hinder. Studies suggesting that high-high leaders are often effective may thus tell us more about the work people do than about good leadership: if most people are working in ill-structured, uninteresting jobs, high-high leaders may be effective; structuring helps productivity and consideration introduces at least some payoffs interpersonally, which increases satisfaction.[28]

TABLE 7–2 Ohio State Leadership Styles

Initiating Structure (IS)	Consideration (C)	
	Low	High
High	High IS, low C	High IS, high C
Low	Low IS, low C	Low IS, high C

Hersey and Blanchard life-cycle theory. Hersey and Blanchard[29] proposed an extension of the model above (under the name "life-cycle" or "maturity" theory) in which the most effective leadership progresses with time through the four quadrants of Table 7-2. For example, in teaching a child a simple task such as tying a shoe, a parent initially concentrates on details of the task (high IS, low C). Then, while continuing to correct task errors, the parent praises the child for successes (high IS, high C), continuing praise after the task has been learned (low IS, high C). Finally, when the task has been ingrained, it no longer requires the attention of the parent (low IS, low C), and attention can be shifted to more advanced tasks.

Situational Approaches

The leadership models described in the two-dimensional approaches above (except for Hersey and Blanchard's) imply that only two factors (one dealing somehow with people and the other with task or production) are important, and that there is one best combination of the two for effective leadership. In 1958, Tannenbaum and Schmidt argued that there really was a continuum of available leadership styles, and one's choice within this continuum should be contingent on the situation. Boone and Bowen assess the significance of this work:

> With the appearance of this article the perspective of *contingency theory,* the dominant theme in management and organizational theory for the next twenty years, was introduced.
>
> Contingency theory basically argues that there is no one right way to manage. The manager must develop a reward system, a leadership style, or an organizational structure to be appropriate for the unique combination of such factors as the nature of the subordinates, the technology of the business and the tasks that result, the rate of change in the organization, the degree of integration of functions required, the amount of time the manager has to accomplish the assignment, the quality of the manager's relationship with subordinates, and so forth.[30]

Leadership continuum. Tannenbaum and Schmidt proposed[31] a continuum of leadership style extending from complete retention of power by the manager to complete freedom for subordinates (they now prefer the term "nonmanagers" to "subordinates"). Although they identify seven "styles of leadership" along the continuum, others have emphasized these four:

1. *Autocratic ("Telling").* Manager makes decisions with little or no involvement of nonmanagers.
2. *Diplomatic ("Selling).* Manager makes decisions without consultation but tries to persuade nonmanagers to accept them (and might even modify them if they object strongly).
3. *Consultative ("Consulting").* Manager obtains nonmanagers' ideas and uses them in decision making.
4. *Participative ("Joining").* Manager involves nonmanagers heavily in the decision (and may even delegate it to them completely).

Tannenbaum and Schmidt proposed that a manager should consider three types of forces before deciding what management style to employ:

1. *Forces in the manager.* His [their pronoun] value system regarding leadership and

his own leadership inclinations, his confidence in the nonmanagers, and his feelings of security (or "tolerance for ambiguity") in an uncertain situation.

2. *Forces in the subordinate (or nonmanager).* Greater delegation can be provided when nonmanagers have a need for independence, are ready to assume responsibility, can tolerate ambiguity, are interested in the problem, understand and relate to the goals of the organization, have the necessary knowledge and experience, and have learned to expect a share in decision making.

3. *Forces in the situation.* The type of organization and the amount of delegation common in it, the experience and success the (subordinate) nonmanagers have had in working together as a group, the nature and complexity of the problem, and the pressure of time.[32]

The Hersey and Blanchard model discussed previously is really a situational model in which leadership styles are selected from a 2×2 matrix rather than a linear continuum, based on "forces in the subordinate" (specifically, maturity). With a new (or antagonistic or lethargic) worker the manager will first emphasize the task (Telling; high IS/low C); next use Selling (high/high), then Consulting (low/high), and then Joining (low/low). In essence the subordinates can choose the leadership style: if they want to participate in decision making, they need only demonstrate the necessary maturity and the skilled leader will encourage them to do so.

Other viewpoints. Other authors have tried to characterize leaders in more complex ways. Cribbin has identified 14 types of executives by their behavior.[33] Eight of these types (Table 7-3) he concludes, are "merely successful" (and could be more effective were it not for some serious weaknesses). The other six types—the entrepreneur, corporateur, developer, craftsman, integrator, and gamesman (Table 7-4 on page 138)—are effective leaders in appropriate situations.

For brevity in describing effective leadership, it is difficult to top the following, attributed to the Chinese philosopher Lao Tsu about 600 B.C., and etched in copper in the office of Jack Smith, CEO of General Motors:

> *A leader is best*
> *when people barely know he exists.*
> *Not so good when people obey and acclaim him.*
> *Worse when they despise him.*
> *But of a good leader, who talks little,*
> *when his work is done and his aim fulfilled,*
> *they will say, "We did it ourselves."*[34]

MOTIVATING AND LEADING TECHNICAL PROFESSIONALS

Now that we have studied the general theories of human motivation and of leadership, let us try to apply them to the technical professional. We will first recall something of the nature of the professional, discuss what motivates scientists and engineers, and finally consider the significance of these factors in the effective leadership of technical professionals.

TABLE 7–3   Leaders Who Are Merely Successful

Executive	Motto	Characteristics	Typical Behavior
Bureaucrat	"We go by the book."	Rational, formal, impersonal, politely proper, disciplined. May be slow-moving and/or jealous of his or her function, rights, and prerogatives. Well versed in the organizational "rocks and shoals."	Follows the letter of the law. Stickler for rules and procedures. Task-oriented, less concerned with people. Logical strategist but may be politically astute and/or a nitpicker.
Zealot	"We do things my way, in spite of the organization."	A loner. Impatient, outspoken, overly independent, extremely competent. Jumps the traces, a nuisance to the bureaucrats. Insensitive to the feelings of others. Modest political skills. Fair but demanding.	Devoted to the good of the organization, *as he or she sees it.* Excessively task-oriented but has little concern for people. Aggressive and domineering. Is insistent but fiercely supports all who are on his or her side.
Machiavellian	"We depersonalize and use you."	Self-oriented, shrewd, devious, calculating, amoral, manipulative. Excellent insight into people's weaknesses. Extremely opportunistic. Flexible, ranges from seeming collaboration to pitiless aggression. Cold but can be charming.	Treats people as things to be exploited and outwitted. Cooperates only when it is to his or her advantage. Personal considerations do not enter into thinking. Must win at any price and in any way possible.
Missionary	"We love one another."	Much too concerned with people and what they think of him or her. Subjective in orientation. Likable but tries too hard to be liked. Excellent interpersonal skills but does not win respect. Insists that conflict and friction be smoothed over.	A soft manager who prizes harmony above all else. Low task orientation. Gets emotionally involved. Acts on a personal basis. Tends to do what is popular or will make him or her liked. Inclined to ignore harder organizational requirements.
Climber	"I vault over anyone I can."	Striving, driving, energetic, self-oriented. Often smooth and polished but always aggressive. Usually opportunistic, always plotting next move or maneuver. No loyalty to the organization or to anyone in it. Often quite competent. Constantly fronts self.	High political skills. Excellent at maneuvering into the limelight. Predatory toward weaker managers. Welcomes and initiates self-propelling change. May have high task orientation but for self-serving purposes, not for the good of the firm. Adroit with people but has no interest in them.

TABLE 7–3 (cont'd)

Executive	Motto	Characteristics	Typical Behavior
Exploiter	"When I bark, they jump."	Arrogant, insistent, abusive. Demeaning, coercive, vindictive, domineering. Often quite competent. Rigid, prejudiced, given to snap judgments. Exploits others' weaknesses.	Exerts constrictive and personal controls. Flogs anyone who is vulnerable. Uses pressure and fear to get things done. Demands subservience. High task orientation. Sees people as minions.
Temporizer	"We bend to the strongest pressure."	Procrastinating, compromising, vacillating. Earns contempt. Feels a helpless sense of being put upon. Survival instincts may be superior. May be politically aware.	Low task orientation, low people concern. Reacts to the strongest immediate pressure. Reactive, not active. Behavior varies with pressures.
Glad-Hander	"We sell the sizzle, not the steak."	Ebullient, superficial, effusive, deceptively friendly, extroverted. Excellent interpersonal skills. Lacks depth, minimally competent. May be an excellent politician. Survival instincts superior. Talkative, humorous, lacks substance.	Sells himself or herself very well. Low or modest task orientation. Unconcerned with people but excellent in dealing with them. Gets by on "personality." Always seeks to impress and to improve his or her position. May use people but rarely threatens them.

Source: Reprinted by permission of the publisher, from *Leadership: Strategies for Organizational Effectiveness*, by James J. Cribbin, pages 36–37 © 1981 AMACOM, a division of American Management Association, New York. All rights reserved.

General Nature of the Technical Professional

A number of authors (for example, Kerr et al.[35] and Rosenbaum[36]) have examined the special characteristics of technical professionals (without distinguishing between scientists and engineers). They are typically described as:

- Having a *high need for achievement* and deriving their motivation primarily from the work itself. As such they are most productive when they can achieve their professional goals in the process of pursuing organizational goals.
- Desiring *autonomy* (independence) over the conditions, pace, and content of their work. To achieve this, they need to participate in goal setting and decision making as it affects their work.
- Tending to *identify first with their profession* and secondarily with their company. As professionals, they look to their peers (whether inside or outside the organization) for recognition, ethical standards, and collegial support and stimulation.
- Seeking to *maintain their expertise*, gained through long and arduous study, and stave off obsolescence (see Chapter 16) through continuing education, reading the

TABLE 7–4 Leaders Who Are Effective

Executive	Motto	Characteristics	Typical Behavior
Entrepreneur	"We do it my way. Only risk-taking achievers need apply."	Extremely competent, forceful, individualistic, egocentric, dominant, self-confident. Extraordinary achievement drive. Innovative, very firm-minded and strong-willed. Something of a loner. Not only listens to his or her own drummer but composes his or her own music. Can be very loyal, protective, and generous to team.	Unable to work well in a subordinate position for very long. Must be prime mover and binds small team to him or her with great loyalty. Offers challenges, opportunities to succeed, and great returns on risks taken. Does not develop subordinates. Is not open to ideas that differ from his or her own. Gets involved in all aspects of the organization. Exercises very tight control. Motivates by example, rewards, and fear.
Corporateur	"I call the shots, but we all work together on my team."	Dominant but not domineering. Quite directive but gives people considerable freedom. Consultative but not really participative. Sizes up people well but relates to them on a surface level. Cordial to people but keeps them at arm's length.	Concerned about the good of the organization. Wins respect. High task orientation. Polished and professional manager. Makes people feel needed. Delegates and consults but keeps effective control. Supportive but not emotionally involved with subordinates.
Developer	"People are our most important resource."	Trustful of subordinates. Intent on helping them actualize their potential. Excellent human relations skills. Wins personal loyalty, builds a supportive and achieving climate. Fine coach and counselor.	Very high people orientation. Although productivity is superior, at times people considerations take precedence. People feel needed. Delegates and consults but keeps effective control. Supportive and emotionally involved with subordinates.
Craftsman	"We do important work as perfectly as we can."	Amiable, conservative, extremely conscientious. Principled, very knowledgeable and skilled, self-reliant. Highly task-oriented. Proud of competence. Work- and family-oriented. Self-contented, honest, straight-forward, perfectionistic, independent, analytical, mild-mannered.	Likes to innovate, build, and tinker with quality products. Not overly concerned with status or politics. Motivated by a desire for excellence. Self-demanding but supportive of subordinates. Competes with projects, not people. Restive with organizational red tape. Likes to solve problems alone or in a small group.

TABLE 7–4 (cont'd.)

Executive	Motto	Characteristics	Typical Behavior
Integrator	"We build consensus and commitment."	Egalitarian, supportive, participative. Excellent interpersonal skills. Superior people insight. A team builder, catalyst, adept at unifying different inputs. A subtle leader, prefers group decision making.	Shares the leadership. Thinks in terms of associates rather than subordinates. Gives great freedom and authority. Welcomes the ideas of others. Geared to win-win interaction. Acts as a synergistic catalyst.
Gamesman	"We win together, but I must win more than you."	Fast-moving, flexible, upwardly mobile. Very knowledgeable and skilled. Autonomous, risk-taking, assertive, and intent on winning but not petty or vindictive. Innovative. Takes no great pleasure in another's loss or defeat. Opportunistic but not unethical, not depressed by defeat.	Wants to be respected as a strategist who builds a winning team. Enjoys the game of winning within the organization's rules. Enjoys competition, jockeying, and maneuvering. Sharp, skilled, unbiased, and tough manager who challenges and rewards contribution. Impersonally eliminates the weak and non-achievers.

Source: Reprinted by permission of the publisher, from *Leadership: Strategies for Organizational Effectiveness*, by James J. Cribbin, pages 44–45 © 1981 AMACOM, a division of American Management Association, New York. All rights reserved.

literature, professional society activity, and especially through work assignments that keep them working at the state of the art.

Differences Among Technical Professionals

Scientists versus engineers. Scientists and engineers often differ in significant ways. Allen[37] and others have identified some of these differences:

- Even as undergraduates, science students place higher value on independence and learning for its own sake; engineers are more concerned with professional preparation, success, and family life.
- The "true scientist" is commonly assumed to have a doctorate; the typical engineer generally begins professional practice with a B.S. degree, and he or she typically earns a master's degree later.
- The scientist puts a high value on professional autonomy and publication of results; the engineer is a team worker and places little value on publication.
- Although both groups desire career development and advancement, the scientist depends heavily on reputation with peers outside the company; the engineer's advancement is tied more to activities within the company. The engineer therefore is motivated more by organizational goals, is comfortable with more applied assignments, and is more likely to seek tangible rewards within the organization.

- Science grows through evolutionary additions to the literature, to which the scientist wants to be free to add; "the technologist's principal legacy to posterity is encoded in physical, not verbal, structure."[38] Further, the engineer is more likely to be working with developments that are considered proprietary information by the organization and thus has less opportunity to publish results.

Field of technical employment. Raudsepp[39] authored a 1960 survey in which 3000 engineers, classified by their occupational field, weighed the relative importance of factors in choosing a new position (Table 7-5). All groups reported that interesting, diversified work was more important to them than salary, but it was especially important to R&D engineers. Similarly, R&D engineers placed a higher value on opportunity to learn, while engineers seeking management, sales, and production positions put a higher value on opportunity for promotion.

Leading Technical Professionals

Dimensions of technical leadership. Rosenbaum[40] believed that, to facilitate achievement of individual and group goals, successful technical leaders should master "five strategic dimensions":

1. *Coach for peak performance*—"listen, ask, facilitate, integrate, provide administrative support"; act as a sounding board and supportive critic; help the professional manage change.
2. *Run organizational interference*—obtain resources, act as advocate for the professional and his/her ideas, and minimize the demands of the bureaucracy (time and paperwork) on the professional.
3. *Orchestrate professional development*—facilitate career development through challenging assignments; foster a business perspective in professionals; find sources where new areas of knowledge are required.
4. *Expand individual productivity through teamwork*—make sure teams are well oriented regarding goals and roles, and that they get the resources and support they need.
5. *Facilitate self-management*—assure that technical professionals are empowered to make their own decisions by encouraging free two-way information flow, delegating enough authority, and providing material and psychological support.

Leading as orchestration. McCall[41] has evaluated a number of studies of the relationship between a formal leader and a follower group of professionals, mostly in R&D settings. He concludes that in such groups "effective supervisory leadership is more orchestration than direct application of authority. It seems a matter of creating and/or maintaining (or at least not destroying) conditions that foster scientific productivity." While the supervisor is not the only factor determining group effectiveness, McCall identifies four general areas where the leader can make a difference:

1. *Technical competence.* "The supervisor's technical competence is related both to scientific productivity and the scientists' willingness to comply with management directives. . . . Leaders of productive groups serve many roles that depend on tech-

TABLE 7-5 Motivating Factors by Fields of Engineering (Percent)

Factors (In Order of Importance)	Total	Fields of Engineering								
		Research	Development	Design	Operation	Production	Admin. Mgt.	Sales	Other	No Answer
Type of work, interesting, diversified	45.0	52.1	45.7	45.0	42.8	41.8	41.1	43.9	46.5	41.7
Salary	33.9	31.3	39.5	37.8	25.2	32.9	34.0	31.1	23.2	16.7
Location, good place to live, family	31.2	33.6	37.3	30.2	27.7	33.6	27.1	13.5	24.5	50.0
Opportunity for advancement	29.8	22.6	25.6	27.7	27.0	32.2	39.5	36.5	22.6	8.3
Challenge, more responsibility, chance to use creative ability	16.9	18.5	15.5	17.0	20.1	19.2	18.7	22.3	25.8	25.0
Reputation, prestige of company	13.7	9.1	13.5	13.6	13.8	15.8	14.9	12.8	11.6	8.3
Working conditions, personnel policies	11.7	14.0	12.1	11.9	13.2	10.3	9.6	14.9	7.7	8.3
Growing organization, growing field	6.9	6.4	9.3	5.8	5.7	7.5	8.4	6.1	5.8	—
Security, retirement plan, benefits	6.8	7.2	7.3	6.3	12.6	3.4	6.9	5.4	3.2	8.3
Opportunity to learn, broaden experience, training programs	6.6	11.3	7.9	6.0	5.0	7.5	5.9	2.7	6.4	—
Small company	4.1	2.6	3.3	4.3	4.4	6.2	4.3	4.1	3.9	8.3
Job was available	3.4	2.6	2.6	4.9	3.1	4.8	2.8	3.4	3.2	—
Progressive research and development program	2.8	8.7	3.5	3.6	3.8	2.1	2.6	3.4	1.9	—
Own business, partnership, independence	2.7	2.6	2.0	4.0	4.4	2.1	3.5	6.8	3.9	—
Type of product	2.5	0.8	2.6	2.7	1.3	0.7	1.6	4.1	2.6	8.3
Previous association with company	1.8	2.6	1.5	1.6	1.9	3.4	1.0	1.4	1.9	8.3
Public service, humanistic reasons	1.3	1.5	0.7	0.9	2.5	2.1	1.2	2.7	1.9	8.3
Opportunity to travel	1.3	—	0.4	1.8	1.9	—	1.0	4.1	1.9	—
Regular salary increases	0.4	0.8	0.4	0.7	—	0.7	0.6	—	1.9	—
All others	1.9	1.9	2.2	1.3	1.3	0.7	2.8	2.0	1.9	—
No answer	2.7	1.5	1.5	2.7	4.4	3.4	2.4	0.7	2.6	8.3

Source: Eugene Raudsepp, "Why Engineers Work," *Machine Design*, February 4, 1960.

nical expertise, including: recognizing good ideas emerging inside and outside the group; defining the significant problems, influencing work goals on the basis of expertise; and providing technical stimulation."

2. *Controlled freedom.* "In general, leaders of productive groups create controlled freedom, a condition in which decision making is shared but not given away, and autonomy is partially preserved."

3. *Leader as metronome.* McCall views this image as "perhaps the best statement of the subtlety of leadership in professional groups," and quotes Sayles and Chandler[42] as describing the job of project manager as one that "widens or narrows limits, adds or subtracts weights where tradeoffs are to be made, speeds up or slows down actions, increases emphasis on some activities and decreases emphasis on others."

4. *Work challenge.* Since challenging work is one of the most important things to a professional, the technical manager is measured by the extent to which he or she can provide challenging assignments. The professional's view of what is challenging must be reconciled with the needs of the organization, and the challenge to the supervisor is not just making wise assignments, but structuring them as much as possible to provide the desired challenge and then persuading the individual of their importance.

Breakpoint leadership. McCall confined himself above largely to direct supervision of a group of technical professionals, especially in R&D. Then he added:

> At some point on the way up the managerial ladder, a different kind of leadership demand occurs. When influencing other parts of the organization is as important, or more important, than influencing a subordinate group, leadership is a breakpoint. Effectiveness is no longer measured simply as group productivity, but involves such things as impact on organizational direction, influence across organizational and even hierarchical boundaries, and securing and protecting organizational (and external) resources and support. . . . For many professionals the first breakpoint leadership role is that of project manager.[43]

We study the special leadership challenges of project management in Chapter 15.

Use of Motivational Theories by Engineers

Utley and Westerbrook[44] conducted a survey of 408 engineering managers in Tennessee to determine which motivational theories they felt familiar with, and which theories they were actually using. As Table 7-6 shows, they were most familiar with three related concepts not discussed in this chapter: management by objectives (MBO, discussed in Chapter 3), quality circles (discussed in Chapter 12), and the findings of Peters and Waterman (*In Search of Excellence,* introduced toward the end of Chapter 2). Of the theories discussed in this chapter, they were most familiar (in descending order) with Maslow's hierarchy, McGregor's Theory X and Theory Y, Herzberg's two-factor theory, the Managerial Grid® (now the Leadership Grid®), and the Tannenbaum and Schmidt leadership continuum; they thought they were making use of them in about the same order. Engineering managers in government agencies favored use of MBO and quality circles more then did managers in government contractor organizations and private industry; other management concepts were used with about the same frequency in the three types of organizations. Top-level managers were more

TABLE 7–6 Familiarity and Use of Motivational Theories by Engineering Managers[a]

Theory	High Tech n = 229 F	U	Med Tech n = 179 F	U	Top Level n = 95 F	U	Mid Level n = 162 F	U	First Level n = 151 F	U	Total n = 408 F	U	Ranking n = 17 F	U
Herzberg	50	28	39	16	58	31	45	21	37	21	45	23	6	4
Maslow	69	40	59	34	76	49	67	40	55	26	64	37	4	2
McGregor	60	26	51	18	64	28	59	22	48	19	56	23	5	4
Managerial Grid	44	17	40	12	52	20	38	14	40	13	42	15	7	7
Likert System IV[b]	9	1	11	1	8	1	10	1	11	2	10	1	*	*
In Search of Excellence[c]	66	20	64	15	87	41	63	18	54	15	65	22	3	6
McClelland	21	7	15	3	16	6	19	3	19	7	18	5	*	*
Porter and Lawler	31	15	26	9	32	15	28	14	28	10	29	13	11*	8
Likert linking pin[d]	17	7	13	4	13	5	15	4	18	7	16	6	*	11*
Vroom[e]	10	0	8	2	9	0	9	1	10	1	10	1	*	*
Argyris[f]	16	4	11	2	14	3	10	2	17	4	13	3	*	*
MBO[g]	87	60	85	50	96	68	89	59	77	43	86	55	1	1
Quality circles[h]	86	38	78	34	85	46	83	40	81	26	83	36	2	3
Hersey and Blanchard	19	9	17	5	19	5	17	7	19	9	18	7	*	10
Tannenbaum and Schmidt	36	12	27	8	35	16	33	8	28	9	32	10	8	9
Ouchi[i]	31	5	30	8	45	15	27	4	25	3	30	6	10*	11*
Drucker*	33	6	28	5	41	11	28	4	27	4	31	6	9*	11*

[a]F, familiar with theory; U, use theory.

[b–i]Theories not discussed in the chapter are (b) Likert's four systems varied from System I (very autocratic) to System IV (very participative); (c) *In Search of Excellence* by Peters and Waterman was introduced in Chapter 2; (d) Likert's linking pin concept involves forming teams, starting at the top, for good decision making and communication; (e) Expectancy theory; (f) Argyris's study of personality development concluded that classical management concepts such as chain of command and span of control tended to encourage childlike, dependent behavior; (g) management by objectives, introduced in Chapter 3; (h) quality circles are introduced in Chapter 12; (i) Ouchi's Theory Z applied Japanese management systems to American companies.

*"Drucker's survival principles—modern businessperson's attitude" was the last "motivational principle" in the survey. Although Peter Drucker is perhaps the most widely read contemporary management theorist and author, his survival principles are not related to motivation. The number checking this concept may relate to the approximate error level in the survey, and lower scores on the other theories are not considered significant.

Source: Dawn R. Utley and Jerry D. Westerbrook, "A Survey of Management Concepts in Technical Organizations," *Proceedings of the Ninth Annual Conference, American Society for Engineering Management*, Knoxville, TN, October 2–4, 1988, ASEM, 1988, pp. 345–351.

familiar with these motivational concepts than were lower-level managers. Managers at all levels in high-technology companies were more likely to use motivational concepts than were managers in lower-technology companies.

DISCUSSION QUESTIONS

7-1. Suggest your own definitions for **(a)** motivation and **(b)** leadership.

7-2. For what kind of worker, in what type of environment, does McGregor's Theory X make some sense?

7-3. If you were structuring Maslow's need hierarchy specifically for (**a**) scientists and (**b**) engineers, would you do it any differently?

7-4. Herzberg specifically classed *salary* as a hygiene factor, not a motivator. How would you classify it? Discuss.

7-5. *Job enrichment* seeks to make work more meaningful and give employees more control over their work. Discuss the negative response of many blue-collar production workers toward this initiative. Why do you think workers have this attitude?

7-6. Would you expect the need for affiliation among managers to be somewhat dependent on the culture in which they grew up and/or work? If so, give an example.

7-7. Provide another example of the validity of equity theory as a motivator of human performance at work or elsewhere.

7-8. How might American management increase the "instrumentality" (performance-to-outcome expectancy) of workers?

7-9. If you employ Skinner's positive reinforcement to encourage favorable behavior in employees, should this be done after every instance of proper behavior? At regular intervals? At increasing intervals?

7-10. From among the leadership traits suggested in the text and other sources, plus your own ideas, identify and defend your view of the most important half dozen traits needed by the engineering manager.

7-11. From your analysis of the findings of Harris (Table 7-1), why do you think engineers look for different qualities in their managers as they (the engineers) grow in experience?

7-12. Provide an example of the application of Hersey and Blanchard life-cycle theory to engineering practice.

7-13. Tannenbaum and Schmidt assume that a manager can choose among alternative leadership styles rather than being confined to a single "inborn" style. Do you agree? Discuss.

7-14. Would you regard engineers in an "applied" field such as production management as being any more or less "professional" than a research scientist? Explain your viewpoint.

7-15. Raudsepp reported (Table 7-5) on the parameters that motivated engineers in different fields to accept a position. How, in your view, might their motivations to perform effectively once they were in that position differ?

7-16. McCall's observations on leading small professional groups was based largely on studies of research scientists. Would you expect any difference in leading a group of design engineers? Of production staff specialists such as industrial engineers?

7-17. Which of the 16 concepts of Utley and Westerbrook (Table 7-6, ignoring Drucker) were you familiar with before reading this book? Which do you now feel would be useful to you as an engineering manager? Discuss why.

7-18. Would you expect the factors motivating an engineer to change as he or she proceeds through a career? In what ways? How can the engineering manager make use of these changes?

NOTES

1. B. Berelson and G. A. Steiner, *Human Behavior: An Inventory of Scientific Findings* (New York: Harcourt, Brace, & World, 1964), p. 240.

2. Stephen P. Robbins, *Management,* 4th ed. (Englewood Cliffs, NJ: Prentice-Hall, Inc., 1994), p. 465.

3. John P. Campbell, Marvin D. Dunnette, Edward E. Lawler III, and Karl E. Weick Jr., *Managerial Behavior, Performance, and Effectiveness* (New York: McGraw-Hill Book Company, 1970), p. 340.

4. Robert E. Shannon, *Engineering Management* (New York: John Wiley & Sons, Inc., 1980), p. 173.

5. Douglas M. McGregor, "The Human Side of Enterprise," *Management Review,* November 1957, reprinted by permission of publisher, © 1957 American Management Association, New York. All rights reserved.

6. McGregor, "Human Side of Enterprise," p. 133.

7. Peter F. Drucker, "Beyond Stick and Carrot: Hysteria over the Work Ethic," *Psychology Today,* November 1973, pp. 89, 91–92.

8. Abraham H. Maslow, "A Theory of Human Motivation," *Psychological Review,* 50, 1943, pp. 370–396.

9. Lyman W. Porter, Edward E. Lawler III, and J. Richard Hackman, *Behavior in Organizations* (New York: McGraw-Hill Book Company, 1975), p. 43.

10. Frederick Herzberg, "One More Time: How Do You Motivate Employees?" *Harvard Business Review,* 46:1, January–February 1968, p. 57.

11. "Worthington's Human Resources: Building Quality on the Strength of Its People," *Quality,* February 1988, p. 22.

12. M. Scott Myers, "Who Are Your Motivated Workers?" *Harvard Business Review,* 42:1, January–February 1964, pp. 73–88.

13. Mitchell Fein, "Job Enrichment: A Reevaluation," *Sloan Management Review,* 15:2, Winter 1974, pp. 69–88.

14. David C. McClelland, *The Achieving Society* (Princeton, NJ: D. Van Nostrand Company, 1961), *Power: The Inner Experience* (New York: Irvington Publishers, Inc., 1975), and *Human Motivation* (Glenview, IL: Scott, Foresman and Company, 1985).

15. David C. McClelland, "Achievement Motivation Can Be Learned," *Harvard Business Review,* November–December 1965, pp. 6–24.

16. McClelland, *The Achieving Society.*

17. J. Stacey Adams, "Towards an Understanding of Equity," *Journal of Abnormal and Social Psychology,* November 1963, pp. 422–436.

18. Victor H. Vroom, *Work and Motivation* (New York: John Wiley & Sons, Inc., 1964).

19. Pat Choate, "Where Does Quality Fit in with the Competitiveness Debate?" *Quality Progress,* February 1988, p. 26.

20. Lyman W. Porter and Edward E. Lawler III, *Managerial Attitudes and Performance* (Homewood, IL: Richard D. Irwin, Inc., 1968), p. 165.

21. B. F. Skinner, *Science and Human Behavior* (New York: The Free Press, 1953) and *Contingencies of Reinforcement* (New York: Appleton-Century-Crofts, 1969).

22. Shannon, *Engineering Management,* p. 207.

23. E. Peterson and E. G. Plowman, *Business Organization and Management* (Homewood, IL: Richard D. Irwin, Inc., 1957), pp. 50–62.

24. E. Douglas Harris, "Leadership Characteristics: Engineers Want More from Their Leaders," *Proceedings of the Ninth Annual Conference, American Society for Engineering Management,* Rolla, MO, October 2–4, 1988, pp. 209–216.

25. Terry Connolly, *Scientists, Engineers, and Organizations* (Monterey, CA: Brooks/Cole Engineering Division, a Division of Wadsworth, Inc. of Belmont, CA, 1983), p. 129.

26. Information on the MBTI® can be obtained from Consulting Psychologists Press Inc., 3803 E. Bayshore Road, Palo Alto, CA 94303; most college and employer career or personnel offices can provide Myers–Briggs evaluations.

27. Robert R. Blake and Jane S. Mouton, *The Managerial Grid III: The Key to Leadership Excellence* (Houston, TX: Gulf Publishing Company, © 1985).

28. Connolly, *Scientists, Engineers, and Organizations,* p. 141.

29. Paul Hersey and Kenneth H. Blanchard, *Management of Organizational Behavior: Utilizing Human Resources,* 4th ed. (Englewood Cliffs, NJ: Prentice-Hall, Inc., 1982), pp. 150–175.

30. Louis E. Boone and Donald D. Bowen, *The Great Writings in Management and Organizational Behavior,* 2nd ed. (New York: Random House, Inc., Business Division, 1987), p. 124.

31. Robert Tannenbaum and Warren H. Schmidt, "How to Choose a Leadership Pattern," *Harvard Business Review,* 36:2, March–April 1958, pp. 95–101, reprinted (with "retrospective commentary") in *Harvard Business Review,* 51:3, May–June 1973, pp. 162–180.

32. Tannenbaum and Schmidt, "How to Choose a Leadership Pattern," pp. 162–180.

33. James J. Cribbin, *Leadership: Strategies for Organizational Effectiveness* (New York: American Management Associations, Inc., 1981), pp. 36–37.

34. "Can Jack Smith Fix GM?" *Business Week,* November 1, 1993, p. 126.

35. S. Kerr, M. A. Von Glinow, and J. Schriesheim, "Issues in the Study of Professionals' in Organizations: The Case of Scientists and Engineers," *Organizational Behavior and Human Performance,* 18, 1977, pp. 329–345.

36. Bernard L. Rosenbaum, "Leading Today's Professional," *Research Technology Management,* 34:2, March–April 1991, p. 30.

37. Thomas J. Allen, *Managing the Flow of Technology: Technology Transfer and Dissemination of Technological Information Within the R&D Organization* (Cambridge, MA: The MIT Press, 1977), pp. 35–41.

38. Allen, *Managing the Flow,* p. 40.

39. Eugene Raudsepp, "Why Engineers Work," *Machine Design,* February 4, 1960.

40. Rosenbaum, "Leading Today's Professional," pp. 31–35.

41. Morgan W. McCall, Jr., "Leadership and the Professional," in Connolly, *Scientists, Engineers, and Organizations,* pp. 332–335.

42. L. R. Sayles and M. K. Chandler, *Managing Large Systems* (New York: Harper and Row Publishers, Inc., 1971).

43. McCall, "Leadership and the Professional," pp. 337–338.

44. Dawn R. Utley and Jerry D. Westerbrook, "A Survey of Management Concepts in Technical Organizations," *Proceedings of the Ninth Annual Conference, American Society for Engineering Management,* Rolla, MO, October 2–4, 1988, pp. 345–351.

CHAPTER 8

Controlling

PREVIEW

We begin this chapter by introducing the steps in the classical control process, three types of control, and the characteristics of effective control systems. Most of the chapter deals with financial controls: financial statements (especially the balance sheet and income statement), financial ratios used in ratio analysis, financial and operating budgets and the nature of the budgeting process, and financial audits. Human resource controls such as management audits, human resource accounting, and social controls are discussed briefly. Finally, we introduce other nonfinancial controls that will be discussed in later chapters.

THE PROCESS OF CONTROL

Steps in the Control Process

Perhaps the simplest definition of controlling, attributed to B. E. Goetz, is "compelling events to conform to plans." Another author[1] states that "control techniques and actions are intended to insure, as far as possible, that the organization does what management wants it to do." Control is a *process* that pervades not only management, but technology and our everyday lives. Effective control must begin in planning; as shown in Figure 8-1, planning and control are inseparable.

The steps in the control process are simple. The first step, *establishing standards of*

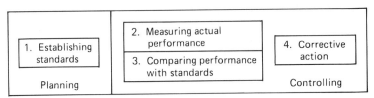

Figure 8-1. The control process.

performance, is an essential part of effective planning. Standards should be measurable, verifiable, and tangible to the extent possible. Examples of standards are a standard rate of production established by work measurement, a budgeted cost of computer usage, a targeted value for product reliability, or a desired room temperature. The second step (and the start of the actual control process) is *measurement of the actual* level of performance achieved. The third step is *comparison of the two,* measurement of the variance (deviation between them), and *communicating this deviation* promptly to the entity responsible for control of this performance. The final step is taking *corrective action* as required to "compel events to conform to plans."

Closed-Loop Versus Open-Loop Control

Closed-loop control, also known as *automatic* or *cybernetic* control, monitors and manages a process by means of a self-regulating system. The essential feature of cybernetic control is a strong feedback system. The common home thermostat provides a simple example of an automatic control process. A desired (standard) temperature is set by adjusting a lever or wheel on the thermostat. A mechanism such as a bimetallic strip or bellows converts the actual temperature surrounding the thermostat into physical movement. When the variance between desired and actual temperature exceeds some design maximum, sensor movement creates an electrical contact that communicates a signal to the correcting entity, in this case the control of a furnace or air conditioner, and the variance is automatically corrected. A more complex application is the automatic control of a nuclear reactor, designed to shut down the reactor under conditions of power surges that could become catastrophic long before a human operator could react.

Open-loop, or *noncybernetic,* control requires an external monitoring system and/or an external agent to complete the control loop. Frequently, the automatic part of the control system provides a warning of a variance from planned values, but then human judgment is required to identify the reason for the variance and to determine corrective action. Even systems that are automated (cybernetic) in the short run are ultimately open loop, because they permit an external agent to adjust the standard (or set point). Cruise control on an automobile, for example, operates automatically, but it may be turned off or set to a different speed by manual control.

In engineering management the last step in the control process, corrective action, usually requires human judgment. Consider the action required when a machining process fails to maintain a specified tolerance of ±0.01 centimeter about some specified (planned) dimension. The problem (and its resolution) might include any of the following:

- The machine used is too worn to maintain such a tolerance (and should be fixed or replaced).
- The operator is not skilled enough to achieve the desired result (and needs training).
- The tolerance specified is more than can be reasonably achieved in the material being machined (and the designer should be asked to relax the specification or choose a more tolerant design or material).

The choice among these solutions and others requires thought and decision making; the control system has done its job when it brings the problem and information surrounding it to the appropriate decision maker.

Three Perspectives on the Timing of Control

Feedback control. Engineers are usually comfortable with the idea of feed*back* systems, in which the output of a system can be measured and the variance between measured and desired output used to adjust the system. Thus the rotational speed of a machine can be measured by the effect of centrifugal force on rotating balls (the traditional "governor"), and the difference between this physical movement and the desired (standard) value can be used to adjust the speed. The thermostat example above is also a feedback system. Such feedback control (also called *post-action* or *output control*) is quite effective for continuing processes or for repetitive actions. For example, the lessons learned in building past McDonald's restaurants have certainly been used to make the next thousand restaurants more efficient. But for many applications managers cannot afford to wait until an activity or product is complete before examining it, because the cost and schedule consequences of late discovery of failure are unacceptable.

Screening or concurrent control. Controls may also be applied concurrently with the effort being controlled. A new engineer may be given an unfamiliar assignment one step at a time, with review by the supervisor after each step. A production schedule may include several in-process inspection points so that further investment in defective parts can be avoided. A baseball coach will observe the effectiveness of a pitcher literally one pitch at a time, prepared at any point to start warming up a replacement in the bullpen. However, concurrent control can be expensive, stifling of initiative, and lead to inactivity while awaiting the next inspection.

Feedforward (or preliminary or steering) control. The essence of feedforward control is a system that can predict the impact of current actions or events on future outcomes, so that current decisions can be adjusted to assure that future goals will be met. Engineers and managers have many applications where controls must be applied in the early phases of a project or program. A nuclear power reactor may take 10 years to produce, and the construction project manager needs management tools that will predict, as the project progresses, whether it is likely to be completed on time and within budget. In Chapter 14 we discuss management tools, such as work breakdown structures and network systems (PERT or CPM), that enable us to identify the longest "critical path" of tasks that must be completed in sequence to complete the project. As the project continues, control over the early tasks in this system gives us "feed*forward*" control over the total project duration. The *earned value* methods of Chapter 15 provide the same sort of feedforward control of costs.

Examples of feedforward control in manufacturing include careful screening of sequences for machine operations, inspection of raw materials, and *preventive maintenance* of machines, all in an attempt to reduce (control) later production problems. The prudent taxpayer does not wait until April 15 to discover his or her tax liability for the previous year; he or she tries to estimate it before the year ends in order to manage cash contributions, sales of stock, and other actions before December 31 to reduce or defer tax. Similarly, the comptroller of a corporation will try to forecast the next period's revenue and sales so that cash will be ready when needed (and effectively invested when not). These also are examples of feedforward control.

Characteristics of Effective Control Systems

An effective control system should satisfy most of the following criteria:

- *Effective.* Control systems should measure what needs to be measured and controlled.
- *Efficient.* Control systems should be economical and worth their cost.
- *Timely.* Control systems should provide the manager with information in time to take corrective action. A tax accounting system is expected to show costs to the nearest dollar, but it does not need to do so for the year ending December 31 until the following April 15. A control system for monthly expenses, however, might be satisfied with ±5% accuracy but demand information within a week after the end of the month measured.
- *Flexible.* Control systems should be tools, not straitjackets, and should be adjustable to changing conditions.
- *Understandable.* Control systems should be easy to understand and use, and they should provide information in the format desired by the users.
- *Tailored.* Where possible, control systems should deliver to each level of manager the information they need for decisions, at the level of detail appropriate for that level.
- *Highlight deviations.* Good control systems will "flag" parameters that deviate from planned values by more than a specified percentage or amount for special management attention.
- *Lead to corrective action.* Control systems should either incorporate automatic corrective action or communicate effectively to an agent that will provide effective action; this is why the control system exists.

Delegation and Control

In Chapter 6 the concept of delegating authority was introduced. Such delegation requires effective control systems to assure that delegated power is not used unwisely. Drucker offers as a topical example "Irangate," in which arms were supplied to Iran despite clearly stated government policy to the contrary. He asserts that (among other things) the Reagan administration missed this basic principle of control and accountability.

First—in one of the most common but also most unforgivable management mistakes—it confused delegation of authority with abdication of responsibility. A chief executive officer must delegate. Otherwise, he'll end up like Gulliver in Lilliput, ineffectual and ensnared in details, as were Lyndon Johnson and Jimmy Carter. But delegation requires greater accountability and tighter control. Delegation requires clear assignments of a specific task, clear definition of the expected results and a deadline. Above all it requires that the subordinate to whom a task is delegated keep the boss fully informed. It is the subordinate's job to alert the boss immediately to any possible "surprise"—rather than to try to "protect" the boss against surprises, as Mr. Reagan's subordinates apparently did. If they keep surprises away from the boss, they invariably will end up making him look incompetent or not in control or a liar—or all three.[2]

FINANCIAL CONTROLS

Financial Statements

Financial statements provide the basic information for the control of cash and credit, which are essential to the survival of a company. There are three major types of financial statements: the balance sheet, the income statement, and the cash flow statement. The *balance sheet* shows the firm's financial position at a particular instant in time—a financial "snapshot," as it were. Table 8-1 gives an example. *Assets* are what the company "owns," and consist principally of *current assets* (assets that can be converted into cash within a year) and *fixed assets* (property, plant, and equipment at original cost, less the cumulative *depreciation* of plant and equipment (but not land) and *depletion* of natural resources since they were purchased). *Liabilities* are what the firm "owes," and consist of *current liabilities* that must be paid within a year and long-term debt. The difference between assets and liabilities is the *net worth* or *equity* of the stockholders, and it consists of the original investment (what was paid in for common and preferred stock) plus the *retained earnings* (the cumulative profits over the years after dividends are paid).

An *income statement* (see Table 8-2), also called a *profit and loss* or *revenue and expense* statement, shows the financial performance of the firm over a period of time (usually three months or a year). Sterling Chemicals had 1995 net sales of about $3.1 million. Production costs (materials, labor, and production overhead costs) were $2.0 million, and the depreciation and depletion related to 1995 production were $0.26 millon. Selling, advertising, and shipping cost $0.10 million, and "general and administrative" expenses ("G&A," the cost of general management, R&D, and miscellaneous activities not chargeable elsewhere) were $0.18 million. Subtracting the total expense of about $2.55 million from net sales leaves an *operating profit* of $0.55 million. After adjusting for interest and other nonoperating income and expense, the pretax income is found to be $0.58 million, and the net income (after taxes) is $0.32 million. The board of directors decided to return part ($0.22 million) of net income to the stockholder-owners as dividends, and to reinvest the rest on their behalf as an addition to retained earnings.

The *cash flow,* or *sources and uses of funds,* statement shows where funds come from (net profit plus depreciation, increased debt, sale of stock, sale of assets) and what they are used for (plant and equipment, debt reduction, stock repurchase, and dividends). Like the income statement, it concerns financial activities over time. Note the special nature of

TABLE 8–1   Balance Sheet, Sterling Chemicals, Inc., December 31, 1995

Assets		
Current assets		
Cash	$150,724	
Securities (at cost)	99,866	$ 250,590
Accounts receivable		416,304
Inventories (at lower of cost or market)		
Raw materials and supplies	208,046	
Work in progress	182,702	
Finished goods	289,610	680,358
Prepaid expenses		29,498
Total current assets		1,376,750
Property, plant, and equipment	4,461,150	
Less accumulated depreciation and depletion	2,402,024	
Net property, plant, and equipment		2,059,126
Total Assets		$3,435,876

Liabilities and Stockholders' Equity		
Current liabilities		
Accounts payable	$ 105,056	
Installments due within one year on debt	26,836	
Federal income and other taxes	239,194	
Other accrued liabilities	120,768	
Total current liabilities		491,854
Long-term debt		968,664
Total Liabilities		$1,460,518
Stockholders' equity		
Capital stock	505,130	
Retained earnings	1,470,228	1,975,358
Total liabilities and equity		$3,435,876

depreciation and depletion. They represent an expense in that they permit recovery over time of earlier capital investment as a deduction from taxable income. Unlike other expenses, they are only *allocations* and do not represent money expended in the current period. The portion of revenue allocated as depreciation or depletion is therefore available without penalty of taxation for reinvestment in replacement assets or in entirely different assets.

Ratio Analysis

Financial ratios are ratios of two financial numbers taken from the balance sheet and/or the income statement. These ratios can be compared with average values for the industry the firm is in to evaluate relative financial health, and they can be compared with earlier values from the same firm to evaluate trends. Four types of ratios are ordinarily calculated: (1) liquidity, (2) leverage, (3) activity, and (4) profitability ratios. Each is discussed below and calculated for the mythical Sterling Chemical Company in Table 8-3.

Liquidity ratios measure the ability to meet short-term obligations. The most commonly used ratio is that of current assets to current liabilities (*current ratio*). Many analysts use a current ratio of 2.0 as a prudent minimum, but this will vary between industries. A high current ratio (such as 10.0) may simply indicate that assets are not being efficiently employed. Since quickly liquidating (converting to cash) the firm's inventories might prove difficult, analysts also use the *acid test ratio* or *quick assets ratio* of "quick assets" (current assets minus inventories) to current liabilities. An acid test ratio over 1.0 is prudent, and Sterling Chemical satisfies both liquidity tests.

Leverage ratios identify the relative importance of stockholders and outside creditors as a source of the enterprise's capital. A simple measure is the ratio of total debt to total assets (debt as a fraction of the sum of debt and stockholders' equity). A common alternative, which can be derived from this one, is the debt-to-equity ratio. Leverage ratios vary significantly by industry. For example, an electric utility might well have a debt/assets ratio of 0.5 (debt/equity ratio of 1.0), while retail firms might have much lower debt ratios. The rapid growth of Japanese firms has led to quite high leverage ratios, which are accepted in that culture and economy. In the United States, the "leveraged buyout" (LBO) mania of the late 1980s resulted in the floating of billions of dollars of high-interest-rate "junk bonds" to finance company takeovers. This resulted in high debt loads (high leverage ratios) for the purchased companies, making it difficult for them to pay interest on their debt when the economy slowed down.

Activity ratios (also known as *operating ratios*) show how effectively the firm is using its resources. One common measure is *inventory turnover,* measured in Table 8-3 by dividing the cost of goods sold by total inventory (both valued at the manufacturing cost invested in them). Many textbooks define inventory turnover instead as sales divided by inventory (which would give us 4.55 rather than 2.94 in our example), but it seems more prudent to measure both by the same standard. Another activity ratio is *asset turnover,* or sales/assets,

TABLE 8–2 Statement of Income and Retained Earnings, Sterling Chemicals, Inc., Year Ended 12/31/95

Gross sales	$3,246,386	
Less returns and allowances	150,050	
Net sales		$3,096,336
Less expenses and costs of goods sold		
Cost of goods sold	2,002,376	
Depreciation and depletion	258,502	
Selling expenses	104,500	
General and administrative expenses	180,076	2,545,454
Operating profit		550,882
Plus interest and other income		59,480
Gross income		610,362
Less interest expense		33,260
Income before taxes		577,102
Provision for income taxes		261,142
Net income		315,960
Retained earnings January 1, 1995		1,370,988
		1,686,948
Dividends paid		216,720
Retained earnings December 31, 1995		1,470,228

TABLE 8–3   Financial Ratios for Sterling Chemicals, Inc., 1995

Ratio	Formula	Calculation
Liquidity ratios Current ratio	$\dfrac{\text{Current assets}}{\text{Current liabilities}}$	$\dfrac{\$1,376,750}{491,854} = 2.80$
Acid test ratio	$\dfrac{\text{Current assets} - \text{inventory}}{\text{Current liabilities}}$	$\dfrac{\$\,696,392}{491,854} = 1.42$
Leverage ratios Debt-to-assets ratio	$\dfrac{\text{Total debt}}{\text{Total assets}}$	$\dfrac{\$1,460,518}{3,435,876} = 0.425$
Activity ratios Inventory turnover	$\dfrac{\text{Cost of goods sold}}{\text{Inventory}}$	$\dfrac{\$2,002,376}{680,358} = 2.94$
Accounts receivable turnover	$\dfrac{\text{Net sales}}{\text{Accounts receivable}}$	$\dfrac{\$3,096,336}{416,304} = 7.44$
Asset turnover	$\dfrac{\text{Net sales}}{\text{Total assets}}$	$\dfrac{\$3,096,336}{3,435,876} = 0.901$
Profitability ratios Profit margin	$\dfrac{\text{Net income}}{\text{Net sales}}$	$\dfrac{\$\,315,960}{3,096,336} = 10.2\%$
Return on total assets	$\dfrac{\text{Net income}}{\text{Total assets}}$	$\dfrac{\$\,315,960}{3,435,876} = 9.20\%$

a measure of how well the firm is using its assets to produce sales. A third is the *accounts receivable turnover,* the ratio of net sales to accounts receivable. This ratio is often, in turn, divided into the traditional (but inaccurate) measure of 360 days per year to calculate the average collection period. In the case of our chemical company this would be 360/7.44, or 48.4 days required to receive payment after billing.

Profitability ratios describe the organization's profit. Two common measures are the profit as a percentage of sales, and the profit as a percentage of total assets (a measure of the return to the investor). A third measure is the earnings per share of common stock (the net income less preferred stock dividends, divided by the shares of common stock outstanding), which the stock market investor can compare with the current market price. We reiterate that desirable levels of financial ratios vary with the industry, economy, culture, and recent company history. Used with care, however, they are invaluable tools for financial management.

Budgets

Budgets are perhaps the most common and universally used of control techniques. They are plans for the future allocation and use of resources (usually but not always financial ones) over a fixed period of time. The budgeting process forces managers to think through future operations in quantitative terms and obtain approval of the planned scope of operations, and it provides a standard of comparison for judging actual performance in the control process.

Financial budgets describe where the firm intends to get its cash for the coming period and how it intends to use it. There are three common types. *Cash budgets* estimate future revenues and expenditures and their timing during the budgeting period, telling the manager when cash must be borrowed and when excess cash will be available for temporary investment. *Capital expenditure budgets* describe future investments in plant and equipment. Because expenditures for fixed assets require their use for an extended period to recover the investment, capital expenditures usually are scrutinized more carefully by upper management than are operating expenditures. Finally, a *balance sheet budget* uses the previous two estimates to predict what the balance sheet will look like at the end of the budgeting period.

For closer control, organizations are divided into *responsibility centers. Expense or cost centers* are those (such as manufacturing units or staff offices) where the manager's primary financial concern is control of costs. In a *revenue center,* such as sales or marketing, the manager has revenue targets to meet. Where an organization can be divided into business units containing both production and sales of a distinct product, so that *profit centers* are created, the manager has more freedom to manipulate costs in order to increase profit.

Where one unit of a company has as its primary customer another unit of the same company, the "transfer price" credited to one profit center and debited to the other must be established with care, especially where no accurate market price for the product exists. Not only does this price establish which unit makes the most apparent profit, but (where the units are in different states or countries) it determines the amounts and beneficiaries of tax receipts on these profits.

Operating budgets can be created for each of these responsibility centers. These also are of three (corresponding) types: the *expense budget,* the *revenue budget,* and the *profit budget,* which is a combination of the other two for profit centers.

Budgeting process. Budgets can be prepared by a central staff group and imposed on everyone by top management (the "top-down approach"), but this approach is usually unwise. It does not take advantage of information from lower management levels that would improve the budget process, and it does not foster commitment from lower managers to conform to the budget. Alternatively, budgets could be prepared at the responsibility center level and then just added up, but such budgets tend to be inflated and often do not consider adequately upper management's goals and objectives for the coming period.

Many organizations employ a combination of these two approaches. Top management first provides guidelines for the budgeting process, including estimates of future sales and production levels and changes in priorities to meet new objectives. After middle management has provided more detail, the various responsibility centers prepare proposed revenue and expense budgets. These are merged, "massaged" (modified), and negotiated at each middle management level, approved at the top, and then passed back down as operating guidelines for the coming period.

Budgets are frequently proposed and approved as percentage increases or decreases in current levels, which makes it difficult to change priorities in resource use quickly to meet new priorities. The technique of *zero-base budgeting* was developed to overcome this problem. Each responsibility center develops a budget package with a core of resource expenditure that is absolutely necessary to meet next year's objectives, and one or more supplemental additions required to do the job more effectively or to carry out "nice to have" functions. Packages and supplements are then ranked on a cost–benefit basis at each management level, and top management then allocates resources to meet organizational goals, which may require expansion of some units and shrinking or elimination of others.

Budgets should be a tool, and management should be flexible in adapting them as conditions change. Many budgets are valid only for the level of production and sales on which they were based; thus, when the level of output can vary substantially, a *variable budget* is needed. In such a budget, costs for labor, materials, and certain overhead and sales costs are set up as functions of output, while others are kept fixed. For a given month, for example, budget expenditures might be authorized at the level corresponding to 60% of capacity.

Cost Accounting

The financial budgets just discussed are plans for the future in quantitative (dollar) terms. Before effective decisions for the future (plans) can be made, the costs of alternative decisions must be understood. Historical accounting systems that determine the profitability of past operations are needed to determine income tax liability and produce quarterly and annual reports for stockholders, but they are often not adequate for determining if particular products, whether produced in the past or proposed for the future, have been/will be profitable. To find that out, costs must be divided among (allocated to) specific products, and this is the arena of *cost accounting*.

For example, assume that a plant produces 4000 units of product A and 1000 units of product B, and that each unit (whether A or B) requires one hour of direct labor at $10 per hour. Total labor cost is therefore $10(4000 + 1000), or $50,000. Now if supervisory effort costing $5000 is required to coordinate this production, it might be reasonably assumed that each hour of direct labor requires a proportional amount of supervision, resulting in an *overhead* or *burden* charge of $1 a direct labor hour, and a total cost for labor and supervision of $11 per unit (whether A or B).

Now, assume that costs of setting up the production line for products A and B total $8000. If we allocate this overhead cost in proportion to direct labor it will amount to $8000/5000, or $1.60, and we will now have a unit cost of $12.60 for both product A and B. However, this setup cost may represent one $4000 setup activity for each of products A and B, so that a fairer representation of setup cost would be $4000/4000 or $1.00 per unit of A and $4000/1000 or $4.00 per unit of B. Now we find that the unit costs for direct labor, supervision, and setup total $12 for product A and $15 for product B. Knowing this, we may try to get a higher price for product B, or we may want to quit making it.

Historically, direct labor formed the major part of manufacturing costs, and distribution of overhead costs in proportion to direct labor hours or direct labor dollars was often an acceptable estimate. With modern automation, direct labor costs are often reduced to less than 10% of total costs, and allocation of overhead costs by *activity-based costing*, as illustrated in our simple example of setup costs, becomes essential for making good decisions.

Audits of Financial Data

Audits are investigations of an organization's activities to verify their correctness and identify any need for improvement. Audits of accounting and financial systems and records are the most common type, and these may be either internal or external. *External audits* are required at least annually for any publicly held organization and are performed by independent accounting firms. They determine if financial records are accurate and reflect generally accepted accounting practices, and provide stockholders and creditors with greater confidence in the firm's financial statements. Most large firms also have *internal auditing* staffs, who spend their time auditing the several subunits of the organization. These staffs often have a more intimate knowledge of the firm's accounting systems, and they may be charged by management to evaluate organizational efficiency as well as just the accuracy of financial data; this does not replace the legal need for periodic external audit.

NONFINANCIAL CONTROLS

Human Resource Controls

Just as essential as assuring that actual financial performance conforms to plans is assuring that human and organizational performance conform to expectations. On an individual basis this is accomplished using the tools of performance appraisal discussed in Chapter 6, especially management by objectives (MBO), which is by its very nature a control system. Two tools used to evaluate collective human and organizational performance are the management audit and human resource accounting. Finally, one should consider social controls through group values and self-control.

Management audits. The definition of an audit provided under "Audits of Financial Data" above can be applied equally well to other areas. One area of increasing importance is the audit of the entire system of managing an enterprise. A number of the major accounting firms have developed management services staffs that are prepared to conduct management audits, or firms may seek many of the same objectives through an *enterprise self-audit.* Some of the questions on administrative effectiveness that might be asked in such an audit appear in Figure 8-2. A related set of questions from the same source that might be used to evaluate the technical effectiveness of an engineering design organization appears in Chapter 10.

Human resource accounting. Conventional financial accounting deals with the prudent handling of revenue and expenses and with investments in tangible items that appear as assets on the balance sheet. Increasingly, however, the biggest assets of an enterprise are its people. Investments in acquiring outstanding people and in extensive training programs for them represent capital investments in the future as much as does the purchase of new machinery. Quantifying the value of human resource investment is difficult, but a number of approaches are being tested. (Similarly, costs for R&D and in-process improvement are written off as current expense, but these might more appropriately be recognized as capital investments.)

Organizational Effectiveness Review Administrative Factors Worksheet		Administrative Factors Worksheet	
Factor	Rating	Current Strengths, Weaknesses, Needs	Future Objectives and Plans
Planning. *Does the organization:* —Develop realistic, time-phased plans for the long, medium, and short term? —Analyze risks and provide for contingencies? —Integrate plans and objectives with interfacing organizations? —Produce valid and timely proposals and accurate cost estimates? —Forecast funding and labor requirements accurately?	—High —Avg. —Low		
Organizing and staffing. *Does the organization:* —Establish clear definitions of function, authority, and accountability? —Select the most qualified personnel to fill its needs? —Assign personnel so as to best utilize their capabilities and potential? —Assess its strengths and weaknesses and promptly correct deficiencies?	—High —Avg. —Low		
Directing. *Does the organization:* —Maintain high performance standards? —Stress people-oriented leadership and the importance of personal example? —Delegate work effectively, encouraging maximum employee involvement and responsibility? —Recognize achievement and distribute rewards equitably? —Encourage employee development and growth?	—High —Avg. —Low		
Control. *Does the organization:* —Monitor operational progress and promptly correct deficiencies? —Control expenditures as required to assure achievement of profit objectives? —Adhere to schedules? —Assess its productivity and continually strive to improve it?	—High —Avg. —Low		
Communication. *Does the organization:* —Maintain good intra- and interorganizational communications? —Keep management informed of key operations and problems? —Keep employees informed and solicit their ideas and opinions? —Encourage the exchange of technical information?	—High —Avg. —Low		
Procurement/subcontracting. *Does the organization:* —Act promptly on procurement matters? —Establish effective time-phased plans for procurement? —Assume an active role in "make or buy" decisions? —Assist in developing subcontract sources? —Ensure an adequate definition of work on all subcontracted efforts? —Maintain an effective interface with subcontractors and monitor subcontractor progress?	—High —Avg. —Low		
Space and Facilities. *Does the organization:* —Accurately predict its space and facilities needs? —Make optimal use of available space and facilities? —Ensure proper maintenance and calibration of all instruments and equipment? —Maintain required accountability records of all property? —Maintain high standards of housekeeping?	—High —Avg. —Low		

Figure 8-2. Management audit worksheet for administrative activities.
(From *R&D Productivity: Study Report*, 2nd ed., Hughes Aircraft Company, El Segundo, CA, 1978, pp. 26–27.)

Social controls. No organization that relies on formal controls only will be truly effective. Peters and Waterman found in *In Search of Excellence*[3] the central importance of the underlying values imbued in their corporate cultures and inculcated into all employees:

> The excellent companies live their commitment to people, as they do their preference for action—any action—over countless standing committees and endless 500-page studies, their fetish about quality and service standards that others, using optimization techniques, would consider pipe dreams, and their insistence on regular initiative (practical autonomy) from tens of thousands, not just 200 designated $75,000-a-year thinkers. . . .
>
> The excellent companies seem to have developed cultures that have incorporated the values and practices of the great leaders and thus those shared values can be seen to survive for decades after the passing of the original guru. Second, . . . it appears that the real role of the chief executive is to manage the values of the organization.[4]

For values imbued in the corporate culture to be effective requires that employees in general exercise self-control over their actions. Like other control systems, self-control requires:

- The existence of standards (knowledge by the general worker of the organization's objectives and values)
- Comparison with actual outcomes (which implies feedback of performance to the individual, not just to management or a "quality control" group)
- Corrective action (which requires that the individual have the tools, the autonomy, and the motivation to make corrections)

Obviously, an emphasis on self-control is a "Theory Y" approach to leadership. It will not work with every person, and it requires careful selection and training of personnel, but carries with it handsome payoffs for success.

Other Nonfinancial Controls

The control process we've described pervades all the functions and applications of management, and we will meet it in a number of later chapters. In Chapter 9, on research, we consider methods of evaluating the effectiveness of research activities. In Chapter 10, we discuss control systems for drawing release and for engineering design changes (configuration management). Effective production management (Chapters 11 and 12) requires inventory control and quality control, among other control systems. Project management (Chapters 14 and 15) requires control systems monitoring all three of its key variables: schedule, cost, and the performance of the resulting product.

DISCUSSION QUESTIONS

8-1. Provide two additional examples of (**a**) feedback, (**b**) screening (concurrent), and (**c**) feedforward control. In each example identify the four steps of the control process.

8-2. Which, in your opinion, are the most important characteristics of an effective control system? Justify your choices. What other important characteristic(s) might one add?

8-3. Hytek Corporation ended 1995 with cash of $50,000, accounts receivable of $100,000, and inventory of $300,000. Property, plant, and equipment were valued at their original cost of $470,000, less accumulated depreciation of $170,000. Current liabilities other than income taxes owed (see below) were $120,000, and long-term debt was $250,000. Stockholders' equity consisted of (**i**) $90,000 capital stock investment and (**ii**) accumulated retained earnings, which had totaled $130,000 at the end of *1994*. Net sales for 1995 were $900,000. Expenses included $500,000 cost of goods sold, $50,000 allowance for depreciation, $85,000 selling expense, and $65,000 G&A expense. Interest income and expense were $5000 and $25,000, respectively, and income taxes for the year (unpaid at year's end) were $80,000. Dividends of $20,000 had been declared but not paid at year's end. Prepare a balance sheet and an income statement reflecting these figures.

8-4. Use the output of question 8-3 to calculate the eight financial ratios of Table 8-3. Comment on the values you obtain.

8-5. Excelsior Corporation reported the following status (in thousands of dollars) as of December 31, 1995: accounts payable of $150; accounts receivable of $250; cash of $150; inventory of $200; long-term debt of $260; net plant and equipment of $500; notes payable during 1996 of $250; and stockholders' equity of $440. (**a**) Prepare a balance sheet as of 12/31/95, and (**b**) calculate as many financial ratios as you can with the information provided.

8-6. For 1995 a company reported (in millions of dollars) net sales of $10.0, cost of goods sold of $4.4, other (sales, G&A, and interest) expense of $1.2, and income taxes of $1.6. As of December 31, 1995, the company had $1.0 cash and securities, $1.4 accounts receivable, and $2.0 inventory; it owed $2.0 in current liabilities (including unpaid taxes) and $2.5 in long-term debt. Calculate as many financial ratios as you can with the information provided.

8-7. Suggest some characteristics that distinguish an effective budgeting system from an ineffective one.

8-8. How does the existence of *profit centers* assist top executives in doing their job?

8-9. Discuss how allocation of overhead costs on the basis of direct labor might distort product pricing where some products are produced by automated machining centers and others by more labor-intensive methods.

8-10. ABC Corporation produces 50,000 units of product X and 5000 units of product Y at a direct materials cost of $3 per unit. Product X requires 3 minutes and product Y 30 minutes direct labor per unit (at $20 per hour). Other costs (tooling, setup, and equipment depreciation and maintenance) for this period amount to $60,000. (**a**) If these "other costs" are allocated on the basis of direct labor hours, what is the apparent unit cost of each product? (**b**) Production of product X is highly automated to reduce direct labor cost; it is responsible for $55,000 of this "other cost," and product Y only $5000. Using *activity-based costing,* what do the unit costs now become? (**c**) What difference might this make in ABC Corporation's actions?

8-11. If you were preparing to lead a management audit of a large automobile assembly plant, what are some questions you might seek to answer in the investigation?

8-12. What steps might you take to try to inculcate the essential values of your organization into new employees and to keep them in the minds of other employees? (For example, "Quality Is Job One" or "The customer comes first" might represent one such value.)

8-13. Recalling what you learned about motivation in Chapter 7, how might you encourage your technical employees to support corporate goals and values through self-control?

NOTES

1. Robert E. Shannon, *Engineering Management* (New York: John Wiley & Sons, Inc., 1980), p. 261.
2. Peter F. Drucker, "Management Lessons of Irangate," *Wall Street Journal,* March 24, 1987, editorial page.
3. Thomas J. Peters and Robert H. Waterman Jr., *In Search of Excellence: Lessons from America's Best-Run Companies* (New York: Harper & Row, Publishers, Inc., 1982).
4. Peters and Waterman, *In Search of Excellence,* pp. 16, 26.

Managing Technology Through the Product Life Cycle

CHAPTER 9

Managing the Research Function

PREVIEW

We begin this chapter by introducing the product and technology life cycles, in which research is the first active step. We define and describe research and development. We discuss briefly new product strategies and organization for research. We introduce the sequential process of winnowing the many ideas for product research and development to an affordable level, based on technical, market, and organizational considerations. Next follows a contributed section on the important topic of protecting ideas through patents, trade secrets, and other means. Creativity, which is essential to effective research, is then considered carefully. Finally, we introduce some additional methods of making R&D organizations successful.

PRODUCT AND TECHNOLOGY LIFE CYCLES

A new product begins as an idea for the solution of a problem or the satisfaction of a need. In nature only a few out of a hundred tadpoles survive to become frogs; in research only a few out of many research ideas will be vigorous enough to survive and will reach the right environment to mature into a successful product. Like the buggy whip, our product will have its day and will then be replaced by newer ideas that satisfy newer needs. This cradle-to-grave sequence is known as the *product life cycle* (Figure 9-1).

This product life cycle begins with an *identification of need* or suggestion of a prod-

Product life cycle	Consumer	Identification of need	"Wants or desires" for products (because of obvious deficiencies/problems are made evident through basic research results)
	Producer	Product planning function	Marketing analysis; feasibility study; advanced product planning (product selection, specifications and plans, acquisition plan–research/design/production, evaluation plan, product use and logistic support plan); planning review; proposal
		Product research function	Basic research; applied research ("need" oriented); research methods; results of research; evolution from basic research to product design and development
		Product design function	Design requirements; conceptual design; preliminary system design; detailed design; design support; engineering model/prototype development; transition from design to production
		Production and/or construction function	Production and/or construction requirements; industrial engineering and operations analysis (plant engineering, manufacturing engineering, methods engineering, production control); quality control; production operations
		Product evaluation function	Evaluation requirements; categories of test and evaluation; test preparation phase (planning, resource requirements, etc.); formal test and evaluation; data collection, analysis, reporting, and corrective action; retesting
	Consumer*	Product use and logistic support function	Product distribution and operational use; elements of logistics and life cycle maintenance support; product evaluation; modifications, product phase-out; material disposal, reclamation, and/or recycling

*Some of the specific supporting functions indicated may be accomplished by the producer throughout and/or at various stages in the product life cycle.

Figure 9-1. Steps or functions and typical activities in the product life cycle.
(From Benjamin S. Blanchard, *Engineering Organization and Management,* © 1976, p. 16. Reprinted by permission of Prentice-Hall, Inc., Englewood Cliffs, NJ.)

uct opportunity, which might come from researchers, salespeople, or customers, from observation of a competitor, or (for military goods) from fear of a potential enemy. The product idea must then be subjected to a screening process to select from the many ideas available those that are technically and economically feasible, and to propose a program for their successful design and development. These preliminary steps (the *product planning and research* functions in Figure 9-1) are the subject of this chapter.

Proposed products that appear attractive at this point are approved for the *product design* function, itself a process of several steps discussed as *systems engineering phases* or *engineering stages of new product development* in Chapter 10. Products that still appear desirable after the design process then go to the *production* (and/or construction) function, which is treated in Chapters 11 and 12. Finally, the products are put into use, and if they are at all complex, they will require continuing technical effort to support their operation and maintenance (the *product use and logistic support* function in Figure 9-1), as discussed in Chapter 13. The *product evaluation* function is spread throughout the design, production, and early system use phases and is discussed under each of these topics. Finally, in a step not shown in Figure 9-1, the product undergoes phase-out, disposal, reclamation, and/or recycling.

The preceding model of the product life cycle fits the construction of a building or a ship or the design and development of an aerospace system well. For a product line (or family of products) based on a technology that is developed and improved over a period of years of product manufacture, the model of the *technology life cycle* portrayed by Betz[1] (Figure 9-2) is more appropriate. Betz illustrates this model using the automobile as an example:

> When a new industry (based on new technology) is begun, there will come a point in time that one can mark as the inception point of the technology. In the case of the automobile, that was 1896, when Duryea made and sold those first 13 cars from the same design.
>
> Then the first technological phase of the industry will be one of rapid development of

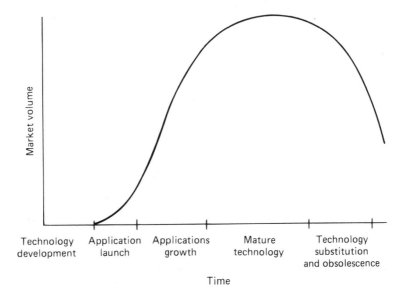

Figure 9-2. Technology life cycle.
(From Frederick Betz, *Managing Technology: Competing Through New Ventures, Innovation, and Corporate Research,* © 1987, pp. 72–74. Reprinted by permission of Prentice-Hall, Inc., Englewood Cliffs, NJ.)

the new technology—technology development. For the automobile this lasted from 1896 to 1902, as experiments in steam-, electric-, and gasoline-engine-powered vehicles were tried. . . .

In any new technology, the early new products are created in a wild variety of configurations and with differing features. . . . Finally, when enough experimentation has occurred to map out the general boundaries of possibilities of the product line, some managerial genius usually puts all the best features together in one design and creates the model which then becomes the standard design for the industry. Thereafter all product models generally follow the standard design. This makes possible large market volume growth. For the automobile, this occurred [in 1908] with Ford's Model T design.

After the applications launch, there occurs a rapid growth in the penetration of technology into markets (or in creating new markets). After some time, however, the innovation rate slows and market creation will peak. This is the phase of technology maturity. Finally, . . . when competing or substituting technologies emerge, the mature technology begins to degrade in competition with the competing technologies.[2]

NATURE OF RESEARCH AND DEVELOPMENT

R&D Defined

Research and development are commonly lumped together under the catchall term "R&D." To distinguish between them, let us adopt the definitions commonly used by the National Science Foundation:

Research, both basic and applied, is systematic, intensive study directed toward fuller scientific knowledge of the subject studied.

Basic research is . . . research devoted to achieving a fuller knowledge or understanding, rather than a practical application, of the subject under study . . . [although when funded by commercial firms it] may be in fields of present or potential interest to the company.

Applied research is directed toward the *practical application* of knowledge, which for industry means the discovery of *new* scientific knowledge that has specific commercial objectives with respect to either products or processes.

Development is the systematic *use* of scientific knowledge directed toward the production of useful materials, devices, systems, or methods, including design and development of prototypes and processes.[3]

Distribution by Expenditure and Performance

U.S. investment in research and development amounted to $160.7 billion in 1993, as itemized in Table 9-1. Basic (fundamental or pure) research of $26 billion made up 16.3% of the total, primarily paid for by the federal government but performed at universities and industry under contract to the National Science Foundation and other government agencies. Few corporations can afford to invest much of their own funds in the search for fundamental knowledge for its own sake, since it would benefit competitors as much as themselves. Before the breakup of the Bell System and its near monopoly, Bell Telephone Laboratories was an exception, and much of the basic research underlying statistical quality control (for example) was performed there in the 1930s.

TABLE 9-1   R&D Expenditures and Performance in 1993, $ Billions (%)

	Federal Government		Industry		Universities		Other Nonprofit		Total (100%)
Spending on ↓ by →									
Basic research	16.5	(62.7%)	4.6	(17.7%)	3.5	(13.5%)	1.6	(6.1%)	26.2
Applied research	15.5	(38.9%)	21.1	(53.1%)	2.0	(5.1%)	1.1	(2.8%)	39.7
Development	<u>36.1</u>	<u>(38.1%)</u>	<u>57.8</u>	<u>(61.0%)</u>	<u>0.4</u>	<u>(0.4%)</u>	<u>0.5</u>	<u>(0.5%)</u>	<u>94.8</u>
Total	68.0	(42.3%)	83.6	(52.0%)	6.0	(3.7%)	3.2	(2.0%)	160.7
(1995 estimates)*		(36.0%)		(59.0%)		(3.0%)		(2.0%)	
Work done on ↓ by →					***				
Basic research	2.9	(11.1%)	4.7	(17.9%)	16.4	(62.4%)	2.3	(8.7%)	26.2
Applied research	4.9	(12.3%)	26.5	(66.8%)	6.4	(16.0%)	1.9	(4.8%)	39.7
Development	<u>8.8</u>	<u>(9.3%)</u>	<u>81.1</u>	<u>(85.5%)</u>	<u>3.1</u>	<u>(3.3%)</u>	<u>1.8</u>	<u>(1.9%)</u>	<u>94.8</u>
Total	16.6	(10.3%)	112.3	(69.9%)	25.9	(16.1%)	6.0	(3.7%)	160.7
(1995 estimates)*		(10.0%)		(72.0%)		(14.0%)		(4.0%)	

***Work done by universities includes that done by federally funded R&D centers operated by them.

Source: National Science Board, *Science and Engineering Indicators—1993* (Washington, DC: U.S. Government Printing Office, 1993, NSB 93-10) pp. 233–236. *1995 estimates from Batelle and *R&D Management*, as reported in *Technology Transfer Business*, Spring 1995, p. 8.

Applied research of $40 billion made up ¼ (24.7%) of R&D in 1993; industry funded over ½ of this, but performed ⅔ of it, funding the rest from federal contracts. Applied research may be divided into materials research, product research, and equipment and process research.

The bulk of 1993 R&D expenditure was development ($95 billion, or 59.0%), primarily paid for and performed by industry. We defer further discussion of development to Chapter 10. Two other categories of research not included in the R&D discussion above are *operations research* (management science, discussed in Chapter 4) and *marketing research,* which is an essential part of product development and commercialization.

The nature of U.S. government spending for R&D has changed over the years with external military threats and internal political administrations. In 1980 about half of the $30 billion federal expense was for military purposes; by 1989 about ⅔ of the $62.5 billion federal R&D expense was military; considering inflation, federal investment in nondefense R&D had shown no significant increase over the "Reagan years." Further, federal funding for R&D was concentrated in a few defense-related industries, as shown in Table 9-2. About ⅓ of the $92.6 billion R&D carried out by industry in 1989 was paid for by the federal government, but ½ of this $30.8 billion federal investment went to the aerospace industry and another ¼ to electrical and electronic industries.

Although R&D expenditures in the United States, Japan, and Germany had been comparable in the late 1980s as a percentage of gross domestic product (about 3%), in the United States about one-third of the total (and two-thirds of federal) R&D funding had been for military purposes, so that the United States spent less proportionately than these two major competitors on nonmilitary products for the global marketplace (about 3.0% of GDP for Japan, 2.7% for Germany, and 1.9% for the United States in 1990).[4]

TABLE 9–2   Estimated Sources of Funds for R&D Performed by Industry, by Industrial Class, 1989

Industrial Class	Funds (Billions of Current Dollars)		
	Federal	Industry	Total
Aerospace	15.6	3.5	19.2
Electrical machinery and communications	7.9	10.6	18.5
Nonelectric machinery	1.6	10.5	12.1
Chemicals	0.4	11.1	11.5
Autos and other transportation equipment	2.0	9.4	11.4
Professional and scientific instruments	1.0	5.5	6.5
Petroleum products	0.0	2.1	2.1
Rubber products	0.3	0.9	1.2
Food and beverage	0.0	1.2	1.2
Other manufacturing	0.3	4.8	5.1
Total manufacturing	29.2	59.6	88.9
Nonmanufacturing	1.6	2.2	3.8
Total	30.8	61.9	92.6

Source: Battelle, Columbus, OH.

With the collapse of the Soviet Union, the federal part of the U.S. R&D investment decreased from 47% in 1989 to 42% in 1993 and to an estimated 36% by 1995; the portion of federal R&D for nondefense purposes increased from about ⅓ in 1989 to 42% in 1993 and 46% in 1994.

RESEARCH STRATEGY AND ORGANIZATION

New Product Strategies

Within a specific industry, deciding the relative investment a company should make in R&D is a part of strategic planning and should be based on the organization's concept of its fundamental mission and objectives. Ansoff and Stewart suggest four alternative new-product strategies:

> *First-to-market.* This . . . demands major expenditures for research before there is any guarantee of a successful product. It also demands heavy development expenditures and perhaps a large marketing effort to introduce an innovative product. The possibilities of reward from the R&D, however, are tremendous.
>
> *Follow-the-leader.* This strategy does not require a massive research effort, but it demands strong development engineering. As soon as a competitor is found to have had research success that could lead to a product, the firm playing follow-the-leader joins the race and tries to introduce a product to market almost as soon as the innovator.
>
> *Me-too.* A me-too strategy differs from follow-the-leader in that there is no research or development. In its purest form this strategy means copying designs from others, buying or leasing the necessary technology, and then concentrating on being the ab-

solute minimum-cost producer. The firm following this strategy will try to maintain the lowest possible overhead expenses.

Application engineering. . . . This role involves taking an established product and producing it in forms particularly well suited to customers' needs. It requires no research and little development, but a good deal of understanding of customers' needs and flexibility in production.[5]

Corporate Research Organizations

Through the end of the nineteenth century, industrial support of research was unknown. The first corporate research laboratory in the United States began when General Electric Company observed that newer inventions were making its principal product, the carbon filament lamp, technically obsolete, and hired MIT Professor Willis R. Whitney to organize what became the GE R&D Center in 1900. One of his former students at MIT was William Coolidge, who returned to MIT at $1500 a year as a faculty member in 1905 after earning a doctorate at the University of Leipzig in Germany. Whitney hired him away at twice that salary (a phenomenon still a problem for universities in retaining good faculty) and put him to work trying to develop a lamp filament of tungsten to compete with a tantalum filament developed by a German company. It took five years and $100,000, but by 1911 Whitney and Coolidge had saved GE's principal business and "by 1920 two-thirds of GE's $22 million profit came from the new lamps."[6] Research from the laboratory helped keep GE a leader in technology; indeed, "Over the years, GE had consistently led all other companies in obtaining U.S. patents, becoming in 1979 the first firm in history to pass the 50,000th patent milestone."[7] Other early corporate research laboratories that were very successful were those of AT&T (Bell Labs), DuPont, Dow Chemical, and General Motors. Today, most large corporations consider corporate research at some level essential. Although some companies are extremely successful in creating profitable new products from research, others are not. Robert Frosch, in charge of the $200,000,000 effort at General Motors Research Laboratories, recently identified three ways a corporate research laboratory could fail:

> Many research laboratories have been opened with great fanfare, only to fail later because they had the idea that producing great science, whether or not it had anything to do with the business, was why they were there. There *is* a role for the production of good science—provided you can eventually make it relevant to the business. But if a laboratory goes for a long period of time doing nothing relevant to the business, then it probably deserves to fail, because a corporation is not a university.
>
> The second reason laboratories fail is because of what can best be described as rampant short-termism. Financial support for business seems to come increasingly from markets and groups who regard two and a half weeks as an eternity and a quarter [year] as the age of the universe. So, sometimes research and development fail because nobody has enough patience to let them succeed.
>
> The third way in which research efforts fail is that the connection between research and development and the business breaks down. What is done in the research and development laboratories may be applicable to the business. The business may need these useful developments. But somehow the developments never get out of the laboratory and into the business. This is a failure of technology transfer.[8]

Large corporations normally have two kinds of research activities: applied research staffs attached to each of the major business units, and a central laboratory with a broader scope of scientific expertise and a long-range outlook. In General Electric, for example, the central laboratory represents only about 10% of the research effort, but it plays an essential role. Central corporate laboratories also make their special expertise available to the business units to solve current problems, but they must be careful that this does not cripple their basic function.

SELECTING R&D PROJECTS

Need for Selection

Any successful technology-based manufacturing firm will have many more ideas for research projects than it has resources to invest in them. Booz, Allen, and Hamilton, Inc. has suggested approximately the following ratio of raw new product ideas to profitable products (also illustrated in Figure 9-3):

- 60 ideas (from researchers, other employees, customers, and suppliers) need to be screened quickly down to
- 12 ideas worthy of preliminary technical evaluation and analysis of profitability, to produce

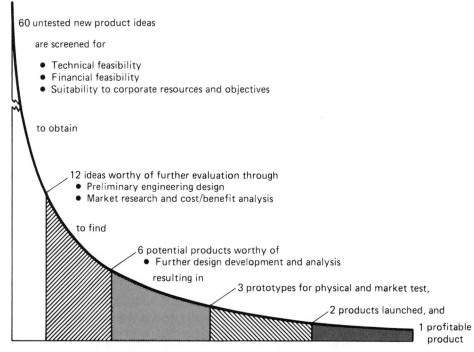

Figure 9-3. Screening of research project ideas.

- 6 defined potential products worth further development, to obtain
- 3 prototypes for detailed physical and market testing, resulting in
- 2 products committed to full-scale production and marketing, of which
- 1 product should be a real market success.

Initial Screening

To slash 60 crude ideas into 12 worthy of any significant evaluation requires a method that is quick and inexpensive. A common method is use of a simple *checklist,* in which the proposed product is given a simple judgmental rating (poor/fair/good/excellent or $-2/-1/+1/+2$, for example) for each of a number of characteristics. Seiler[9] suggests, for example, scoring 10 items:

1. *Technical factors* (availability of needed skills and facilities; probability of technical success)
2. *Research direction and balance* (compatibility with research goals and desired research balance)
3. *Timing* (of R&D and market development relative to the competition)
4. *Stability* (of the potential market to economic changes and difficulty of substitution)
5. *Position factor* (relative to other product lines and raw materials)
6. *Market growth factors* for the product
7. *Marketability* and compatibility with current marketing goals, distribution methods, and customer makeup
8. *Producibility* with current production facilities and manpower
9. *Financial factors* (expected investment need and rate of return from it)
10. *Patentability* and the need for continuing defensive research

Only slightly more sophisticated is the use of a *weighted checklist* or *scoring model* in which each factor is scored on a scale, often from 0.0 to 1.0. A relative *weight* representing the importance of that factor is then used as a multiple, and the weighted scores for all factors are added. Table 9-3 provides an example of such a scoring model. In this example, a potential new product has been given a raw score of 36 (60% of the maximum 60) and a weighted score of 75 (only 50% of the maximum 150). The product was judged very favorably on technical factors and could be developed with some confidence of technical success. However, it was rated poorly on its marketing factors, which had been assigned greater weight in the model, and therefore probably would not be developed.

Quantitative Approaches

Once the large number of ideas for research projects has been screened to a more manageable number, the remaining proposals justify more detailed consideration of their technical and financial merits. The technical evaluation can take place in several stages increasing in depth and detail (such as the conceptual, technical feasibility, development, and commercial validation stages of new product development discussed in Chapter 10), with a decision point at the end of each phase. Hand in hand with evolution of the technology should come

TABLE 9-3   Example of a Weighted Scoring Model

PRODUCT CONCEPT EVALUATION SHEET

Criteria	Weight	Score	Weighted Score
Technical factors			
Compatibility with research objectives	1	9	9
Compatibility with production facilities and capabilities	2	8	16
Probability of technical success	2	9	18
Marketing factors			
Compatibility with marketing goals, distribution, customers	4	4	16
Probability of marketing success	4	2	8
Potential profitability	2	4	8
Totals		36	75

increasingly detailed analysis of costs of producing the proposed product and market estimates of potential sales and profits.

Many mathematical models have been proposed for evaluating the financial suitability of proposed projects (see, for example, Balderston et al.,[10] Dhillon,[11] and Shannon[12]). Typically they involve estimating the relationship between the investment required and the benefits to be gained. Easiest to calculate is the *simple payback time* T_{pb}, which is the ratio of required investment I and mean annual gross profit A:

$$T_{pb} = \frac{I}{A}$$

Simple payback time is often used to justify investments that need to be recovered quickly because of uncertainties, but it is unsuitable for longer-term investments because it ignores profits expected beyond the point of payback and does not consider the time value of money (the fact that a dollar profit returned at some future time has less value than a dollar available today). Many engineers learn these valuable methods of justifying investment in a new project or purchase of new equipment in a course in *engineering economy* and return to tell their teachers that it was one of the most useful, practical courses they took in college. Using the standard engineering economy nomenclature:

P = present worth of future cash flow

A_j = cash flow (revenue less expense) in the jth year

i = discount rate (minimum attractive rate of return) required by the organization to justify investment, expressed as a decimal

n = number of years of future cash flow

Any sum P today, placed at an (annually compounded) interest i would compound to $A_1 = P(1 + i)$ in one year, $A_2 = P(1 + i)^2$ in two years, and $A_j = P(1 + i)^j$ in j years. Therefore, the present worth of any future sum A_j can be calculated as

$$P = \frac{A_j}{(1 + i)^j}$$

The present worth of n years of such cash flow would be

$$P = \sum_{j=1}^{n} \frac{A_j}{(1+i)^j}$$

For example, suppose it was proposed to invest ($P =$) \$400,000 today with the certainty of a return of ($A_1 =$) \$209,000 in one year and ($A_2 =$) \$242,000 in two years. On the surface, the return of \$451,000 for an investment of only \$400,000 seems attractive, and the *payback time* is under two years. However, if the rate of return on corporate investment must be at least 10% ($i = 0.1$), the overall *net present worth* (NPW) of the proposal would be

$$\begin{aligned} \text{NPW} &= \$-400,000 + \frac{\$209,000}{1.1} + \frac{\$242,000}{(1.1)^2} \\ &= \$-400,000 + \$190,000 + \$200,000 \\ &= \$-10,000 \end{aligned}$$

This negative net present worth shows that the project would not earn the required return on investment, and the research proposal would be rejected.

Even if the net present worth were positive, there would normally be no certainty that the projected future earnings would be realized. For this reason it is prudent to calculate a *maximum expenditure justified* E_{mj} based on the estimated probabilities of success:

$$E_{mj} = F_c \times F_t \times P$$

where F_c is the probability of commercial success, F_t is the probability of technical success, and P is the NPW assuming certainty of success (from the previous equation, excluding initial investment). Other quantitative methods that calculate the expected return on investment or the benefit-to-cost ratio achieve much the same result.

PROTECTION OF IDEAS

by Dr. Donald D. Myers[13]
Professor of Engineering Management
University of Missouri–Rolla

Generally, four legal means protect a person's ideas and right to benefit from those ideas. They are patents, copyrights, trade secrets, and trademarks and other marks. Each of these legal protections is discussed in a section below.

Patents

A patent is an exclusive property right to an invention issued by the Commissioner of Patents and Trademarks, U.S. Department of Commerce. The rights granted are limited to the "claims" of the patent. There are three classifications of patents: (1) utility, (2) design, and (3) plant. A *utility* patent may be obtained for a process, machine, article of manufacture, composition of material, or any improvement thereof for 20 years from date of filing. Utility patents cannot be obtained on laws of nature, methods of doing business, scientific principles, or printed matters.

To be patentable, the invention must be (1) new or novel, (2) useful or have utility, and (3) nonobvious. If the invention has been used, sold, or known by others in the United States or patented or disclosed in a printed publication in the United States or a foreign country before the invention was made by the inventor, a patent is barred. It is also barred if the invention was patented or described in a publication or in public use or on sale in the United States more than one year prior to the application for the patent. Useful inventions must advance the useful arts and benefit the public. The test of obviousness is whether it is obvious to those "with ordinary skill in the art involved."

Design patents are granted on new, original, and ornamental design of an article of manufacture for a term of 14 years. The design patent is not concerned with how the article of manufacture was made and how it was constituted, but with how it looks. *Plant patents* are granted for 17 years for plants when asexually reproduced, with the exception of tuber-propagated plants or plants found in the uncultivated state.

Establishing patent rights. The invention process includes (1) conception and (2) reduction to practice. In the United States, if the first to conceive makes a reasonable, diligent effort to reduce the invention to practice, he or she will receive the patent, even if someone else actually reduces it to practice earlier. Accordingly, it has been essential for the American inventor to maintain good records to establish the date of conception and diligence in reduction to practice in case of any later interference. The filing of the patent application satisfies reduction to practice if, from the patent specification, one skilled in the art to which it relates is capable of constructing or carrying out the invention.

A written disclosure of the invention should be made as soon after conception as possible. There is no specific requirement about the form a written disclosure must take to document the conception of an invention. A disclosure's primary purpose is to prove the date of conception where there is question of invention. The disclosure should include sufficient description and sketches to describe fully what has been conceived. The disclosure should be witnessed by at least two persons who fully understand its content.

To demonstrate diligence to "reduce to practice," a written record of developmental activities should be maintained in a bound notebook. Daily entries are encouraged. Each page should be signed and witnessed in proximity to the entries on that page. Each entry should be made in chronological order. Notebook pages should be consecutively numbered, with all entries made in ink. If an error is made in an entry, it should not be erased; it should be crossed out. All entries should be made by the inventor in his own handwriting. Although it is permissible for an inventor to file his or her own application, it is strongly advised that a patent attorney or agent be used to make and prosecute the application.

In almost all other countries, patents are awarded to the first person to file, rather than the first to conceive. However, Commerce Secretary Ron Brown concluded January 24, 1994, that "The first-to-invent system has served us well in the past, and while the United States may move to first-to-file sometime in the future, I am not convinced that enough small inventors and entrepreneurs would benefit if we made a switch at this time."[14] Just over half of U.S. utility patents have been awarded to Americans in recent years; the three companies awarded the most U.S. patents in 1992 were all Japanese; General Electric was 4th, Mitsubishi Electric 5th; four more U.S. companies and Matsushita rounded out the top ten, followed by Bayer of Germany (11th).[15]

Trademarks and Other Marks

The Lanham Act defines a mark as "any word, name, symbol, or device, or any combination thereof." The U.S. Patent and Trademark Office recognizes four types of marks: trademarks, service marks, certification marks, and collective marks. A *trademark* is "used by a manufacturer or merchant to identify his goods and distinguish them from those manufactured or sold by others." A trademark differs from a trade name. IBM may be both a trademark and a trade name, but only the trademark is protected by federal statutes and registered with the Patent and Trademark Office.

A *service mark* is associated with services rather than goods. A *certification mark* indicates that the marked goods or services meet standards or services established by the mark's owner, for example, Good Housekeeping. A *collective mark* identifies members of a group such as an organization, union, or association.

The rights to a mark can be lost, especially if a mark is abandoned or allowed to become a generic word. To avoid losing a mark, vigilance must be exercised even to the point of suing infringers. Under the Trademark Law Revision Act of 1988, beginning November 16, 1989, application for a mark can be made before any use has taken place. Previously a mark had to be used and products bearing the mark sold and shipped to a commercial customer before the mark could be registered. Now the applicant need only indicate a *bona fide* intent to use the mark within the next three years.

Almost all states have their own trademark law. If a mark is to be used entirely within one state, the only protection it has, other than common law, is registration under the state's trademark law. Federal trademark law applies only to marks used in interstate commerce.

A mark does not have to be registered, but the symbol "®" or the notice "Reg. U.S. Pat. and TM Off." should be used with registered trademarks and "™" or "Trademark" with nonregistered marks. For example, the first edition of this book was written on an IBM Personal Computer® using Volkswriter® word processing software for the initial drafts and Total Word™ for subsequent drafts; the second edition was written in WordPerfect® 5.1. A nonregistered mark has common law rights. Official registration, however, provides distinct advantages.

Copyrights

A copyright generally prevents reproduction of a copyrighted work for the life of the author, plus 50 years. Copyright owners can sue anyone who infringes their rights to stop illegal reproduction, impound infringing articles, collect lost profits, court costs, and attorney's fees, and in extreme cases, invoke criminal penalties.

A copyright is the exclusive right to reproduce, publish, and sell an original, creative work in some tangible form. Copyrights can be given for literary works; dramatic works; musical works, including any accompanying music; motion pictures and other audiovisual works; and sound recordings. A copyright protects expressions, not ideas. A potentially patentable idea expressed in a copyrighted text may be used by others.

For works created after 1989, copyright notices are not necessary (although they are recommended). The copyright notice has three elements: (1) the copyright symbol ©, the word "copyright," or the abbreviation "copr."; (2) the year of first publication; and (3) the name of the copyright owner. A copyright notice can appear any place in or on the work as long as it can be readily seen, but in a book such as this the notice is usually on the back of the title page. Copyright registration is not a condition for protection but is a prerequisite for an infringement suit. Copyrighted material is registered with the copyright office at the Library of Congress, which requires one copy of unpublished work and two copies of published work, plus a $20 fee for processing of registration forms.

Trade Secrets

Trade secrets, or confidential technological and commercial information, are the most important assets of many businesses. The law protects trade secrets as alternatives to patents and copyrights. Trade secrets have no precise definition, but to be protected by the courts, they must be secret, substantial, and valuable. The secret can be almost anything as long as it is not generally known in the trade or industry to which it applies. A trade secret provides its owner with a competitive advantage. It may be a formula, process, know-how, specifications, pricing information, customer lists, supply sources, merchandising methods, or other business information. It may or may not be protected by other means.

Unlike patents or copyrights, trade secrets have no time limitations. A trade secret, however, has value only while it remains secret. For instance, a trade secret may lose its privileged status when it is ascertained through "reverse engineering" or when it is discovered independently. A trade secret revealed in these ways can be used without any obligation to the trade secret's originator or owner. If a trade secret is unlawfully obtained—for example, by breach of trust or violation of a confidential relationship—the courts could award the trade secret's owner compensation for damages suffered and forbid the infringer use and further disclosure of the trade secret.

Comparison of Means of Protecting Ideas

Table 9-4 compares the various means of protecting ideas just discussed. Any innovator or author should be familiar with these options so that an intelligent decision can be made on the proper protection needed for each idea. Different options offer very different kinds of protection. For example, the Coca-Cola Company has elected to protect the ingredients, mixing, and brewing of its principal product, Coca-Cola, as trade secrets. This decision does not prevent another company that claims to have discovered these secrets from marketing or patenting a similar product. The trade secret approach, however, protects the Coca-Cola Company's information for as long as it remains secret. Had the company patented these formulas, the knowledge would have been dedicated to the public 17 years after the patent was issued.

Many ideas that are protected as trade secrets cannot be patented. On the other hand, an item that is patentable can theoretically be protected as a trade secret. If the idea can be easily discovered through reverse engineering, however, a patent is the only practical choice for protection.

The law can, however, provide some protection against deliberate theft of trade secrets. For example, the General Electric Company has invested over $200,000,000 in designing better turbine parts, and drawings of these parts are normally marked "GE propri-

TABLE 9-4 Comparison of Means of Protecting Ideas

Category	Utility Patents	Design Patents	Trademarks	Copyrights	Trade Secrets
Idea or subject matter	New and useful processes, machines, articles of manufacture, and compositions of matter	New ornamental designs for articles of manufacture	Words, names, symbols, or other devices that serve to distinguish goods or services	Writings, music, works of art, and the like that have been reduced to a tangible medium of expression	Almost anything that is secret, substantial, and valuable
Sources of protection	U.S. Patent and Trademark Office patent	U.S. Patent and Trademark Office patent	Registration with the U.S. Patent and Trademark Office Registration with the secretary of state Common law protection through courts as long as proper use continues	Federal law protects only a tangible medium of expression Enforceable only when registered with the copyright office	Common law protection through courts
Terms of protection	20 years from filing date; before June 8, 1995 was 17 years from patent date	14 years	10 years from registration with federal office; renewable for additional 10-year terms	Life of author, plus 50 years for works created after Jan. 1, 1978	For as long as it remains a secret
Tests for infringement	Making, using, or selling invention described in patent claim	Making, using, or selling design shown in patent claim	Likelihood of confusion, mistake, or deception	Copying of protected subject matter	Taking of trade secret by breach of trust or violation of a confidential relationship

etary data" and regarded as trade secrets. GE was shocked in 1984 when "AEG, a German manufacturer of power plants and GE licensee, landed a big Soviet turbine order, but didn't follow its usual practice of buying the high-tech spare parts from GE. This time, AEG placed a $6 million parts order with Turbine Services, a company "set up in the early 1970s by Mr. [Charles] Mothon after he resigned as GE's director of gas turbine international sales."[16] When AEG tests found parts from Turbine Services remarkably close to GE's own, GE executives concluded that they must have been made from stolen GE drawings. There followed an intensive investigation over the next several years, involving a dummy corporation, undercover agents, a "sting" operation, and finally a raid on Turbine Services offices accompanied by federal marshals. The matter climaxed in federal court in Albany, NY, on May 6, 1988. The *Wall Street Journal* reported:

> In a tough settlement, GE will get $5.2 million in damages, the return of hundreds of technical drawings and a slew of restrictions on Turbine Services' future activity. . . .
>
> For GE, however, a troubling question lingers. Why were so many current and former employees disloyal? In all, nearly two dozen people in Schenectady, New York and Houston have been implicated in the theft of its trade secrets.[17]

CREATIVITY

Nature of Creativity

Creativity is the ability to produce new and useful ideas through the combination of known principles and components in novel and nonobvious ways. Creativity exists throughout the population, largely independent of age, sex, and education. Yet in any group a few individuals will display creativity completely out of proportion to their number. To have an effective research organization requires understanding the creative process, identifying and acquiring creative people, and maintaining an environment that supports rather than inhibits creativity.

The Creative Process

There are a number of models for problem solving. One method, often inefficient, is simple trial and error. A second is the planning/decision-making process introduced in Chapter 3 (see Figure 3-1), which involves problem definition, identification of alternatives, and evaluating alternatives against objectives. Its major thrust is analytical reasoning, although its success is enhanced by some creativity in selection of alternatives to be evaluated. The creative process uses some of the same steps, but it emphasizes the insight that can occur subconsciously when a perplexing problem is not resolved through the analytical process and is temporarily set aside. Following are the steps usually identified in describing this process.

1. *Preparation.* Shannon describes this step as "a period of conscious, direct, mental effort devoted to the accumulation of information pertinent to the problem. . . . Quite often the problem is solved at this stage as one submerges oneself in the problem while trying to (a) structure the problem, (b) collect all available information, (c) understand relations and effects, (d) solve subproblems, and (e) explore all possible solutions and combinations that may lead to a satisfactory solution."[18]

2. *Frustration and incubation.* Failure to solve the problem satisfactorily by the analytical process above leads to frustration and the decision to set it aside and get on with something else. However, the problem, fortified with all the facts gathered about it, "stews" or incubates in the subconscious mind.

3. *Inspiration or illumination.* A possible solution to the problem may occur as a spontaneous insight, often when the conscious mind is at rest during relaxation or sleep. Raudsepp writes that "creative ideas may and do appear at any hour and under the strangest of circumstances. . . . [An] incident has been reported about Newton, who, during the course of a dinner he was giving to his guests, left the table to get some wine from his cellar. On the way from the cellar he was overcome by an idea, forgot his errand and company, and was soon hard at work in his study."[19] Many creative individuals are never without a notepad and pen, on their person or bedside table, to write down these flashes of insight.

4. *Verification.* Intuition or insight is not always correct, and the solution revealed in a flash of insight must now be tested and evaluated to assure it is, indeed, a satisfactory solution to the problem.

Shannon defends this model:

How do we know this process is true? Because thousands of creative people have described exactly this process when discussing their work. Over and over again we see this interplay between the conscious and the subconscious. For creative work we have this wondrously competent coupling where each part (conscious and subconscious) is indispensable in its own way, but each is helpless without the other. When applied to problem solving, the human mind has two aspects: (1) a judicial, logical, conscious mind that analyzes, compares, and chooses; and (2) an imaginative, creative, subconscious mind that visualizes, foresees, and generates ideas from stored knowledge and experience.[20]

Brainstorming and Other Techniques for Creativity

Dhillon[21] describes eight creativity techniques designed for one, two, or up to a dozen people. Best known is *brainstorming,* a modern method for "organized ideation" first employed in the West by Alex Osborne in 1938, although he reports that a similar procedure had "been used in India for more than 400 years as part of the technique of Hindu teachers" under the name *Prai-Barshana,* literally "outside yourself-question."[22] The essence of brainstorming is a creative conference, ideally of 8 to 12 people meeting for less than an hour to develop a long list of 50 or more ideas. Suggestions are listed without criticism on a blackboard or newsprint as they are offered; one visible idea leads to others. At the end of this session participants are asked how the ideas could be combined or improved. Organizing, weeding, and prioritizing the ideas produced is a separate, subsequent step.

Dhillon next lists two brainstorming techniques that can be used by two people. In one, known as the "tear-down" approach, the first person (person A) must disagree with the existing solution to a problem and suggest another approach; next, person B must disagree with both ideas and suggest a third; then person A must suggest yet another solution; this "cycle continues until a useful idea clicks." In a variant, known as the "and-also" method, person A suggests an improvement on the subject under study; person B agrees, but suggests a further improvement; this sequential improvement "continues until a sound solution is reached."

In a somewhat different group technique developed by W. J. Gordon, a "team explores the underlying concept of the problem. For example, if a new can opener is desired the team would first discuss . . . [the] meanings of the word opening and examples of opening in real life things." The method encourages finding unusual approaches by preventing early closure on the problem. Gordon used a team of six meeting for about a day on a problem.

Dhillon describes two approaches in which individuals are given a description of a problem and required to list solutions in advance of group effort. In the simpler, "each participant has to have a certain number of solution ideas, say 17, to the problem before he is allowed to attend the meeting." In a more complex version known as the "CNB method," each member of a team is given a notebook with a problem statement and supporting material a month in advance. Each day during that month the team member writes one or more ideas in the notebook, and at the end of the month selects the best idea along with "fruitful suggestions for further exploration." A problem coordinator collects and studies notebooks and prepares a detailed summary for distribution; if necessary, all team members then participate in a final meeting.

Finally, Dhillon includes two methods that individuals may use. In an "attribute-listing" approach, a person lists attributes of an idea or item, then concentrates on one attribute at a time to make improvements in the original idea or item. The other method tries to generate new ideas by creating a "forced relationship" between two or more usually unrelated ideas or items. For example, an office equipment manufacturer might consider the relationship between a chair and a desk, "start up a line of free associations," and end up with a combined unit consisting of both desk and chair.

Characteristics of Creative People

There have been many studies comparing more creative with less creative people. Characteristics of creative people can be grouped in the following categories:

Self-confidence and independence. Creative people seem to be self-confident, self-sufficient, emotionally stable, and able to tolerate ambiguity. They are independent in thought and action and tend to reduce group pressures for conformity and rules and regulations that do not make sense.

Curiosity. They have a drive for knowledge about how or why things work, are good observers with good memories, and build a broad knowledge about a wide range of subjects.

Approach to problems. Creative people are open-minded and uncritical in the early stages of problem solving, generating many ideas. They enjoy abstract thinking and employ method, precision, and exactness in their work. They concentrate intensively on problems that interest them and resent interruptions to their concentration.

Some personal attributes. Creative people may be more comfortable with things than people, have fewer close friends, and are not "joiners." They have broad intellectual interests: They enjoy intellectual games, practical jokes, creative writing, and are almost always attracted by complexity.

Providing a Creative Environment

Creative people tend to be independent, nonconformist, and to work intensively for long periods but with a disregard for conventional work hours. They are most effective in an organization that will tolerate idiosyncrasies, remove as much routine regulation and reporting as feasible, provide support personnel and equipment as required, and recognize and reward successes. People doing routine work and those doing creative work should be separated where possible. One of my students [Claire Moore] provided an example in a homework assignment (answering discussion question 9-13) from an earlier co-op work assignment at a nuclear plant:

> In our group was a man who was quite an oddball. He didn't like people, phones, or anyone using the computer when he was. To work "normal hours" unnerved him, so he was allowed to come in anytime he wanted to. Many people would have taken advantage of this, but he worked *longer* [and] more productive hours. Sometimes he would work 5 P.M. until 7 A.M. nonstop. They put a computer in his home and hooked it to the mainframe at work for those sudden brainstorms—the results were great!

The prolific production of ideas in the early phases of problem solving is a hallmark of creativity. "Nothing can inhibit and stifle the creative process more—and on this there is unanimous agreement among all creative individuals and investigators of creativity—than critical judgment applied to the beginning stages of the creative process. . . . Critical attitude, according to several psychologists, seems to be the engineer's most notable personality trait, and it colors all of his perceptions."[23] Engineer managers therefore must be especially careful to withhold criticism until its appropriate place—at the conclusion (verification) of the creative process.

Creative people value working on problems of interest to themselves and working on their own schedule. It is important to explain the problem and its importance fully, agree on a timetable, and stay in contact without close supervision as long as reasonable progress is made.

Creativity and Innovation

Invention (the creative process) only produces ideas. They are not useful until they are reduced to practice and use, which is the process of innovation. Kidder[24] provides an excellent study of motivation and creativity in the development of a 32-bit computer at Data General. Roberts and Wainer[25] have identified five kinds of people who are needed for technological innovation:

Idea generator—the creative individual

Entrepreneur—the person who "carries the ball"

Gatekeepers—discussed below

Program managers—who manage without inhibiting

Sponsor or Champion—the person, often in senior management, who provides financial and moral support

Technological Gatekeepers in R&D Organizations

Allen and Cohen[26] found that only about 15% of the scientific and technical ideas being worked on in industrial laboratories came directly from the scientific and technical literature—most of it reached lab members in a two-step process involving "gatekeepers." These are research staff members who, through their professional work habits, bring essential information into the organization. Gatekeepers (1) are more likely to read the more sophisticated (refereed) journals, (2) are in contact with "outside" specialists, and (3) form a network with other gatekeepers.[27] They often are high technical performers, usually produce more than their share of conference papers and refereed articles, and are likely to be promoted to first- and second-line supervision ahead of their peers. Gatekeepers are not appointed, but the wise research manager recognizes them and their function. Professional staff who are hired away from other organizations or who transfer in from other parts of a corporation provide another important source of new ideas and ways of doing things.

MAKING R&D ORGANIZATIONS SUCCESSFUL

Three topics are treated herein: the relation of R&D strategy to business strategy, evaluating the effectiveness of R&D (both at the organizational and individual level), and providing effective support for researchers.

R&D and Business Strategy

Chapter 3 discussed the importance of strategic planning to the success of any enterprise. In the technology-driven organization, a carefully planned technology strategy must be thought through to support the overall strategy of the enterprise. This strategy should encompass research, product and process development, and manufacturing engineering. Erickson et al. identify "three broad classes of technologies" a typical firm must consider:

- *Base Technologies.* These are the technologies that a firm must master to be an effective competitor in its chosen product-market mix. They are necessary, but not sufficient. . . . The trick for R&D management is to invest enough—but only enough—to maintain competence in these technologies.
- *Key Technologies.* These technologies provide competitive advantage. They may permit the producer to embed differentiating features or functions in the product or to attain greater efficiencies in the production process.
- *Pacing Technologies.* These technologies could become tomorrow's key technologies. Not every participant in an industry can afford to invest in pacing technologies; this is typically what differentiates the leaders (who do) from the followers (who do not). The critical issue in technology management is balancing support of key technologies to sustain current competitive position and support of pacing technologies to create future vitality.[28]

Evaluating R&D Effectiveness

Organizational effectiveness. Balderston et al.[29] suggest the following 11 criteria for business enterprise R&D:

1. Ratio of research costs to profits.
2. Percentage of total earnings due to new products.
3. Share of market due to new products (usually computed as the volume of sales from a firm's new products in a specific product market to the total sales available from that market, which confounds the measure by including marketing proficiency as well).
4. Research costs related to increases in sales.
5. Research costs to ratio of new and old sales.
6. Research costs per employee.
7. Ratio of research costs to overhead expenses such as administrative and selling costs.
8. Cash flows (continuing evaluation of the pattern of outflows for research expense and actual and projected inflows from resulting revenue).
9. Research audits, including "indicators of administrative and technical objectives such as costs, time, completion dates, probability of technical success, probability of commercial success, expected market share, expected profits, expected return on investment, design, and development." Blake[30] provides a checklist of questions to ask in such an audit.
10. Weighted averages of costs and objectives (a measure of the extent the average R&D dollar contributed toward objectives with weights on a scale such as 0.0 equals "project badly missed objectives" to 3.0 equals "project far exceeded objectives").
11. Project profiles (a more complex weighted scoring of each project using criteria such as those in the research audits, item 9); Whelan[31] has described use of such profiles at Union Carbide.

A number of these measures (such as items 1, 4, and 5) are obscured by the lag between research expenditures and the sales and profits that result from them, as well as the contribution of production and marketing to sales and profits. Others (items 6 and 7) are measures of the intensity of research expenditures rather than research effectiveness. The last three are more time consuming and require subjective opinion, but they also may be more effective.

Individual effectiveness. The effectiveness of individual researchers can be evaluated by the normal techniques of performance appraisal introduced in Chapter 6, especially management by objectives (MBO), emphasizing research goals. A few quantitative measures such as the number of patents and publications, and citations by others of those publications, give limited insight into research effectiveness.

Support for R&D

Quality supporting services need to be supplied to make the work of the highly trained scientist and engineer more efficient and productive. A few special types of assistance that are needed in research and engineering are:

1. Technician support to carry out repetitive testing and other functions not requiring a graduate engineer or scientist

2. Shop support of mechanics, glassblowers, and carpenters to produce test and research equipment based on researchers' sketches

3. A technical library with technical information specialists conversant in the fields of the company's interest and willing and able to suggest sources to researchers and structure and run searches in the appropriate data bases for them

4. Technical publication support, including typing, editing, and graphical support to simplify researchers' production of reports, technical papers, and presentations

5. A flexible, responsive system for approving and acquiring equipment as needed by researchers

6. Ample computer facilities conveniently available to researchers, and programming assistance to provide consultation and programming to those researchers not wishing to do it themselves

DISCUSSION QUESTIONS

9-1. Contrast the application of Blanchard's product life cycle with that of Betz's technology life cycle.

9-2. Would the same kind of leader be suitable throughout Betz's technology life cycle? If not, what kind of leader would be effective in each portion of it?

9-3. Summarize the principal contributions to U.S. R&D activity by each of (**a**) the federal government; (**b**) industry; and (**c**) universities.

9-4. In an industry with which you are familiar, give an example of one or more firms that appear to have chosen each of Ansoff and Stewart's new product strategies.

9-5. Discuss the relationship between the central corporate research laboratory and divisional research in a corporation you know or have found described in the literature.

9-6. Why are simple checklists used as a first screening of ideas in research projects by many companies?

9-7. An engineer proposes to buy a machine for $100,000 today that will save $60,000 in labor costs at the end of each of the next two years. If the company demands a 15% return on investments such as this, what is the net present worth (NPW) of the proposal? Should it be funded?

9-8. Your company has two alternative opportunities, each requiring your entire capital investment budget of $325,000. Alternative A will return $390,000 at the end of one year; alternative B will return $216,000 at the end of each of the first two years. Which (if either) alternative should you recommend on the basis of (**a**) simple payback time? (**b**) net present worth?

9-9. If you have been exposed to capital investment analysis and/or engineering economy, comment on the proposal to invest $1,000,000 in a new product now that is projected to generate $200,000 profit at the end of each year for eight years, assuming that your company requires 15% return on investment before taxes.

9-10. As an R&D manager, what actions might you take or programs might you implement to assure your organization got maximum benefit from patentable ideas?

9-11. How do the kinds of ideas best protected by patent differ from those best protected by keeping them a trade secret?

9-12. The General Electric Company was troubled to find that almost two dozen employees were implicated in the theft of turbine trade secrets. What actions or defensive programs might a firm institute to reduce the likelihood of this happening?

9-13. What are some of the steps a manager can take to encourage creativity in his or her technical employees?

9-14. Why is it so difficult to measure the effectiveness and productivity of research organizations?

9-15. How would you try to evaluate the effectiveness of researchers if you were their research manager?

9-16. What are some of the support services an organization might provide to make the work of researchers and design engineers more effective?

NOTES

1. Frederick Betz, *Managing Technology: Competing Through New Ventures, Innovation, and Corporate Research* (Englewood Cliffs, NJ: Prentice-Hall, Inc., 1987).

2. Betz, *Managing Technology,* pp. 72–74.

3. National Science Foundation, *Patterns of R&D Resources,* Report 74-304 (Washington, DC: U.S. Government Printing Office, 1974), p. 17.

4. "Reinventing America 1992," *Business Week,* October 23, 1992, p. 169.

5. H. Igor Ansoff and John M. Stewart, "Strategies for a Technology-Based Business," *Harvard Business Review,* 45:6, November–December 1967, pp. 71–83.

6. Betz, *Managing Technology,* pp. 153–154, 187–189.

7. Descriptive material on the GE Corporate Research Laboratories (Fairfield, CT: General Electric, 1980), p. 22.

8. Robert A. Frosch, "GM's Healthy Pain," *Mechanical Engineering,* December 1987, pp. 23–25.

9. Robert E. Seiler, *Improving the Effectiveness of Research and Development: Special Report to Management* (New York: McGraw-Hill Book Company, 1965).

10. Jack Balderston, Philip Birnbaum, Richard Goodman, and Michael Stahl, *Modern Management Techniques in Engineering and R&D* (New York: Van Nostrand Reinhold Company, Inc., 1984), pp. 34–58.

11. B. S. Dhillon, *Engineering Management* (Lancaster, PA: Technomic Publishing Company, Inc., 1987), pp. 79–104.

12. Robert E. Shannon, *Engineering Management* (New York: John Wiley & Sons, Inc., 1980), pp. 235–257.

13. Some of Dr. Myers's remarks have appeared in John M. Amos and Bernard R. Sarchet, *Management for Engineers* (Englewood Cliffs, NJ: Prentice-Hall, 1981).

14. Lucy Reilly Fitch, "'First to Invent' Spared . . . For the Moment," *Technology Transfer Business,* Spring 1994, p. 82.

15. "The Global Patent Race Picks Up Speed," *Business Week,* August 9, 1993, p. 57.

16. William M. Carley, "Keeping Secrets: GE Presses Campaign to Halt Rivals' Misuse of Turbine-Parts Data," *Wall Street Journal,* August 16, 1988, pp. A1, A12.

17. Carley, "Keeping Secrets," pp. A1, A12.

18. Shannon, *Engineering Management,* p. 146.

19. Eugene Raudsepp, "A New Look at the Creative Process," Part 2, *Creative Computing,* September 1980, pp. 82–90.

20. Shannon, *Engineering Management,* p. 147.

21. Dhillon, *Engineering Management,* pp. 59–62.

22. Alex F. Osborne, *Applied Imagination,* 3rd ed. (New York: Charles Scribner's Sons, 1963), p. 151.

23. Eugene Raudsepp, "A New Look at the Creative Process," Part 1, *Creative Computing,* August 1980, pp. 46–51.

24. Tracy Kidder, *The Soul of a New Machine* (Boston: Little, Brown, 1981).

25. E. B. Roberts and H. A. Wainer, *IEEE Trans. Engineering Management,* 18:3, 1971, pp. 100–109, summarized in George E. Dieter, *Engineering Design: A Materials and Processing Approach* (New York: McGraw-Hill Book Company, 1983), p. 25.

26. T. J. Allen and D. I. Cohen, "Information Flow in Research and Development Laboratories," *Administrative Science Quarterly,* 14, 1969, pp. 12–19.

27. Thomas J. Allen, *Managing the Flow of Technology: Technology Transfer and the Dissemination of Technological Information Within the R&D Organization* (Cambridge, MA: The MIT Press, 1977), pp. 144–149, 163–173.

28. Tamara J. Erickson, John F. Magee, Philip A. Roussel, and Kamal N. Saad, "Managing Technology as a Business Strategy," *Sloan Management Review,* 31:3, Spring 1990, pp. 73–78.

29. Balderston et al., *Modern Management Techniques,* pp. 179–185.

30. Stewart P. Blake, *Managing for Responsive Research and Development* (San Francisco: W.H. Freeman and Company, 1978), pp. 250–261.

31. J. M. Whelan, "Project Profile Reports Measure R&D Effectiveness," *Research Management,* 14, September 1976, pp. 14–16.

CHAPTER 10

Managing Engineering Design

PREVIEW

We begin this chapter by considering the nature of engineering design and the phases or stages in the systems engineering and new product development processes. We next discuss the modern emphases on concurrent (simultaneous) engineering and on Computer-aided Acquisition and Logistics Support (CALS). Special control systems in engineering design—drawing/design release, configuration management, and design review—are considered. We then introduce product liability and safety, which require special precautions in design. Reliability, the assurance function most associated with design, is discussed from several aspects: its significance, some simple models of reliability, and the development of increased reliability over the product life cycle. Finally, we discuss six more considerations important to design: maintainability, availability, human factors engineering, standardization, producibility, and value engineering.

NATURE OF ENGINEERING DESIGN

Design is the activity that best describes the engineer. To design is to create something that has never existed before, either as a solution to a new problem or as a better solution to a problem solved previously. J. B. Reswick summarizes the process of design well:

> Design is the central purpose of engineering. It begins with the recognition of a need and the conception of an idea to meet that need. It proceeds with the definition of the problem, continues

with a program of directed research and development, and leads to the construction and evaluation of a prototype.[1]

Essentially, design is the process of creating a *model,* usually described in terms of drawings and specifications (whether on paper or in computer memory), of a system that will meet an identified need. The model can then be reproduced by some suitable manufacturing process and distributed for use, as described in the succeeding chapters.

Engineering design is a process of transforming information, as illustrated in Figure 10-1. Information provides the input to the process: a statement of the problem to be solved, design standards, design methods, and the methods of engineering science. Within the box labeled "Engineering design process" the engineer performs some logical sequence of activities, decisions, and analyses to develop a solution to the problem. However, this solution is of little use until the engineer communicates the solution in the form of drawings, specifications, financial estimates, written reports, and oral presentations to explain and promote the solution. Unfortunately, many engineers do not realize the importance of this vital last step of communication, without which the rest of the work done is fruitless.

SYSTEMS ENGINEERING/NEW PRODUCT DEVELOPMENT

The design of a complex engineered system, from the realization of a need (for a new system or improvement of an existing system) through production to engineering support in use is known as *systems engineering* (especially with military or space systems) or as *new product development* (with commercial systems). The (U.S.) National Aeronautics and Space Administration (NASA) offers a good definition of the first in their (draft) *NASA Systems Engineering Handbook:*

> *Systems engineering* is a robust approach to the design, creation, and operation of systems. In simple terms, the approach consists of identification and quantification of system goals, creation of alternative system design concepts, performance of design trades, selection and implementation of the best design, verification that the design is actually built and properly integrated, and post-implementation assessment of how well the system meets (or met) the goals. The approach is usually applied repeatedly and recursively, with several increases in the resolution of the system baselines (which contain requirements, design details, verification procedures and standards, cost and performance estimates, and so on).[2]

Whether called systems engineering or new product development, the engineering of complex systems is carried out in a series of sequential phases or stages. A task group formed from U.S. engineering societies, coordinated by the (U.S.) National Society of Professional Engineers (NSPE) in cooperation with the (U.S.) National Institute of Science and Technology (NIST) agreed to define such a stage as follows:

Figure 10-1. The engineering design process.

An Engineering Stage of Development is a period during which designated engineering activities are completed. The end of a stage is an optimum decision point in the progression toward realization of a product, process or service. The objective of each stage is to establish the engineering information (technical, economic and risk assessment) necessary to make the decision to proceed or not.[3]

As you learned in Chapter 9, many research concepts enter the initial phases of evaluation and development for each one that becomes a successful product. Each proposed idea is beset with many unknowns and uncertainties. It should be the function of the early phases of systems engineering/new product development to resolve these uncertainties and to determine at minimum cost whether the proposed system or product is feasible and is worth the resources required to develop it further and produce it. In each development phase it is necessary to carry out a *systems engineering process* not unlike the planning/decision-making process introduced in Chapter 3. The (U.S.) Department of Defense (DOD) visualized this process in their draft MIL-STD-499B (proposed but never issued) as including the following major activities:[4]

- *Requirements Analysis.* Analyze customer needs, objectives, and constraints to determine the functions that must be performed (functional requirements) by the system to meet objectives.
- *Functional Analysis/Allocation.* Identify lower level functions needed to meet these functional requirements, and translate them into design requirements suitable as design criteria.
- *Synthesis.* "Define the system concept, configuration item (component), or system element [design] alternatives and select the preferred set of product or process solutions" to the level of detail required in the phase being conducted.
- *System Analysis and Control.* " . . . provide the progress measurement, assessment, and decision mechanisms required to evaluate design capabilities and document the design and decision data." These may include *trade-off (trade) studies* (to optimize choices among alternatives); *risk management* (to identify and quantify sources of technical risks and their program impacts and find ways to avoid or control them); *configuration management* (discussed later); *interface management* (to assure that components designed and built by different organizations will fit and perform together); a *systems engineering master schedule (SEMS)* (based on major events, tasks needed to complete these events, and measurable criteria that define task completion); *technical performance measurement (TPM)* (to assure the designs as they evolve will meet performance requirements); and such technical (design) reviews as are needed to demonstrate that the requirements for each phase have been met.

Phases/Stages in Systems Engineering or New Product Development

The phases of the *system life cycle* (which extends from original concept through systems engineering to product disposal) are given slightly different names by the U.S. DOD[5] and NASA,[6] but they cover the same functions. A third approach, perhaps more applicable to nonmilitary products, is described in the *Engineering Stages of New Product Development*[7] developed by the NSPE/NIST task force. The sequence of phases or stages in these three approaches appear in Table 10-1.

Each phase begins with approval to expend the resources that phase will require and agreement on the work to be accomplished in that phase. Next comes accomplishment of the work of the phase, which may be modest or enormous. The results of that phase are then compiled: designs and specifications, analyses and reports, and a proposed plan for conducting the following phase if one is recommended. At this point there should be a conscious, and often formal, review to decide whether the expense of the next phase, which will usually represent a substantial increase in resource commitment, is justified. Typically, one of three types of decisions should come out of this review: (1) to cancel the development, if study to that point has shown that further development cannot be justified; (2) to go back (recycle) and do more work in the present phase if too many uncertainties still exist; or (3) to proceed with the next phase and its increased resource expenditure with confidence.

NSPE's titles for their "Engineering Stages of New Product Development" are used in the discussion that follows. NSPE uses the term *product* "in the more global sense to refer to structures, machines, systems, processes, software, or technical services." Details discussed under each stage (or phase) are a composite of those outlined by NSPE and those discussed in DOD or NASA publications for equivalent systems engineering phases, with additions from other sources, and therefore represent a more generic treatment.

Conceptual stage. This stage should begin with a careful statement of the design problem, clearly defining what the desired product is intended to accomplish, the functions it must perform, the performance characteristics it must meet, any constraints limiting the solution, and the criteria that will be used to judge the quality of the design solution. Wilson and Wilson[8] suggest identifying four categories of objectives and goals for the solution:

TABLE 10–1 Comparison of Systems Engineering/New Product Development Phases or Stages

Department of Defense Draft MIL-STD-499B	NASA Systems Engineering Handbook	NSPE Stages of New Product Development
Pre-milestone zero studies	A. Conceptual design studies	1. Conceptual
Concept exploration and definition	B. Concept definition	2. Technical feasibility
Demonstration and validation	C. Design and development	3. Development
Engineering and manufacturing development	D. Fabrication, integration, test, and certification	4. Commercial validation and production preparation
Production and deployment	E. Pre-operations	5. Full-scale production
Operations and support	F. Operations and disposal	6. Product support

1. *Musts:* requirements that must be met
2. *Must nots:* constraints defining what the system must not be or do
3. *Wants:* features that would significantly enhance the value of the solution but are not mandatory (to which an additional, even less compelling category of "*nice to have*" is often added)
4. *Don't wants:* characteristics that reduce the value of the solution

These criteria lead to a set of *functional* (or *operational*) *requirements* defining what the required design solution must and/or should be able to achieve. Once the problem is well understood, the synthesis of possible design concepts can begin. Essential at this stage is "preliminary identification of the potential barriers to development, manufacturing, and marketing"[9] the proposed product. Before proceeding to the next phase, it may be desirable to reduce the uncertainty created by these potential barriers by devising and testing a simple and inexpensive *test-of-principle* model to reduce technical uncertainties, and to carry out order-of-magnitude economic analyses and preliminary market surveys to reduce financial uncertainty.

The outcomes of these activities may then be subjected to a *conceptual design review* (by whatever name) to decide whether proceeding to the next phase is justified. Doing the work of the first design stages right is essential. Zangwill states[10] that, although the concept development phase [conceptual stage] only *spends* about one percent of the cost of the product, the decisions made therein *determine* about 70 percent of the life-cycle cost of the product; similarly, at the end of the advanced development phase [technical feasibility stage, discussed next], about 7 percent of the development cost has been spent but about 85 percent of the life-cycle costs are determined.

Technical feasibility stage. According to NSPE, "The objective of this stage is to confirm the target performance of the new product through experimentation and/or accepted engineering analysis and to ascertain that there are no technical or economic barriers to implementation that cannot be overcome by development."[11] The purpose of the corresponding DOD/NAS *Definition phase* is to select a single system design concept, identify the building blocks needed, develop design requirements for them, and reconfirm the desirability of the system. Following are some typical steps:

- Subsystem identification: transform the functions required of the system into a set of building blocks (subsystems and components) with defined functions or subfunctions that will make up the system.
- Trade-off studies: examine alternative concepts for subsystem designs and "trade off" the advantages (benefits) and disadvantages (costs) of these alternatives to arrive at a consistent set of subsystem design concepts.
- System integration: develop *layout* drawings showing how these subsystems will fit together.
- Interface definition: define the *interfaces* between subsystems in terms of the space *envelope* surrounding each subsystem and the physical, electrical, thermal, and other interactions passing through each subsystem boundary.
- Preliminary *breadboard-level* testing of only those components or concepts for

which the uncertainty of successful resolution threatens overall system success. A breadboard model is one level of sophistication beyond the test-of-principle model of the conceptual stage, and it may be supplemented by materials tests and computer simulations.

- Establishment of subsystem and system *design requirements* for use in the next phase (including such requirements as reliability, safety, maintainability, and environmental impact).
- Development of preliminary test plans, production methods, maintenance and logistic concepts, and marketing plans.
- Preliminary estimation of the life-cycle cost of the system (including development, production, and use).
- Preparation of a proposal for the next (development) stage, including tasks to be performed, cost, and schedule. This proposal, along with results of the other activities above, should then be subjected to a formal design review to decide whether to proceed with the next phase, to do additional work to reduce uncertainties, or to cancel the project.

Development stage. "The objective of this stage is to make the needed improvements in materials, designs and processes and to confirm that the product will perform as specified by constructing and testing *engineering prototypes* or pilot processes."[12] Effective development normally requires that design and testing proceed in parallel in an iterative manner. This may begin by identifying and testing critical materials and developing and testing components and process steps, then continue by integrating components into subsystems, which are then subjected to tests of increasing complexity. At each level the design is optimized using build–test–fix–retest sequences, supplemented by computer analyses where appropriate. Physical models (*mockups*) or three-dimensional computer models can be crudely prepared at first, then updated as the design proceeds, to better visualize problems of manufacture, operation, and maintenance. Optimized subsystems and components are then integrated into engineering prototypes and subjected to extensive system tests simulating as much as possible the operating and environmental conditions of actual use, and demonstrating the performance, safety, and reliability expected of the product.

In the process, most of the activities of the previous (technical feasibility) stage are repeated in much greater detail. Preliminary test plans evolve into detailed test plans and test setups, then to test results. Preliminary concepts of production methods, maintenance, logistics, marketing approaches, and environmental impact are developed much more completely. Now that the components and performance of the system are far better understood, cost estimates for production, marketing, and product support can be developed in much greater detail. Results of this stage and estimates for the remaining stages should be the subject of another formal review before the next stage is authorized.

Commercial validation and production preparation stage. NSPE defines the objective of this stage "to develop the manufacturing techniques and establish test market validity of the new product." The manufacturing development portion includes selecting manufacturing procedures, production tools and technology, installation and start-up plans for the manufacturing process, and vendors for purchased materials, components, and subsystems. Market validity may require producing *preproduction prototypes,* using production processes as close to those planned for full-scale production as possible, both for final systems tests and for market trials (often with cooperating "early adopter" customers). The equivalent to "market validity" in development for the U.S. Department of Defense or NASA (left two columns of Table 10-1) is obtained by delivering the first systems produced to the government customer for systems testing under field conditions of use.

At this point, a very detailed analysis of the proposed methods and estimated costs of producing, marketing, supporting in use, and disposing of the system is prepared, and revenues throughout the system's effective marketing life are forecast. This analysis provides the basis for a final system review, often by top management, since the decision to release to production may require massive expenditures before the first revenues are realized. For example, developing a new automobile model and preparing production plant(s), suppliers, and dealers for it may cost several billion dollars.

Full-scale production stage. During the initial phase of this stage, the final design drawings, specifications, flow charts, and procedures are completed for manufacture and assembly of all components and subsystems of the product, as well as for the production facility. Quality control procedures and reliability standards are established, contracts made with suppliers, and procedures established for product distribution and support. Manufacturing facilities are constructed, and trial runs made to "shake down" the new plant and adjust the process until a quality product is being economically produced.

As production continues, minor evolutionary changes should be undertaken to enhance the product's value to the customer (quality, reliability, function), and to reduce production cost so that profit can be maintained as increasing competition makes price reductions necessary. The Japanese excel in continuous process improvement (*kaizen*), and American firms must likewise excel to remain competitive.

Chapters 11 and 12 provide details on planning and managing production activity and maintaining product quality.

Product support stage. Complex engineered systems require continuing support by the manufacturer throughout product life. Technical manuals for product installation, operation, and maintenance must be prepared and maintained, and training programs for customer personnel may be needed. A cadre of engineers and technicians to resolve customer problems with the product may be necessary, and warranty services to support product guarantees must be provided. Repair parts and replacement "consumables" must be manufactured and distributed. Product deficiencies discovered in use may require providing the customer new procedures for operation and maintenance, improved parts for retrofit (whether by customer or manufacturer's personnel), or even notification of product recall for safety reasons.

Staying close to the customer may lead to improvements in materials, components, subsystems, and software either in subsequent production or in upgrades offered to existing customers, and experience with the product may lead to new product spin-offs or major product design changes that involve returning to earlier stages to create a new model or new product.

Disposal stage. Although this is not listed as a separate stage in Table 10-1, every product causes waste during manufacture, while in use, and at the end of useful life that can create disposal problems. The time to begin asking "how do we get rid of this" is in the early stages of product or process design. The U.S. nuclear weapons program leaves us with many billions of dollars in costs to mitigate radioactive waste, much of which could have been saved had disposal been considered from the beginning. Chemical, petroleum, steel, and other "smokestack" industries send enormous amounts of waste into our air and water, much of which can be eliminated (sometimes even at a profit in material recovered) by improving production processes.

The simplest example of waste problems created during product use is the automobile. Used tires, discarded lead acid batteries, fluorocarbons from air conditioners, and gasoline tank fumes each present an environmental hazard and require a complex recovery network. Packaging gives us mountains of glass, plastic, steel cans, and aluminum, each of which requires a different process to recycle. An Alcoa executive provides a good example of the need to consider disposal in the early stages of design:

> Perhaps the quintessential case study on the value of recyclability is the aluminum beer and beverage container. Even though it made its debut in the mid-1960s, it took about 10 years in the marketplace, the 1973–74 oil embargo and an environmental concern over detachable pull-tabs to focus the public as well as the aluminum and can manufacturing industries on the inherent value of used beverage containers as a recyclable end-product.
>
> I would love to report that we in Alcoa knew all along this would happen and were shrewd enough to design a can lending itself to convenient remelting and refabrication. The fact of the matter is that we probably could not have designed a more difficult package to recycle. We were confronted with the need to separate two different alloys, one for the lids and the other for the can bodies, with problems around removing organic coatings, and with substantial difficulties pertaining to the most efficient way to prepare the scrap for remelting. Our first recycling attempts routinely resulted in melt losses in the 65–70 percent range. Through continuous improvement of our technologies, we have been able to reduce melt losses to between 7 and 8 percent and today see the path to virtually eliminating them.[13]

CONCURRENT ENGINEERING AND CALS

Concurrent (Simultaneous) Engineering

Traditionally, in industries such as automobile manufacturing, a new product proposal originated in research or in marketing, and then it was transferred to engineering design to carry out the technical feasibility and development stages described above. The resulting design drawings and specifications were then "thrown over the wall" to a separate manufacturing engineering department and to the engineering departments of suppliers, who would then figure out how to make what had been designed. Often, manufacturing difficulties would surface that would make the product excessively expensive, or would result in compromises that reduced quality and made maintenance and repair difficult and expensive. Not only was this sequential process costly, but the complete process to come up with a new model or new product took much longer than it should have, and the resulting product was often launched too late to be competitive. A 1988 article in the *Wall Street Journal* gives examples of companies that have faced—and solved—this problem:

> A few years ago, Xerox Corp. executives were stunned to learn that Japanese competitors were developing new copier models twice as fast as Xerox and at half the cost. Its market share eroding, Xerox faced a painful choice: either slash its traditional four- to five-year product development cycle or be overtaken by more nimble competitors. Today, after a sweeping reorganization and millions of dollars of investment, Xerox can produce a new copier in two years. . . .
>
> Now, U.S. manufacturers of all stripes are scrambling to shorten their development cycles and be the first to market at home and abroad. To do that, they are attempting to break out of old, stratified way of developing new products—methods that, up to now, have left them uncompetitive. The Big Three auto makers, for instance, all recently formed task forces to cut ponderously bureaucratic development cycles that have swollen to nearly five years. The Japanese, by comparison, can design and build a new car in a little over 3½ years. . . .
>
> It is often desperation, not enlightened planning, that drives U.S. industry to shorten development cycles. Until recently, Honeywell required four years to design and build a new thermostat. Then a customer, worried about the delay, threatened to take its request for a new climate-control device to a competitor. In response, Honeywell set up a special "tiger team" of marketing, design, and engineering employees and gave it carte blanche. "We told them to break all the rules, but get it done in 12 months," said John Bailey, a Honeywell vice president and general manager. The team did.[14]

This approach to reduce time-to-market had become widely adopted, under the name *concurrent* (or *simultaneous*) *engineering,* for development of both industrial and military systems. Lake[15] quotes a 1989 definition of concurrent engineering by the Pymatuning Group, Inc.:

> The set of methods, techniques, and practices that:
>
> 1. cause significant consideration within the design phases of factors from later in the life cycle;
> 2. produce, along with the product design, the design of processes to be employed later in the life of the product;
> 3. facilitate the reduction of the time required to translate the design into distributed products; and
> 4. enhance the ability of products to satisfy users' expectations and needs.

How, then, can the sequential systems engineering/new product development multistage process described earlier be retained while gaining the benefits of concurrent engineering? Representatives of future stages must be present in earlier stages to provide guidance. For example, Honda Motor Co. began brainstorming for the 1994 Honda Accord in September 1989, just as the 1990 model was being launched. By August 1990 the first of "nearly 60 American production engineers and their families began moving to Japan for two- to three-year stints [at a cost of $500,000 each] working with development engineers at Honda's Sayama assembly and Wako engine plants. They're to make sure each part can be easily and cheaply manufactured at Honda's plants in Ohio."[16]

Figure 10-2 shows the interplay of technical specialties (and change in team leader-

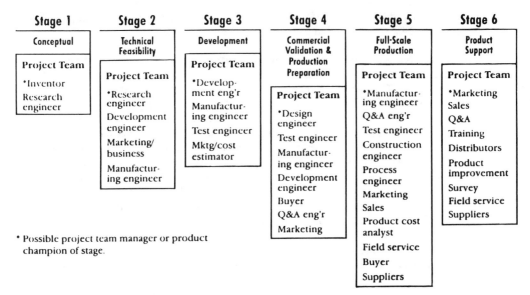

Figure 10-2. Formulation of a multispecialty project team designed to accomplish concurrent engineering in the stages of new product development.
(From National Society of Professional Engineers, *Engineering Stages of New Product Development,* NSPE Publication #3018, Figure 2, p. 10.)

ship) that can take place as a concurrent engineering team carries a system from its first conception through to postproduction product support.

CALS

CALS is an acronym for a set of international initiatives essential to the success of concurrent engineering and to modern design techniques. [CALS initially meant "Computer Aided Logistics Support," then "Computer-aided Acquisition and Logistics Support," which remains its commercial interpretation in Europe; in 1993 the U.S. Department of Defense redefined CALS to mean "Continuous Acquisition and Life-Cycle Support," and NATO military offices have adopted this; more recently, the interpretation "Commerce At Light Speed" is being used in U.S. industry.[17]] According to the U.S. National Institute of Standards and Technology (NIST):

> The CALS Initiative is an industry and government strategy to enable more effective generation, management, and use of digital data supporting the life cycle of a product through the use of international standards, business process change, and advanced technology application. The CALS initiative was started in September 1985 by the U.S. Department of Defense with the goal of enabling the integration of enterprises on a worldwide basis through the development, implementation, and integration of digital information standards for product design, manufacture, and support. The vision is for all parts of a single enterprise to be able to work from a common digital database, in real time, on the design, development, manufacturing, distribution, and servicing of products.[18]

To achieve the objectives of CALS and of concurrent engineering, the cumbersome process of committing ideas to paper and sending them back and forth must be replaced by electronic storage, transmission, and retrieval (a) between engineers representing the several design stages, (b) between organization functions such as marketing, design, manufacturing, and product support, and (c) between cooperating organizations such as customer and supplier. To accomplish this in an international marketplace, international (ISO) and American (ANSI) commercial standards have been developed; related [U.S.] Federal Information Processing Standards (FIST) and [U.S.] Department of Defense and NATO standards also exist. Examples of the more important commercial standards are:

- *Computer Graphics Metafile (CGM)* (ISO-8632): A standard means of representing line drawings in a device-independent way [regardless of the computer hardware or application programs being used for Computer-Aided Design/Manufacture (CAD/CAM)].
- *Electronic Data Interchange for Administration, Commerce, and Transport (EDIFACT)* (ISO 9735, ANSI X12): An international standard means for communicating commercial (trade) information.
- *Initial Graphics Exchange Specification (IGES)* (ANSI Y14.26M): A standard means of representing product data in a device-independent way.
- *Standard Generalized Markup Language (SGML)* (ISO 8879): A standardized language to indicate the structure of the information content of a document, especially text, in accordance with predetermined rules.[19]

The ultimate goal of CALS is to structure integrated data bases so that people working in different parts of the world and for different organizations can form a *virtual organization*[20] working together without mistrust, confusion, or delay on a common problem or product.

CONTROL SYSTEMS IN DESIGN

In creating a complex system, hundreds or thousands of engineers, technicians, and other workers may be involved in creating designs, reviewing them, manufacturing or constructing in accordance with them, or inspecting to assure that what has been made agrees with what was specified. However, no complex system is specified perfectly to begin with and the needs of the user may change during the design phase, so design changes are inevitable. Control systems for drawing/design release and configuration management are essential to assure that everyone knows what the official design (configuration) is at any instant, yet change can be managed effectively. Design review, introduced earlier in this chapter, is also an essential control system in design.

Drawing/Design Release

Drawing release is the process of identifying when a particular design drawing or change has been officially accepted. When a designer finishes with a drawing, it must go through a series of checks and/or analyses by appropriate specialists. At some point the drawing or design (which might consist of a number of drawings) needs to be officially *released,* so that others may work with it. Depending on the design phase, this release might be to further development, to validation testing, or to production. In the past, a drawing made the rounds through designated analysts and checkers; was signed off by the appropriate supervisor(s) and/or project engineer(s); and then was released by rubber-stamping the reproducible master at an official "release desk," from which copies were sent to an official distribution list with instructions to destroy any previous version—those not on the list were left in the dark. With modern team management concepts, the needed specialists are brought together in integrated product teams empowered to create, review, and approve designs concurrently; design release is then effected by changing the design detail recorded in a common electronic data base accessible to all who need it (including, increasingly, suppliers and even customers).

Configuration Management

As the design of a complex system evolves, design specialists often find some need for or advantage in changing the design in one way or another. However, the designer often cannot tell what the impact of the change will be on other parts of the complex system being developed. A design change in one area may require engineers in a second discipline to provide more electrical energy, those in a third to find a way to carry off more heat, test engineers to modify their test equipment, and training manual writers to discard what they have written and start over. Moreover, unless everyone knows the design criteria their design neighbors are working toward, they may be investing substantial effort in designing a system that no longer exists. Therefore, there must be agreement as to the current design criteria (or *configuration*), and there must be a control system to define this configuration and to control changes to it rigorously.

In systems design programs for the U.S. Department of Defense, *baseline* documents are required at several points to specify the current design criteria. At the end of the concept exploration and definition phase, a *functional baseline* is developed to identify the functional characteristics and design constraints that must be satisfied by the design. At the end of the demonstration and validation phase, an *allocated baseline* is prepared, which describes the performance characteristics each subsystem and component must meet. Finally, at the end of the engineering and management development phase, a *product baseline* is established consisting of all the detailed specifications required for production. Baselines are therefore part of the material submitted at the end of a phase for approval in the design review (described next), and they form the basis for beginning work in the next phase.

Changes to these baselines during a phase of the design process are governed by a system known as *configuration management* (or *control*). This system usually involves a committee known as a *configuration control board (CCB),* made up of members from the major design branches and other functions (reliability, production planning, training, etc.) that are affected by change. If, during the design process, a need or desire arises to change a significant part of the design, the exact change proposed is identified to all CCB members. They then analyze the impact in dollars and time delay of the change to their organizations. The CCB then discusses the total impact and compares it with the benefit afforded by the change. The executive appointing the CCB or, where the system is being created under contract to someone else, the customer or client then makes a decision based on the CCB estimate. Up to that decision point, no work is done on the proposed new configuration; as soon as it is adopted, all parts of the organization are immediately notified and everyone begins incorporating the change. Where change in a configuration item affects its *interface* with other systems (its space envelope or the physical, energy, or other interactions that pass through it), an interface control working group (ICWG) of some sort is needed to coordinate these changes.

Design Review

Earlier in this chapter the phased nature of engineering design was introduced. At the end of a phase it is desirable to come to an explicit decision as to whether (1) the work of that phase has progressed far enough to justify going on to the next phase at greatly increased resource cost; (2) the project should be scrapped at that point; or (3) further work needs to be done in that phase before proceeding. This is accomplished by a *design review board (DRB)* convened by the executive responsible for the overall system being created. The DRB often consists of engineering managers not directly involved in the design but familiar with the general type of system, and a client representative if there is one. All major design groups present their design concept. Other affected "ilities" (such as reliability, maintainability, producibility, and safety engineering) assess the design from their viewpoints. The DRB convening authority then makes a decision to proceed with the next phase of the system design (often with instructions to fix problems identified in the review), to delay decision pending remedial work, or to cancel the project. For more information on design reviews, see Bronikowski[21] and Fisher and Lake[22].

Management Audit of the Design Function

The concept of a management audit as a control system was introduced in Chapter 8. This approach can easily be tailored to evaluate the quality of the design-related functions of an organization. Figure 10-3 illustrates a worksheet suggested by Hughes Aircraft to facilitate such an audit.

PRODUCT LIABILITY AND SAFETY

Development of Product Liability

Through the centuries a relationship of buyer and seller stemming from the Roman philosophy of *caveat emptor* (let the buyer beware) persisted in Western thinking. Although English law permitted recovery by a plaintiff based on the negligence of a defendant, this was possible only where a direct contractual relationship ("privity of contract") existed. In the famous 1842 case *Winterbottom* v. *Wright,* for example, the injured driver of a defective mail coach could not sue the maker of the coach, because there was no such privity. This changed in U.S. law in New York in 1916, when a man named MacPherson was awarded damages from Buick Motor Company for harm done by a defect in his car, even though his contract for the car was with a dealer (who in turn had purchased the car from Buick).

Organizational Effectiveness Review — Technical Factors Worksheet		Technical Factors Worksheet	
Factor	Rating	Current Strengths, Weaknesses, Needs	Future Objectives and Plans
Marketing and contracting. *Does the organization:* – Work closely with the marketing and contracts organizations? – Remain alert and responsive to customer needs, market trends, and competition? – Aggressively seek new applications for current products? – Develop new products and prepare effective sales proposals? – Assist in contract negotiation, administration, and closure?	– High – Avg. – Low		
Conceptual design. *Does the organization:* – Keep abreast of the technology? – Originate creative concepts that fulfill customer needs? – Stress simplicity of design? – Conduct performance/cost/risk trade-off studies to seek optimal design approaches?	– High – Avg. – Low		
Development engineering. *Does the organization:* – Develop creative designs that make use of advanced techniques? – Emphasize practicality of design? – Perform trade-off studies involving product performance, reliability, maintainability, etc.? – Effectively use failure mode analyses, tolerance studies, etc., to optimize designs? – Give special attention to quality, safety, producibility, and cost reduction?	– High – Avg. – Low		
Design–assist resources. *Does the organization:* – Effectively use design–assist capabilities, such as computer–aided analysis and design, automatic drafting, etc.? – Apply engineering standards effectively in the design process? – Make optimal use of design reviews? – Solicit and use feedback from test, production, and field support activities? – Take full advantage of experience gained on earlier design efforts?	– High – Avg. – Low		
Test and evaluation. *Does the organization:* – Plan and integrate effective laboratory, demonstration, and field tests? – Develop test specifications, test methods, and simulation techniques to accurately evaluate the product? – Ensure timely availability of needed test facilities and equipment? – Complete test programs on schedule? – Analyze test results and initiate action to correct deficiencies noted?	– High – Avg. – Low		
Production and field support. *Does the organization:* – Establish, early in a design effort, effective interfaces with production and field support activities? – Ensure production and field support participation in engineering design reviews? – Provide complete, on–schedule release of engineering data to manufacturing? – Implement an effective system of configuration management? – Support production and field support activities with engineering assistance as required?	– High – Avg. – Low		

Figure 10-3. Management audit worksheet for design activities.
(From *R&D Productivity: Study Report*, 2nd ed., Hughes Aircraft Company, El Segundo, CA, 1978 pp. 28–29.)

Thus began the era of *product liability,* which has had far-reaching effects on how companies make and describe their products, and which requires great care on the part of their engineers and managers. Initially, the plaintiff (the injured party) had to prove *negligence;* that is, to show that the manufacturer omitted doing something that a "reasonable man," guided by the "ordinary considerations" that regulate human affairs, would have done, or did something that a "reasonable and prudent man" would not have done. The manufacturer, on the other hand, could defend by showing that the plaintiff did not use the product as a "reasonable person" would (and was therefore guilty of *contributory negligence*). However, in the 1960 California case of *Hennington* v. *Bloomfield Motors,* another auto manufacturer was found liable when the steering mechanism on a new car failed at only 20 mph, causing the car to swerve and hit a wall. Although there was no privity and negligence could not be proven, the court concluded that there had been a breach of an *implied warranty* of merchantability and fitness for use.

More constraining was the case of *Greenman* v. *Yuba Power Products,* in which Mr. Greenman was injured when a piece of wood he was turning on his combination lathe/saw/drill press flew out and struck his head. The California Supreme Court ruled that "a manufacturer is *strictly liable* in tort when an article he places on the market, knowing it is to be used without inspection for defects, proves to have a defect that causes injury to a human being." Still worse is the threat of *absolute liability,* where "a manufacturer could be held strictly liable for failure to warn of a product hazard, even if the hazard was scientifically unknowable at the time of the manufacture and sale of the product."[23]

Liability problems must be attacked prudently and rapidly, for they can destroy even large companies. Manville Corporation was forced into bankruptcy in 1982 because of the claims of tens of thousands of workers exposed to asbestos. A. H. Robins Co. faced 200,000 claims of injuries (about 20 resulting in death), because of a plastic intrauterine device (IUD) the size of a nickel; they were forced into bankruptcy in 1985 and have established, with their insurers, a $2.38 billion trust fund to meet claims. A 1994 newspaper "Important Legal Notice [to] . . . Women With Breast Implants" invites them to request "the detailed Breast Implant Settlement Notice describing the proposed settlement under which certain manufacturers of breast implants and their suppliers have agreed to pay up to $4,225,070,000 . . . to resolve current and future claims against them."[24] Also in 1994, a plaintiff who claimed his Remington Model 700 bolt-action rifle went off without his touching the trigger, causing loss of a foot, was awarded $17 million (mostly punitive) damages when his lawyer unearthed Remington records of development work beginning in 1981 on a "new bolt action rifle"; the court concluded these records provided evidence of "Remington's knowledge of defects and of its ability to implement safer alternative designs."[25] Johnson and Johnson, on the other hand, withdrew their profitable Tylenol product from retail shelves everywhere immediately on learning that poison had been inserted in some bottles of product; they retained consumer confidence and regained most of their market share when they resumed deliveries in new tamper-resistant containers.

Liability awards in American courts reach unreasonable levels, fueled by plaintiffs' attorneys who receive a third or more of the award on a contingency basis, juries who fail to apply any test of reason to awards, and legislators who fear to bite the bullet and put some limit to the millions that can be awarded for pain and suffering or as punitive damages. The

implications of excessive awards pervade our society. The author of this book is also trea-
surer for a fraternity building corporation that in 1989 paid $4000 for liability insurance an-
nually for about a dozen fraternity members. Consulting engineers paid an average 4.2% of
their gross billings in 1989 for professional liability insurance or "went bare" (without insurance); rates for small firms, for structural engineers, and for some specific states were
much higher.[26] Entire counties find no doctor left willing to deliver babies. Companies at risk
find it difficult to attract capable outside directors, and firm after firm abandons such essential products as vaccines or contraceptive devices.

As of this writing (May 1995), the incoming Republican congressional leadership proposes to fulfill its "Contract with America" by debating and voting on (among other items)
a "Common Sense Legal Reform Act" to include "loser pays" laws, reasonable limits on
punitive damages, and reform of product liability laws to stem the endless tide of litigation.
The reader will know how much of this has taken place.

Reducing Liability

To protect against product liability, designers must foresee even unlikely conditions. A
manufacturer producing wooden doors with a window at the top packaged the doors in a
stack with windows aligned and a cardboard cover over the stack to protect the windows.
A stevedore walked across a stack of these doors in a ship hold, fell through the glass windows, and sued for injuries. The manufacturer was held liable on the theory that he should
have known that this is the way stevedores behave.

There are many more examples of manufacturers' liability. A 3-year-old child was
awarded damages for burns suffered after knocking over a vaporizer that had a loose-fitting
cap. A high school football player was awarded more than $5,000,000 when the rear edge
of a football helmet pressed against the back of his neck when he was tackled, causing permanent paralysis.

How can the designer reduce the threat of product liability? According to Bass, he or
she must "design for the jury." Bass believes that a design presents a reasonable risk and is
not "defective" if:

- The product meets the user's reasonable expectations of safety.
- The risks are reduced to the greatest extent possible.
- The product contains adequate warnings.
- It is not economically or technologically feasible to reduce the likelihood or the
  severity of injury.
- The benefits obtained by the use of the product outweigh the risks in evaluating the
  product as a *whole.*
- The benefits cannot be obtained in less risky ways.[27]

Some of the things that designers and managers can do to anticipate liability problems are:

- Include safety as a primary specification for product design.
- Use standard, proven materials and components.
- Subject the design to thorough analysis and testing.

- Employ a formal design review process in which safety is emphasized.
- Specify proven manufacturing methods.
- Assure an effective, independent quality control and inspection process.
- Be sure that there are warning labels on the product where necessary.
- Supply clear and unambiguous instructions for installation and use.
- Establish a traceable system of distribution, with warranty cards, against the possibility of product recall.
- Institute an effective failure reporting and analysis system, with timely redesign and retrofit as appropriate.
- Document all product safety precautions, actions, and decisions through the product life cycle.

An increasing number of engineers will be involved in product liability work, supported by organizations such as the System Safety Society, and seeking certification as Certified Safety Professionals. Some will work for manufacturers, some for governments or consumer groups, and some will serve as "expert witnesses" before courts of law. Ethical codes of the engineering professional require placing the public interest and safety paramount, and careful attention to product safety is one way the engineer can meet his or her professional obligations.

DESIGNING FOR RELIABILITY

Significance of Reliability

The color television set is present in almost every American home—often several of them. Color television originated in the West (United States and Europe), and Western manufacturers maintained market leadership by emphasizing picture quality and, through frequent model changes, various innovations and gadgetry. Quality control guru J. M. Juran contrasted the approach of the Japanese where, beginning with the early 1970s:

> Consumer emphasis has been on reliability and function. . . . In response to this emphasis, Japanese manufacturers attained a clear leadership in reliability—a leadership which they hold to this day. . . . During the middle 1970s the Western color TV sets were failing in service at a rate of about five times that prevailing in Japanese sets. . . .
>
> There are hard data available to suggest that as to internal failure (discovered in the factory) the Japanese have an overwhelming superiority over the West. Prior to coming under Japanese management, the (U.S.) Motorola factory ran at a "fall-off" rate of 150 to 180 per 100 sets packed. This means that 150 to 180 defects were found for every 100 sets packed, or 1.5 to 1.8 per set. Three years later [after being taken over by a Japanese company] the fall-off rate at Quasar (the new name of the factory) had gone down to a level of about 3 or 4 per 100 sets, or only about one-fortieth of the previous level. (In Japan the fall-off rates are about 0.5 per 100 sets, which is more than two orders of magnitude lower than the [original] Motorola performance.)[28]

Western automobiles have experienced a similar problem. *Consumer Reports* annually publishes frequency of repair statistics for automobiles, taken from surveys of the magazine's many readers. Over the past decade readers have reported least repairs needed with Toyota, Nissan, Honda, and other Japanese cars—with almost no American names among them. Consumers bought millions of imported cars because they have the reputation of reliability (Japanese cars are no longer cheaper), and each million cars we import represents perhaps $15 billion added to our trade deficit—a further blow to the U.S. economy and our standard of living. Foreign cars assembled in the United States are only a partial solution, since so many of the components used in them are imported (and the profits exported).

Fortunately, there is some progress to report. In 1981 Ford, GM, and Chrysler averaged 6.6 to 8.0 defects per new car versus 2.0 for Japanese; American manufacturers reduced defects below 2.0 per car by 1987 to an average 1.5 by 1991, versus 1.2 for the average Japanese car.[29] Nevertheless, based on member experiences with 1992–94 models, *Consumer Reports* still finds that Japanese nameplates have the lowest frequency of repair, followed by some of the more expensive European models. Among American products, Saturn is the most reliable nameplate; Ford has the best overall record of American companies, and Buick and Oldsmobile models (except the Roadmaster) have shown average to above average reliability.[30]

Reliability and Risk Defined

To this point we have used the term *reliability* without defining it. The rigorous definition of reliability has four parts:

1. Reliability is the *probability* that a system
2. will demonstrate *specified performance*
3. for a *stated period of time*
4. when operated under *specified conditions.*

If the required function or the duration or the environment in which a system operates changes, so does the probability of success (reliability). As an example, the *Challenger* space shuttle solid rocket motor was designed and qualified to operate in the range of 50 to 90°F, and it could not be expected to have the same reliability after the cold night of January 27–28, 1986, for which temperatures at the launch site of 18°F were predicted. The political decision to launch anyway cost seven lives and a delay of over 30 months in our space program. (The ethical considerations of this decision are discussed in Chapter 18.)

TABLE 10–2　Some Reliability Measures

Measure	Symbol	Definition
Reliability (four-part definition)	$R(t)$	$\dfrac{\text{Number surviving at time } t}{\text{Number existing at } t = 0}$
Failure CDF (cumulative distribution function)	$F(t)$	$\dfrac{\text{Cumulative failures by time } t}{\text{Number existing at } t = 0}$
Failure PDF (probability density function)	$f(t)$	$\dfrac{\text{Number failing/unit time at time } t}{\text{Number existing at } t = 0}$
Failure or hazard rate	$\lambda(t)$	$\dfrac{\text{Number failing/unit time at time } t}{\text{Number existing at time } t}$

Several different measures of reliability and its complement, failure, are in common usage. Four of the more common ones, which are used in the rest of this section, are defined in Table 10-2.

Risk may be defined as the chance (i.e., the probability) of injury, damage, or loss. We need a basic feeling for probability and statistics to be able to make good decisions in our daily lives as citizens of a democracy in a technological age. For example, many people have a fear of flying, but it has been calculated[31] that the same increased chance of death (1 in a million) produced by traveling 1000 miles by jet is produced by traveling 300 miles by car, 10 miles by bicycle, or 6 minutes by canoe!

Others feel that nuclear power plants present too great a radiation hazard to their surroundings, but this same source reports that the same 0.000001 increased chance of death from cancer caused by radiation from living five years in the open air at the boundary of a nuclear reactor plant site can be caused by (1) one chest x-ray (even in a good hospital), (2) living two months in an average stone or brick building, (3) vacationing two months in the mile-high city of Denver (if you normally live at sea level), or (4) flying 6000 miles in a jet. [The author also believes any voter today that has not had at least high school physics is functionally illiterate, unable to judge such techo-political problems objectively. Most people (including the author's late wife) would not go that far!]

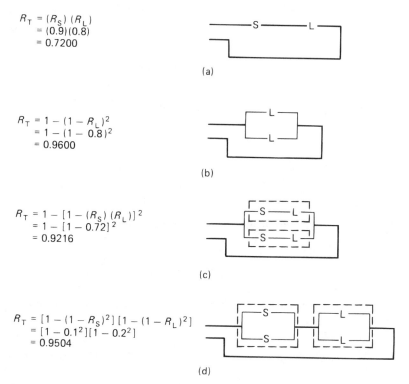

$$R_T = (R_S)(R_L)$$
$$= (0.9)(0.8)$$
$$= 0.7200$$

(a)

$$R_T = 1 - (1 - R_L)^2$$
$$= 1 - (1 - 0.8)^2$$
$$= 0.9600$$

(b)

$$R_T = 1 - [1 - (R_S)(R_L)]^2$$
$$= 1 - [1 - 0.72]^2$$
$$= 0.9216$$

(c)

$$R_T = [1 - (1 - R_S)^2][1 - (1 - R_L)^2]$$
$$= [1 - 0.1^2][1 - 0.2^2]$$
$$= 0.9504$$

(d)

Figure 10-4. Simple reliability models: (a) series; (b) simple parallel; (c) series in parallel; (d) parallel in series. S, switch; L, lamp; R_T, total system reliability; dashed "boxes" indicate subsystems analyzed within [ ] in calculation.

Simple Reliability Models

When we are designing systems we can often obtain a good estimate of the reliability of the individual components we plan to use under approximately the conditions of our application. We need to combine these known reliabilities to estimate the overall reliability of our system, and we use reliability models for this estimation.

For example, consider a system whose purpose is to turn on an electric light on demand over a period of a year under household conditions. Our components are two: lamps (incandescent light bulbs) with a reliability over that period of $R_L = 0.8$, and switches with a reliability $R_S = 0.9$. (We are assuming that failures of the power source and of the wire and connections themselves may be ignored.) There are several ways in which we might connect these components, as shown in Figure 10-4.

Simple series model. If we place one switch and one lamp in series, such that *both must work* for the system to work, the total reliability R_T of the system is the product of the reliabilities of the components. In our example (Figure 10-4a),

$$R_T = R_L \times R_S = 0.8 \times 0.9 = 0.72$$

Even if components are reasonably reliable individually, when a large number are placed in series in a complex system the system reliability can be unacceptable. For example, a system consisting of 14 components in series, each 95% reliable, will have a systems reliability of 0.95^{14}, or 0.488. Modern complex systems will have hundreds or thousands of components.

Simple parallel model. If we place two components in parallel so that *both must fail* for the system to fail (providing *redundancy*), the probability of failure F_T and the reliability R_T of a system consisting of two switches in parallel, of which only one must work, are

$$F_T = F_S \times F_S$$
$$R_T = 1 - F_T = 1 - F_S^2 = 1 - (1 - R_S)^2$$
$$R_T = 1 - (1 - 0.9)^2 = 1 - 0.01 = 0.99$$

Similarly, the reliability of a pair of lamps in parallel (Figure 10-4b) is 0.96.

Series–parallel models. Systems of any complexity consist of a combination of series and parallel arrangements of components. Consider the use of two switches and two lamps, with the requirement that one switch and one lamp in series must work for the system to work. Figure 10-4c calculates a reliability of 92.16% for two series systems placed in parallel; Figure 10-4d a reliability of 95.04% for two parallel systems placed in series.

All of the preceding assumes that failures of components are *statistically independent* of each other—that is, that the failure of one component has no effect on the probability of failure of another component. Further, only one type of failure was considered for each component. If failure of the switch to open (and therefore turn off the light when desired) was considered as well as the failure to close, our reliability calculations would have been more complex.

"Bathtub curve" model. Figure 10-5 shows the pattern of hazard rate (instantaneous failure rate assuming no previous failure) versus time, looking somewhat like the cross section of a bathtub, that is true of many components and systems. During the early "infant mortality" period, numerous failures due to substandard or defective parts or assembly take place—a phenomenon all too familiar to buyers of new cars. Following this early period on many systems is a "constant failure rate" period, where only a low level of random failures occur. Finally comes the "wearout" period, when important parts of the system come to the end of their useful life.

The (approximately) constant failure rate period is the preferred "useful life" of the system. During this period the system can be modeled as having a constant hazard (instantaneous failure) rate lambda (λ) in failures per unit time. The inverse of this rate ($1/\lambda$) is the *mean time between failures (MTBF),* a common figure of merit for reliability.

Developing Reliability over the Product Life Cycle

Planning and apportionment. Reliability is a continuing concern throughout design, manufacture, and use of a complex system. The first step in planning for reliability is establishing a *reliability goal* (the desired probability of successful operation) and its complement, the acceptable failure rate, for the system. This system failure rate is then divided into acceptable failure rates for each subsystem and component (*reliability apportionment*). The component failure rates, in turn, become the design targets for component designers.

Designing for reliability. A number of techniques are used in designing components and subsystems to meet reliability objectives. One is to "start with the best"—to specify and use parts of known high quality. In electronics this may mean specifying HI-REL (high reliability) parts that have been produced and tested in accordance with military specifications—they are much more expensive, but worth it where the cost of failure is high. U.S. industries and government agencies pool their information on (principally electronic) parts reliability through the Government-Industry Data Exchange Program (GIDEP). Designers can use this source to obtain industry experience on the reliability of specific components under specific conditions of temperature and other parameters.

A second design approach is providing redundancy, using components in parallel as previously modeled. Thus a jet airliner will have two, three, or four independent hydraulic lines or electrical wires to control a critical function—routed through different paths, so a

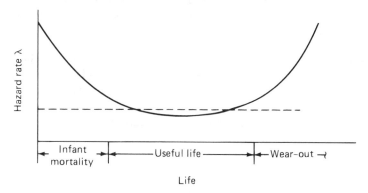

Figure 10-5. The "bathtub curve."

single incident will not affect them all. Even very early automobiles featured "emergency brakes" for redundancy. Redundancy can often be enhanced by having nonoperating "standby" spares that are not turned on unless the primary unit fails, and therefore do not wear out as fast; the weakness in such systems is often unreliability in the added sensor and switching units needed to activate the standby.

Reliability is enhanced by assuring a comfortable *factor of safety,* which is the ratio of the minimum strength provided by the design to the maximum stress anticipated in use. An electrical analog is *derating,* in which electrical components of higher than necessary strength (or "rating") are specified to assure high reliability and durability in the actual service expected. Another approach is *fail-safe* design, in which if failure does occur it leaves the system in a safe (although perhaps inoperable) condition.

Oliver Wendell Holmes long ago recognized the need to "design for reliability" in his delightful poem "The Deacon's Masterpiece; Or, the Wonderful 'One-Hoss Shay'":

> *Now in building of chaises, I tell you what,*
> *There is always somewhere a weakest spot —*
> *In hub, tire, felloe, in spring or thill,*
> *In panel, or crossbar, or floor, or sill,*
> *In screw, bolt, thoroughbrace—lurking still,*
> *Find it somewhere you must and will —*
> *And that's the reason, beyond a doubt*
> *That a chaise breaks down, but doesn't wear out.*
> *But the Deacon swore (as Deacons do,*
> *With an "I dew vum," or an "I tell yeou,")*
> *He would build one shay to beat the taown*
> *'N' the Keounty 'n' all the kentry raoun';*
> *It should be built so that it couldn't break daown;*
> *"Fur," said the Deacon, "it's mighty plain*
> *That the weakes' place mus' stan' the strain*
> *'N' the way t' fix it, uz I maintain, is only jest*
> *T' make that place uz strong uz the rest."*

So the Deacon sought out the finest materials, including "Boot, top, dasher from tough old hide Found in the pit when the tanner died" and finished his chaise. One hundred years later to the day and hour "it went to pieces all at once—All at once and nothing first—Just as bubbles do when they burst."

Flattening the bathtub curve. We have seen in Figure 10-5 that component and system failures are often higher during the early "infant mortality" period due to defective parts or assembly. These failures can be reduced by careful quality control, and by the "burn-in" of electronic systems or the "run-in" of an expensive car motor to discover and replace defective parts. Real systems do not age like Holmes's "one-hoss shay"; different parts will wear out at different rates. Useful life may be extended by replacing those parts that wear out quickly (such as the brake linings on a car), but sooner or later there comes a point where it is cheaper to replace a system than maintain it.

Reliability growth. Reliability is evaluated and improved throughout the system life cycle. Early breadboard tests of critical systems are used to give a first indication. Subsystem tests may be run at several points as parts and components are defined and prototypes become available. Subsystems, and then the system itself, will be "qualified" by conducting rigorous qualification tests under a range of expected environmental conditions (temperature, shock, vibration, and others). At each step, failure modes exhibit themselves, and reliability is enhanced by redesign to eliminate them. In manufacture, each unit or batch is tested to provide further reliability information. In field use, still another set of problems arise, and systems may have to be retrofitted in the field to correct problems not found before. Throughout this process, system reliability improves. Still, the best and most economical place to minimize failures is in the design phase, using such techniques of reliability engineering as failure modes and effects analysis (FMEA).

The Reliability Profession

Reliability engineering has become an established profession, involving a number of professional societies. The Reliability Division of the American Society for Quality Control (ASQC) publishes the quarterly *Reliability Review;* ASQC also has a Certified Reliability Engineer (CRE) designation awarded after a rigorous, day-long examination. The Institute of Electrical and Electronic Engineers (IEEE) has a Reliability Division, and it publishes the *IEEE Transactions on Reliability.* The Society of Reliability Engineers (SRE) also has a journal. These societies and a half dozen others jointly sponsor an annual Reliability and Maintainability Symposium that is extremely well attended.

OTHER "ILITIES" IN DESIGN

"Ilities" is a casual term in aerospace industries for a loosely defined set of specialist disciplines, many ending in the suffix "–ility," that enhance effective design in some aspect. In addition to reliability (above) we will look briefly at maintainability, availability, human factors engineering, standardization, producibility, and value engineering.

Maintainability

Maintainability "is an inherent design characteristic of a system or product [and] it pertains to the ease, accuracy, safety, and economy in the performance of maintenance actions."[32] One can create a four-part definition for maintainability by adding and ~~striking out~~ words in the definition already given for reliability:

1. Maintainability ~~Reliability~~ is the probability that a failed system
2. will ~~demonstrate~~ be restored to specified performance
3. ~~for~~ within a stated period of time
4. when ~~operated~~ maintained under specified conditions.

Maintenance downtime has three components:

1. *Administrative and preparation time:* processing the repair request, waiting for an available worker, travel, and obtaining tools and test equipment

2. *Logistics time:* delay to obtain parts (or test equipment or transportation) after maintenance personnel are available

3. *Active maintenance time:* actually doing the job (including studying repair charts before repair and verifying and documenting the repair afterward)

Maintenance may be divided into *corrective maintenance,* made necessary by failures, and *preventive maintenance,* designed to prevent failures. The average time between maintenance actions (regardless of type) is the *mean time between maintenance (MTBM),* and the average total time for the three components of maintenance is the *mean downtime (MDT).* Maintainability may alternatively be defined by just the active maintenance time for corrective maintenance *(mean time to repair, MTTR),* since only item 3 above is substantially influenced by the designer (although specification of standard parts, tools, and test equipment can shorten the others). The reliability measure of *mean time between failures (MTBF),* the inverse of the hazard rate λ, is often used with the MTTR. These definitions and others are more rigorously defined in Blanchard and Fabrycky.[33]

The designer can reduce active maintenance time by providing easy access to the system, dividing the system in modules that can be replaced as units, specifying preventive maintenance that will delay deterioration and identify worn parts, and providing clear, comprehensive maintenance manuals. Maintainability can be enhanced by creating realistic system models—physical mockups on which maintenance actions can be at least simulated by typical repair people or, using the output of a computer-aided design process, three-dimensional computer simulations that can be rotated and enlarged to provide visibility and understanding of potential maintenance difficulties.

Another aid to maintenance, especially of electronic equipment, is provision for built-in test (BIT). BIT may consist of simple provision of test points to facilitate a mechanic's diagnosis, or it may include an extensive system of sensors, a computer, and software that periodically checks the condition of avionic systems and provides an automatic printout of potential defects as soon as an aircraft lands. (BIT systems can themselves become so complex that a significant number of the problems they identify are false indications, due instead to defects in the BIT.)

Availability

Many users are more concerned that a system operate satisfactorily when called upon, a condition called "availability," than they are in pursuing some ultimate reliability by making the system so complex that it defies repair. Augustine has observed that as military aircraft become more costly and more complex, the maintenance crew hours per flight hour increases. He therefore proposes (tongue in cheek) the Augustine-Morrison Law of Unidirectional Flight: "Aircraft flight in the 21st century will always be in a westerly direction, preferably supersonic, to provide the additional hours needed each day to maintain all the broken parts."[34]

Two definitions of availability, based on the four measures just identified under maintainability, merit mention. The *inherent availability* A_i of a system considers only corrective maintenance in an ideal support environment (with neither administrative nor logistic delays):

$$A_i = \frac{\text{MTBF}}{\text{MTBF} + \text{MTTR}}$$

Operational availability A_o, on the other hand, considers both preventive and corrective maintenance, conducted in the actual support environment:

$$A_o = \frac{\text{MTBM}}{\text{MTBM} + \text{MDT}}$$

Human Factors Engineering[35]

Human factors engineering, also known as *ergonomics,* is concerned with ways of designing machines, operations, and work environments to match human capacities and limitations. Its origin can be traced to early scientific management studies, such as the tailoring of shovel size to material density by Frederick Taylor in 1898 and the efficient arrangement of bricks on the scaffold by Frank Gilbreth in 1911. Human factors engineering did not really emerge as a discipline, however, until the mid-twentieth century, as a result of World War II experience. For example, the location of three critical controls in three military aircraft in common use in 1945 were as shown in Table 10-3.

The hazard created when a pilot is making a difficult landing in an unfamiliar aircraft is obvious. A similar hazard exists when the typical business traveler arrives at an airport on a rainy night and jumps in an unfamiliar rental car to traverse an unfamiliar city to find a hotel. Automobile manufacturers, using ergonomics, have made great strides in locating critical controls in consistent locations that can be reached without taking hands off the steering wheel, in providing climate control and comfortable seats, and in locating and lighting displays for easy visibility.

Human factors engineers have developed a wide variety of standards for illumination, sound, accessibility, controls, displays, and other factors affecting work. As an example, Table 10-4 summarizes the general suitability of controls for different types of operation. By the time the Apollo spacecraft was being designed in the 1960s, many of these standards were well into development. The author, who then had design responsibility for the Launch Escape System in the Apollo command module, can recall extensive design reviews on the placement of critical control switches and displays so that the astronauts could perform essential functions despite the gravity, vibration, and other forces of launch.

Standardization

A standard is defined as "a set of specifications for parts, materials, or processes intended to achieve uniformity, efficiency, and a specified quality. One of the important purposes of

TABLE 10–3   Control Placement in Military Aircraft

Aircraft	Control Placement on the Throttle Quadrant		
	Left	Center	Right
B-25	Throttle	Propeller	Mixture
C-47 (DC3)	Propeller	Throttle	Mixture
C-82	Mixture	Throttle	Propeller

Source: Adapted from Alphonse Chapanis, *Man-Machine Engineering*, Wadsworth Publishing Company, Inc., Belmont, CA, 1966, p. 95.

TABLE 10-4 General Suitability* of Controls for Different Types of Operation

Control Type	Accuracy	Speed	Force	Displacement Range	Number of Discrete Settings	High Frequency of Operation	Long Duration of Operation	Sequence	Visual Identification	Non-visual Identification	Check Reading of Position	Emergency Action	Compatibility with System Response
Pushbutton	—	H	—	—	2	H	L	H	L	L	L (unless lit)	H	H
Toggle switch	—	H	—	—	2/3	M	L	M	M	H	H	H	L
Rocker switch	—	H	—	—	2/3	M	L	M	M	H	H	H	L
Rotary selector	H	H	—	—	3–24	M	L/M	M	M	M	M	M	H
Joystick selector	H	H	L	—	4–8	H	L	L	M	M	M	M	—
Cranks:													
Small	L	H	L	H	—							H	
Large	H	L	H	H		M	M	—	H	H	L	L	M
Horizontal	M	M	H	H									
Vertical	L	H	M	H									
Handwheels	L	H	L	L	—	M	M/H	—	H	M	L	L	M
Levers:													
Horizontal	L	H	L	L	—	M	M	M	M	M	H	M	L
Vertical (to/fro)	M	H	Short L	L								H	M
(to/fro)	M	H	Long H	L								M	L
(across body)	M	M	M	M									M
Knob	H	—	L	L	—	M/L	M/L	H	M	M	L	—	M
Joystick	M	M	L	L	—	H	H	—	M/H	H	L	L	M
Pedals:													
Leg	M/L	M	H	—	—	M	M	L	—	M	—	H	M
Ankle	H	M/H	L	—	—	H	H	M	—	M	—	M	M
Footswitch	—	H	L	—	—	L	L	—	—	M	—	H	H

*General suitability: H, high; M, medium; L, low; —, unsuitable or not applicable. Note that the high accuracy, high speed, large force, and large displacement are generally incompatible.

Source: T. S. Clark and E. N. Corlett, *The Ergonomics of Workspaces and Machines: A Design Manual*, Taylor & Francis Ltd., London, 1984. p. 53.

a standard is to place a limit on the number of items in the specifications so as to provide a reasonable inventory of tooling, sizes, shapes, and varieties."[36]

At one time there were no standards for bolts, nuts, and screw threads, and a ½-inch nut removed from one bolt would not fit another. The same applied to lamp bases—manufacturers once offered 175 different ones, and now there are only about a half dozen. Electronic standards we are familiar with include the 525-line TV format, 5.25- and 3.5-inch computer diskettes, and several sizes of compact discs. Any large design organization has a standards manual that identifies fasteners, tolerances, processes, and the like that are considered acceptable in that organization; specification of nonstandard alternatives requires strong justification. Standardization can be important for reliability, too. In the Apollo program, essentially all devices in which an electrical signal was translated into a pyrotechnic or explosive pulse, whether on a launch stage or command, service, or lunar excursion module, had to use the same initiator; thousands of these initiators were fired in development, and an outstanding reliability record resulted.

The National Institute of Standards and Technology (NIST, formerly the National Bureau of Standards) is only one of many government agencies involved in standardization. Trade associations, industries, professional societies, and government organizations work together in the American National Standards Institute (ANSI) to coordinate standardizing activities. Unfortunately, the United States is alone among developed countries in not standardizing on the metric system of measurement, although increasing numbers of individual companies and industries are doing so to remove this barrier to international trade in American goods.

Producibility

As a product is being designed, careful attention should be paid to assure that it can be produced economically, using available processes and equipment where possible. Manufacturing engineers familiar with production capabilities should be involved in reviewing parts as they are designed, suggesting tolerances, materials, and shapes that are more producible. Two-way understanding is developed—understanding by designers of manufacturing preferences, and by manufacturing engineers of the performance consequences if certain critical specifications are relaxed. Further, the transfer from design to manufacturing is greatly simplified, and the ultimate product is produced not only at lower cost, but with less transition time. These considerations are an essential part of the modern thrust of *concurrent engineering* discussed earlier in this chapter.

Value Engineering/Analysis

Value engineering or value analysis (VE/A) is a methodical study of all components of a product in order to discover and eliminate unnecessary costs over the product life cycle without interfering with the effectiveness of the product. Fasal[37] would use the term "value *engineering*" in developing new products and "value *analysis*" in reviewing old products, but most people use the terms interchangeably. One of the techniques of VE/A is asking a series of penetrating questions about a product, system, process, or component. Weiss[38] provides a typical set of questions to ask about each item:

1. What is it?
2. What does it do?
3. What does it cost?
4. What is it worth?
5. What else might do the job?
6. What do the alternatives cost?
7. Which alternative is least expensive?
8. Will the alternative meet the requirements?
9. What is needed to implement the alternative?

Value engineering activities are encouraged—often required—of contractors by the U.S. Department of Defense and by NASA. Practitioners share their experiences in the Society of American Value Engineers (with the appropriate acronym SAVE) and can earn the title Certified Value Specialist from them by examination.

DISCUSSION QUESTIONS

10-1. What are the reasons for dividing the systems engineering/new product development process into phases or stages?

10-2. How does the design work done in the technical feasibility stage of new product development differ from that done in the later design stages?

10-3. What are the principal reasons that a configuration control system is necessary?

10-4. For an engineering design or project management system you are familiar with, describe the drawing release and design review processes.

10-5. Summarize the history of gradually increasing liability of industry for damage caused by their products.

10-6. What are potential solutions to the liability crisis that haunts American enterprises and institutions today?

10-7. Select a product line in a specific industry and list actions that can be taken to reduce product liability.

10-8. Identify a company and product (aside from television and automobiles) and tell how good or poor reliability has significantly affected company success.

10-9. Given three components, each with a reliability of 0.9, calculate the reliability of a total system in which the three are arranged in (**a**) three in series, (**b**) three in parallel, and (**c,d**) in two different series/parallel designs each using a *total* of only three components.

10-10. A car contains both hydraulic brakes (reliability 0.95) and mechanical brakes (reliability 0.98). What is the probability of stopping (**a**) rapidly at high speed, assuming both systems must work, and (**b**) at low speed, assuming either system will stop the car?

10-11. An engineered system consists of one each of three components X, Y, and Z with reliabilities R_X, R_Y, and R_Z. (**a**) What is the system reliability assuming one component of each type must work? (**b**) If required system reliability is 0.85, show how you can meet this goal by replacing one of these components with two of that same component in parallel.

10-12. For a component to which the bathtub curve model of reliability applies, describe the provisions you would make to assure a low hazard rate in use of the component.

10-13. An engineered system has a hazard rate of 0.01 failure per hour. (**a**) What is its MTBF? If the same system has a MTBM of 60 hours, a MDT of 20 hours, and a MTTR of 6 hours, what is its (**b**) inherent availability and (**c**) operational availability?

10-14. Describe some mechanisms a designer can use to improve maintainability.

10-15. Give an example of a consumer product with which you are familiar that, through recent re-design, seems to be a greater value (a better ratio of utility to apparent cost).

NOTES

1. J. B. Reswick, foreword to Morris Asimow, *Introduction to Design* (Englewood Cliffs, NJ: Prentice-Hall, Inc., 1962), p. iii.

2. National Aeronautics and Space Agency, *NASA Systems Engineering Handbook,* September 1992 draft (still under review as of April 1994), p. 4.

3. National Society of Professional Engineers, *Engineering Stages of New Product Development,* NSPE Publication #3018 (Alexandria VA: NSPE, 1990).

4. U.S. Dept. of Defense, *MIL-STD-499B, Draft Military Standard, Systems Engineering,* May 6, 1992, pp. 10–14. This draft will never be "signed off" as a result of the decision by Secretary of Defense Perry to reduce the number of Military Standards. An industry task force comprised of representatives from the Electronic Industries Association (EIA), the National Council on Systems Engineering (NCOSE), and others is preparing a commercial version of the proposed 499B, but as of November 1994 it had not been released.

5. U.S. DOD Draft (see note above) *MIL-STD-499B,* p. 33.

6. *NASA Systems Engineering Handbook,* pp. 13–20.

7. NSPE, *Engineering Stages.*

8. I. G. Wilson and M. E. Wilson, *From Idea to Working Model* (New York: John Wiley & Sons, Inc., 1970).

9. NSPE, *Engineering Stages,* p. 16.

10. Willard I. Zangwill, "Concurrent Engineering: Concepts and Implementation," *Engineering Management Review,* Winter 1992, p. 41.

11. NSPE, *Engineering Stages,* p. 17.

12. NSPE, *Engineering Stages,* p. 18.

13. Peter R. Bridenbaugh, "Needed: Partnerships Between Materials Producers, Customers, Customers' Customers," *Research•Technology Management,* 34:3, May–June 1991, p. 7.

14. John Bussey and Douglas R. Sease, "Speeding Up: Manufacturers Strive to Slice Time Needed to Develop Products," *Wall Street Journal,* February 23, 1988, p. 1.

15. Jerome G. Lake, "A Program Design for Concurrent Engineering Education," *Proceedings of the International Conference on Engineering Management,* Eatontown, NJ, October 25–28, 1992, pp. 71–76.

16. "A Car Is Born," *Business Week,* September 13, 1933, p. 65.

17. Sharon J. Kemmerer, CALS Program Manager, [U.S. Dept. of Commerce] National Institute of Standards and Technology, personal communication, December 1994.

18. U.S. Dept. of Commerce, National Institute of Standards and Technology, *Putting the Information Infrastructure to Work: A Report on the Information Infrastructure Task Force Committee on Applications and Technology* (Washington, DC: U.S. Government Printing Office, May 1994), p. 16.

19. Definitions have been taken from *The Executive Guide to CALS,* United Kingdom CALS Industry Council, 1993, p. 20, for their brevity and clarity; very similar definitions are provided by the [U.S.] CALS/CE [Concurrent Engineering] Industry Steering Group and by the [U.S. Dept. of Commerce] National Institute of Standards and Technology.

20. For more on the virtual organization, see William H. Davidow and Michael S. Malone, *The Virtual Corporation: Structuring and Revitalizing the Corporation for the 21st Century* (New York: HarperBusiness, a division of HarperCollins Publishers, 1992).

21. Raymond J. Bronikowski, *Managing the Engineering Design Function* (New York: Van Nostrand Reinhold Company, Inc., 1986), pp. 204–213.

22. Jack Fisher and Jerome G. Lake, "MIL-STD-499B and the Practice of Technical Reviews," *Proceedings of the 3rd International Symposium of the National Council on Systems Engineering,* Washington, DC, August 26–29, 1993.

23. Roger Leroy Miller, "Drawing Limits on Liability," *Wall Street Journal,* April 4, 1984, editorial page.

24. "Important Legal Notice," *St. Louis Post-Dispatch,* April 26, 1994, p. A6.

25. "Remington Faces a Misfiring Squad," *Business Week,* May 23, 1994, p. 90.

26. "Liability Burden Eases Slightly for Engineers," *Engineering Times,* December 1989, p. 3.

27. Lewis Bass, "Designing for the Jury," *System Safety Society Newsletter,* 2:5, October 1986, p. 1.

28. J. M. Juran, "Japanese and Western Quality: A Contrast," *Quality Progress,* December 1978, pp. 10–18.

29. University of Michigan Transportation Research Institute data graphed in *Newsweek,* March 30, 1992.

30. "Picking a Reliable New Car: Which New Models Are Most Likely to Be Trouble-Free?" *Consumer Reports,* April 1995, p. 263.

31. Richard Wilson, "Analyzing the Daily Risks of Life," *Technology Review,* February 1979, pp. 40–46.

32. Benjamin S. Blanchard and Wolter J. Fabrycky, *Systems Engineering and Analysis,* 2nd ed. (Englewood Cliffs, NJ: Prentice-Hall, Inc., 1990), pp. 389–390.

33. Blanchard and Fabrycky, *Systems Engineering and Analysis,* pp. 393–405.

34. Norman R. Augustine, *Augustine's Laws, Revised and Enlarged* (Washington, DC: American Institute of Aeronautics and Astronautics, 1983), p. 74.

35. Portions of this section and the following one on Standardization were abstracted with permission from a term paper by University of Missouri—Rolla graduate student Jim Breitbarth.

36. Joseph Shigley and Larry Mitchell, *Mechanical Engineering Design* (New York: McGraw-Hill Book Company, 1983), p. 14.

37. John H. Fasal, *Practical Value Analysis Methods* (Hasbrouck Heights, NJ: Hayden Book Company, Inc., 1972), pp. 8–9.

38. Gorden E. Weiss, "Value Engineering/Analysis, Part III," *UMR Extension Division Continuing Education Series* (Rolla, MO: University of Missouri—Rolla, 1973), p. 1.

CHAPTER 11

Planning Production Activity

PREVIEW

We begin this chapter by emphasizing the importance of competitive production processes to the United States, and the central position of the engineer in the production organization. We next discuss considerations of plant location, design, and layout in planning manufacturing facilities. Three quantitative production planning tools are then considered: economic order quantity, break-even charts, and learning curves. We introduce production planning and control, and we describe four current methods (materials requirements planning, manufacturing resource planning, just-in-time, and synchronized manufacturing). Finally, we consider the evolution of flexible manufacturing systems from simple stand-alone machines to the "factory of the future."

INTRODUCTION

Vital Nature of Production

Alexander Hamilton is reported to have said: "Not only the wealth, but the independence and security of the country appear to be materially connected to the prosperity of manufacturers." For much of this century U.S. manufacturers have, indeed, been the envy of the world. In the last decade this has changed dramatically. Manufactured goods produced in the United States as a percentage of those consumed varied in the comfortable range of 100 to 105% from 1966 until 1982, when it crossed the 100% line; it then plummeted to only 85% by 1986.[1] Our deficit mounts dramatically because we import and consume more than we produce for export. In 1993 our merchandising trade deficit (manufacturing less our surplus food trade) was $115.7 billion, over half ($59.3 billion) with Japan, and an equal amount with the next five countries (mainland China $22.8, Canada $10.7, Germany $9.6, Taiwan $8.9, and Italy $6.8 billion).[2]

Some recent authors might view a manufacturing trade deficit as natural and acceptable for the United States. After all, are we not a "postindustrial society," and should we not expect to emphasize service industries while lesser nations get their hands dirty in factories? Cohen and Zysman refute this:

> Mastery and control of manufacturing is . . . critical to the nation. This . . . has been obscured by the popular myth that sees economic development as a process of sectoral succession. Economies develop as they shift out of sunset industries into sunrise sectors. Agriculture is followed by industry, which in turn is sloughed off to less developed places as the economy moves on to services and high technology. Simply put, this is incorrect. It is incorrect as history and it is incorrect as policy prescription. America did not shift out of agriculture or move it offshore. We automated it; we shifted labor out and substituted massive amounts of capital, technology, and education to increase output. Critically, many of the high value added service jobs we are told will substitute for industrial activity are not substitutes, they are complements. Lose industry and you will lose, not develop, those service activities. These service activities are tightly linked to production just as the crop duster (in employment statistics a service worker) is tightly linked to agriculture. If the farm moves offshore the crop duster does too, as does the large-animal vet. Similar sets of tight linkages—but at a vastly greater scale—tie "service" jobs to mastery and control of production. Many high value added service activities are functional extensions of an ever more elaborate division of labor in production. The shift we are experiencing is not from an industrial economy to a postindustrial economy, but rather to a new kind of industrial economy.[3]

In this chapter we look not only at existing production methods, but also at the nature of the "factory of the future" and the importance of the engineer's contribution to it.

The Engineer in Production Activity

Types of positions. Production organizations vary tremendously with the industry involved, with the size of the organization, with the type of production (mass production of standard items or small-quantity production of specialty items), and with many other factors. However, it is helpful to create a model of the "typical" manufacturing plant from which we can generalize the functions needed and the way in which engineers and engineering managers might fit into such an organization. Our hypothetical plant is assumed to be one of several at different locations, producing products that are researched, designed, financed, and marketed at a corporate headquarters separate from the plant. The plant organization, all reporting to a single *plant manager,* might look something like Figure 11-1.

An obvious job for an engineer in this organization is the position of plant (or facilities) engineer at the right of the diagram. The engineering design function under the plant engineer will normally be responsible for designing small changes to the plant and its production equipment. For more extensive changes, plant engineers would just specify what is

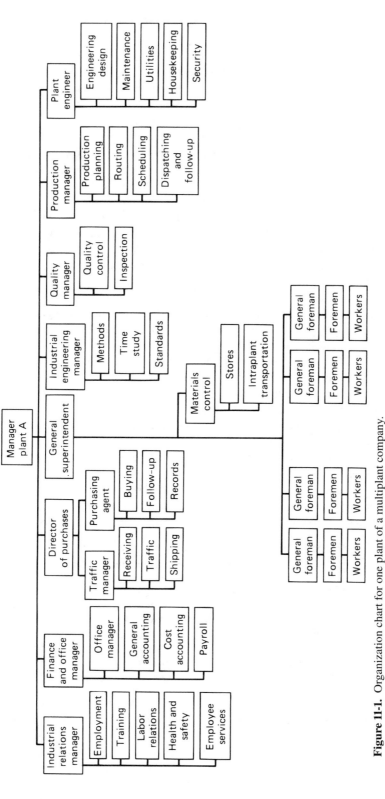

Figure 11-1. Organization chart for one plant of a multiplant company.
(Adapted from James L. Riggs, Lawrence L. Bethel, Franklin S. Atwater, George E. Smith, and Harvey A. Stackman Jr., *Industrial Organization and Management*, 6th ed., McGraw-Hill Book Company, New York, ©1979, p. 53. Reprinted by permission of the publisher.)

needed and monitor construction; the detailed design and construction would then be carried out by an architect/engineer (A/E) firm specializing in that type of manufacturing plant or by the central engineering group at corporate headquarters, since it is not efficient to maintain that level of design capability at each plant.

Maintenance engineering (the design and specification of the criteria for maintenance tasks) commonly might appear under the plant engineer; the routine conduct of maintenance is often under the general superintendent (or plant production manager). The plant engineer is commonly responsible for the utilities (heat, power, steam, water, telephone) throughout the plant and for certain other functions, such as housekeeping and security, that relate to the facility rather than the product.

Another important class of engineering functions in the plant comes under the heading of industrial engineering (IE). Traditional IE functions of plant layout, time-and-motion analysis, and standards setting are performed here. In many metal cutting and chemical processing plants the *process engineer,* often in a separate organizational unit, makes a major contribution. The manager of quality assurance or quality control will be an engineer or scientist in most plants of significant size or product complexity. Two functions are shown under the quality manager: quality control (or quality engineering), responsible for the analysis of quality problems and their prevention (commonly headed by an engineer), and inspection (which often is not).

First-line production management positions such as foreman or assistant foreman provide an excellent opportunity to learn production problems and to test one's wings as a leader. Most such positions will be filled by nonengineers, but they provide a good starting place for the engineer interested in manufacturing management. Positions as general foreman and superintendent or production manager are natural steps up the promotional ladder. Several other positions may call for engineers in larger plants: the safety engineer, whether under the industrial relations (or human resources) department or elsewhere; materials control (where automated storage, retrieval, and transport are extensive); and even purchasing, where the technology of the item being purchased is paramount. More often, such posts and others (technical employment, for example) will be staffed by technical people at the corporate level, with nonengineers implementing their recommendations at the plant level.

Intensity of use of engineers. In an article on postindustrial manufacturing, Jaikumar observes from a study of companies with extensive automation in production that

> the technological literacy of a company's workers is critical. In the Japanese companies I studied, more than 40% of the work force was made up of college-educated engineers, and all had been trained in the use of CNC [computer numerically controlled] machines. In the U.S. companies studied, only 8% of the workers were engineers, and less than 25% had been trained on CNC machines. Training to upgrade skills was 3 times longer in Japan than in the United States. Compared with U.S. plants, Japanese factories had an average of 2½ times as many CNC machines, 4 times as many engineers, and 4 times as many people trained to use the machines.[4]

In a recent (1992–93) academic year, U.S. colleges and universities surveyed[5] granted 65,001 B.S. degrees in engineering. Six years earlier, a larger class of graduates (87,350 in 1986–87) were absorbed, especially into defense industry and related government positions. With the collapse of the Soviet Union and the "downsizing" of American industry, many engineering graduates had to look harder for their first position. In my opinion, this shows that U.S. industry has not yet been ready to employ engineers as intensively as will be necessary if we are to recapture our former leadership in manufacturing.

Future demands on manufacturing engineers. The Society of Manufacturing Engineers commissioned the study, *The Manufacturing Engineer in the 21st Century,* which summarized more than 10,000 opinions from manufacturing practitioners. This report (by A. T. Kearney, Inc.) concluded that:

> The manufacturing engineer of the year 2000 will be faced with new challenges in the form of:
>
> - An environment of exploding scope [increasing product sophistication and variation, a global manufacturing environment, and extensive social and economic changes];
> - Multiple roles [with the manufacturing engineer acting as an operations integrator and manufacturing strategist as well as being a technical specialist];
> - Advanced tools [including more powerful computer hardware, more and larger data bases, a greater choice of software and expert systems, and advanced CAD/CAM (computer aided design/manufacturing) systems]; and
> - Changed work emphasis [focusing on teams, not individuals, with a more human, less technical orientation and the use of more outside services].[6]

John F. Welch Jr., CEO of General Electric Company, has emphasized the need to give higher priority to manufacturing engineering:

> To whom then, is it left to see that American innovation is dynamic enough, and American productivity growth sufficiently rapid, to win in world markets? In large measure it is the engineer, and in that context America needs to see the profession as the bodyguard of its standard of living. If it does; if the country perceives the nexus between a powerful engineering base and our way of life, educational and motivational programs that will preserve and nurture that base will be more forthcoming.
>
> But if engineering should be viewed as a key driver of American competitiveness, is that how the profession views itself? Not quite. The brightest lights are ushered toward the laboratory, the university. The compensation systems, the resources, the glory are largely skewed toward theory, toward elegance, toward self-indulgent innovation; first-of-a-kind solutions. Often the applause for the products of this system dies down when they arrive at manufacturing, and ceases altogether when they reach the marketplace. Making designs work—ensuring that they win in the marketplace—is "someone else's problem," relegated more often than not to the "less gifted" engineers, the "Mister Goodwrenches" on the manufacturing floor. Those whose businesses depend on winning in the global marketplaces are slowly coming to the conclusion that our priorities are exactly wrong.[7]

PLANNING MANUFACTURING FACILITIES

Plant Location

When General Motors Corporation decided to establish its Saturn Division, a billion-dollar investment, it had to make a very important decision on the region of the country in which this plant would be located. Many factors, such as transportation, labor supply and attitude, resource availability, and political climate, had to be considered before the division was finally located in Tennessee. Before the plant was built, corporate executives had to decide the community within that state and the specific site within the community that would provide the best location for the plant. Amrine et al. outline "seven basic steps in locating and building every new plant" followed by one large company:

1. Establish the need for a new plant.
2. Determine the best geographical area for the plant on the basis of the company's business needs.
3. Establish the requirements (i.e., product to be made, equipment and buildings needed, utilities and transportation necessary, number of employees, etc.).
4. Screen many communities within the general area decided upon.
5. Pinpoint a few communities for detailed studies.
6. Select the best location.
7. Build the plant.[8]

Some of the factors affecting the choice of region, community, and site are as follows:

- Transportation (highway, rail, air, water)
- Labor (supply, skill level, local wage rates, union membership and attitudes)
- Geographical location (relative to raw materials, customers, or other company activities)
- Utilities (supply and cost of water, electric power, and fossil fuels)
- Business climate (taxes, pollution controls, community attitudes)
- Amenities (climate, educational facilities, nearby recreation)
- Plant sites (land availability and cost, zoning, space for expansion)

The most important factors for plant location will vary with the industry and its critical factors. For example:

- Kilns used to create charcoal for briquettes from hardwood will be close to the raw material supply to reduce transportation cost, since four-fifths of the mass disappears in charring.
- Aluminum production has traditionally sought a source of cheap electricity, since it is energy intensive.
- High-technology electronic firms have tended to cluster together where technical professionals and educational institutions are available.
- U.S. clothing manufacturers have moved from high-labor-cost areas to lower-labor-cost areas in the United States, and then, increasingly, overseas.

Plant Design

Once the site is selected, engineers must decide on the nature of the plant and its arrangement on the site. *Multistory* plants conserve land area, permit use of gravity flow in moving product along the production line, and are cheaper to heat. However, *single-story* construction is more flexible, permits lighter foundations and columns, and allows higher floor loadings. Most new American plants are now built near major highways on the edge of a city, where available and economical land provides room not only for single-story construction, but also for the ubiquitous parking lot—often larger in area than the plant itself. Materials for plant construction may be steel-reinforced concrete (most expensive but lowest in maintenance cost and most fire resistant), exposed steel beams and trusses, or wood (for low buildings and light loads where fire is not a hazard).

The arrangement of the building on the site will depend on such things as the contours of the site, railroad and truck access, parking-lot provisions, and appearance. Some large companies have their own corporate engineering staffs for plant design, but most companies will call on an architect/engineering (A/E) firm for this specialized service.

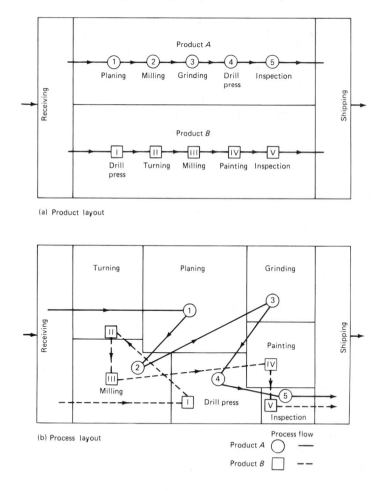

Figure 11-2. Schematic representations of (a) product and (b) process layouts. (From Arthur C. Laufer, *Production and Operations Management,* 3rd ed., South-Western Publishing Company, Cincinnati, OH, 1984, pp. 232–233.)

Plant Layout

Plant layout attempts to achieve the most effective arrangement of the physical facilities and personnel for making a product. The three principal methods of moving the product through the manufacturing steps are product layout, process layout, and group technology. (In a fourth method, *fixed-position layout,* the product remains stationary and the processes are brought to it. This method is largely confined to shipbuilding and other massive construction.)

In *product layout,* machines and personnel are arranged in the sequence of product manufacture so that the product can be moved along the production (assembly) line with a minimum of travel between steps, as shown in Figure 11-2a. This method is especially useful when a large quantity of standardized products are to be produced over a long period of time, and it is the basis for mass production of most automobiles, major household appliances, and the like. Ideally, the assembly line and the plant structure are designed in parallel, since adapting an existing building to a new mass production need can involve undesirable compromises.

In *process layout,* all machines or activity of a particular type are located together. Thus, a plant may have separate departments for turning, planing, grinding, milling, drilling, and painting, as shown in Figure 11-2b. Individual products are transported from department to department in the sequence needed for their production. This layout is particularly useful for the job-shop environment, in which a large number of different products are to be produced using the same equipment and workers. It provides great flexibility in the use of expensive equipment and skilled personnel at the expense of substantial in-plant transportation.

Computer programs have been developed to help in locating departments in relationship with each other so that transportation cost is minimized. CRAFT (Computerized Relative Allocation of Facilities Technique) begins with the expected volume of materials shipment between every pair of departments, the materials handling cost per unit distance, the area needed for each department, and an initial layout. CRAFT then iteratively switches pairs of departments as long as it can reduce materials handling cost, and then it prints out a layout showing the recommended location for each department. Other software programs, with names like CORELAP, ALDEP, PLANET, and COFAD, have been developed to improve on CRAFT's performance. All of these programs have their limitations. Muther, arguably the best-known consultant in plant layout, contributed to the development of several of these but claims that his manual technique (Systematic Layout Planning, SLP) outperforms any of them in the hands of an experienced user.[9]

In *group technology,* a set of products requiring similar processing equipment is identified, and a small group of the machines needed to make this set of similar products is placed together. Transportation between steps in the manufacturing process is therefore minimized, inventory accumulating between steps can be almost eliminated, and products are produced much faster. For example, a General Electric plant reported productivity in making motor frames increased 240%, floor space needed was reduced 30%, and the manufacturing cycle was reduced from 16 days to 16 hours![10] By adding computer control, automated pallet handling of the workpiece, and automatic tool changing one can create a *flexible manufacturing cell* capable of producing this group of related parts with a minimum of human intervention. *Flexible manufacturing systems* are discussed in a later section.

QUANTITATIVE TOOLS IN PRODUCTION PLANNING

Three specific tools are discussed here: the economic order quantity approach to inventory control, break-even charts, and learning curves.

Inventory Control

Types of inventory. Most types of manufacturing processes begin with some type of *raw material* (sheet steel, lumber, leather) that requires processing. They add *purchased parts* (valves, switches, hinges), and consume *supplies* (cutting oils, time cards, drill bits). As work progresses, there will be a considerable investment in *work-in-process* before the *finished goods* are delivered to the warehouse to await sale and shipment. Each of these types of inventory represents an investment of capital, requires storage space, and is subject to loss, so it would seem desirable to make or purchase very small quantities at a time. However, each time a lot of product is made there is a *setup cost,* consisting of the clerical cost of processing and tracking the order and the cost of finding tooling and adjusting machines to make the item; these costs are less when lots are larger. Inventory control is the process of identifying and implementing inventory levels that result in a minimum total cost.

Economic order quantity (EOQ). Consider an inventory item for which the annual requirement is R units. Storing each unit of the item in inventory will cost I dollars per year. These storage costs include interest on the working capital invested in the unit, warehouse expense, and threat of deterioration, theft, and obsolescence while the unit is in storage. If, every time the last item is used, you renew the inventory with a batch of Q units, your average inventory will be $Q/2$ units and you will need R/Q batches per year. Each such batch involves an ordering or setup cost of S dollars. The total annual cost C_T of that inventory item is, therefore,

$$C_T = I\left(\frac{Q}{2}\right) + S\left(\frac{R}{Q}\right)$$

Setting the differential of total cost (with respect to Q) to zero and solving for Q yields the economic order quantity:

$$\text{EOQ} = \sqrt{\frac{2RS}{I}}$$

Determination of the EOQ is shown graphically in Figure 11-3. In this example the minimum annual total cost is attained with an EOQ of about 950. Note the flatness of the total cost curve near the optimum, suggesting that the order quantity can be adjusted over quite a range with little effect on total cost.

Problems with EOQ analysis. This formula has been used for many years, with very few American companies asking the critical question: Why does the setup cost have to be so high? Japanese companies, led by Toyota, developed techniques such as (1) designing dies and tooling so they could be switched quickly and cheaply, and (2) including simple cards (kanban) in each small lot that, when the succeeding process started using the lot, were sent to the pre-

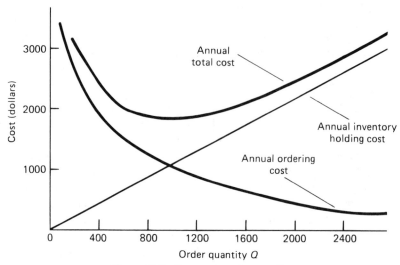

Figure 11-3. Economic order quantity.

ceding step to direct making another small lot with no further paperwork (see the "just-in-time" discussion later in the chapter). This permitted drastic reductions in setup costs.

As an example of the savings possible, consider a Japanese improvement where the time to change tooling was reduced from 2½ hours (150 minutes) to 3 minutes, only a fiftieth (0.02) of the original value. The EOQ equation above calls for a lot size that is only $(0.02)^{0.5}$, or 0.14 of the previous value, so that the total cost of setup and storage, and the floor space required, was cut to about one-seventh of its original level. These savings represented such a significant part of the ability of the Japanese to produce cheaper products that many American firms have instituted similar systems.

Break-Even Charts

Break-even analysis divides costs into their fixed and variable components to estimate the production levels needed for profitable operation. *Fixed costs* are those assumed to be independent of production level, at least in the range of production volume of interest. They include lease payments, insurance costs, executive salaries, plant heating and lighting, and the like. *Variable costs* are those assumed to vary directly with the level of production, such as direct labor, direct materials, and power for production equipment. Some *semivariable* costs may be divisible into fixed and variable components. Selling costs, for example, may consist of both salary (fixed) and commissions (variable).

Consider the example in Figure 11-4, where a plant may produce and sell U units of product up to a plant capacity of 2000 units. Fixed costs F_1 of \$100,000 must be paid regardless of the level of production. The selling price is assumed a constant $S = \$250$, regardless of volume, so that total revenue $R = U \times S$. The unit variable cost V_1 is assumed to be a constant \$150. Each unit sold therefore makes a *contribution* C_1 of

$$C_1 = S - V_1 = \$250 - \$150 = \$100$$

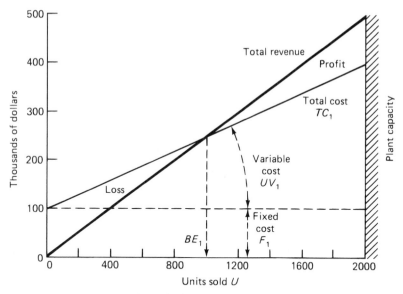

Figure 11-4. Break-even chart.

toward paying the fixed costs and providing a profit. The *break-even point BE₁* is the production level U where total costs TC equal total revenue R:

$$R = U \times S = TC_1 = F_1 + U \times V_1$$

$$BE_1 = U = \frac{F_1}{S - V_1} = \frac{\$100,000}{\$250 - \$150} = 1000 \text{ units}$$

Automation normally involves increasing the (fixed) investment in production equipment in order to make production more efficient (i.e., to reduce the variable cost). Figure 11-5 shows the effect in our example of increasing fixed cost by $80,000 to a total $F_2 = \$180,000$ in order to reduce the variable cost by $50 to $V_2 = \$100$. Our new contribution C_2 is ($250 - $100, or) $150 per unit, and our new break-even BE_2 is (180,000/150, or) 1200 units. This is not the point at which automation is justified, since at 1200 units the plant would make a profit ($20,000) without the added fixed cost of automation. The increased investment will only be justified at the point where the two total costs are equal; in the figure you should be indifferent between fixed cost (automation) levels F_1 and F_2 at a production level I of 1600 units, because the profit will be the same ($60,000) in either case.

Automation will usually increase the break-even point, increasing the vulnerability to low sales levels, but beyond a certain point of production it will also increase profitability. Owners of efficient, highly automated plants will therefore strive to keep their plants busy and may negotiate a lower price (still above their variable cost) for discount chains or foreign shipment if they think this added volume can be achieved without affecting their current sales.

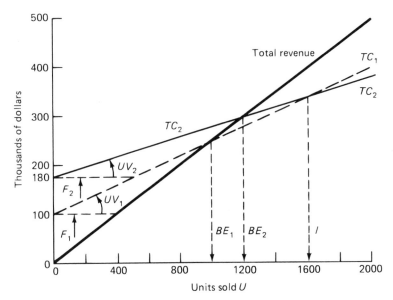

Figure 11-5. Break-even chart showing the effect of automation.

The break-even charts above are idealized; Figure 11-6 represents a more realistic situation. Typically, the revenue line is really curved, since price may have to be reduced to increase the volume of sales. Also, up to some point the incremental cost to produce additional units may decrease because of economies of scale. At some other point, costs may increase as a step function as some additional fixed cost is added (such as the supervision and other overhead for a second shift). As you near plant capacity, incremental costs may increase as less efficient backup equipment and less trained workers are pressed into service. In Figure 11-6, you would not seek to add a second shift until you were confident almost 1300 units could be sold (since 1100 units are more profitable than any higher production quantity less than about 1300), and your most profitable point appears to be at about 80% of plant capacity (1600 units).

Learning Curves

The learning curve concept derives from the observation that, in many repetitive human activities, the time required to produce a unit of output is reduced by a constant factor when the number of units produced is doubled. With a 90% learning curve, for example, if the first unit takes 1000 labor hours to produce, the second will take 900 hours, the fourth 810, the eighth 729, and so on, as shown in Figure 11-7. If it takes Y_1 time periods to make the first unit, the time Y_n to produce the nth unit can be found from

$$Y_n = Y_1 n^{-b}$$
$$\ln Y_n = \ln Y_1 - b \ln n$$

The exponent b can be found for any learning curve rate k by setting $n = 2$:

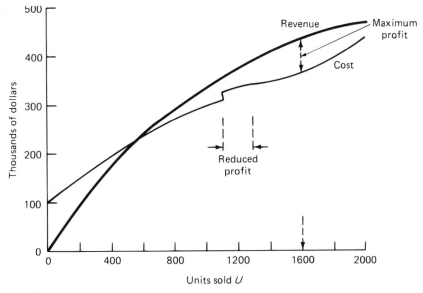

Figure 11-6. A more realistic break-even chart.

$$\frac{Y_2}{Y_1} = k = 2^{-b}$$

so that, for the 90% learning curve, for which $k = 0.9$,

$$\ln(0.9) = -b \ln(2); b = \frac{-\ln(0.9)}{\ln(2)} = 0.152$$

This relationship was developed in the aircraft industry, and its most common use has been there. Other applications are in the automobile industry, electronics assembly, and repetitive construction. Improvements are from a combination of factors, including increased worker skill, better work methods, better tooling and equipment, and organizational improvements, but tasks with greater manual and less mechanical content tend to show a

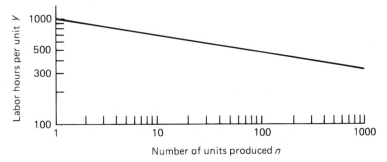

Figure 11-7. A 90% learning curve.

TABLE 11–1 Sample of the Percent Learning Curves Found in Various Industries

Industry	Percent Learning Rate
Volkswagen, 1945 to 1949	60
Volkswagen, 1950 to 1954	80
Twenty light-alloy products	80
Home construction	73–86
Welding of thin steel	70
Airplane production	70–75
Shipbuilding	74–90
Vehicle bodies	70–80
German armament industry	65–82
Railway carriages	75–93

Source: Adapted from J. R. DeJong, "Increasing Skills and Reduction of Work Time—Concluded," *Time and Motion Study*, October 1964, pp. 20–33.

faster reduction in time required (lower percent learning curve). Table 11-1 gives a sample of the percent learning curves found in various industries.

Between competitors in mass production of similar products it appears that the competitor with the largest market share would always enjoy a production cost farther down the learning curve, and therefore it would be able to sell at a lower price or enjoy a higher profit. Fortunately, in many cases the learning curve may end in a plateau, permitting competition on other grounds.

Note that the learning curve applies only for a continuous sequence of activity; if production stops at the end of one batch or lot and resumes later, the time to produce the first unit of the new batch will be greater than that for the last unit of the previous batch, and the learning curve will begin again at that point. This would seem to encourage large production batches, which carry with them large inventory levels. In the factory of the future, however, more and more production will be automated and accomplished by numerically controlled machinery that does not "forget" how to produce a part, so that there will not be the loss of efficiency present when a human worker returns to a job done before but now partially forgotten.

PRODUCTION PLANNING AND CONTROL

Introduction[11]

Any activity whose success is dependent on the coordination and cooperation of many people will benefit from careful planning and control, and the manufacturing environment requires the interaction of many people and machines. Often, there is an exact sequence of operations that must be performed, and any deviation from this sequence will result in a scrapped part. The manager cannot focus solely on number of parts produced or even the cost per part, because quality, due date performance, and efficiency are also scrutinized by upper management. Machines break down, parts are scrapped, raw material arrives late, and salespeople insist on the delivery of unscheduled rush orders.

In manufacturing it is essential to strike a strategic balance between idle resources and idle inventory. If inventories are very low, a worker may be "starved" for parts whenever the preceding work station slows down, breaks down, sets up for a new product, or switches to a different product that does not require processing by the worker in question. The shorter these disruptions are, the shorter the idle period will be. If we wish to prevent idle periods, we must hold enough inventory between stations to keep the worker busy whenever the feed is unexpectedly disrupted.

The costs of idle resources are widely recognized. Management often believes that the wages of an idle worker have been wasted. There is the fear that more workers or more machines will be needed if one sits idle for a while. Worse, there is a belief that if someone has no work to do, it is time to have a layoff, even if that capacity is clearly needed to fill future demand.

The costs of idle inventory are just beginning to be recognized by many people. In the past, "inventory holding costs" were considered to consist primarily of the interest on working capital and the rental of warehouse space. This neglects the increased delay that long production runs of each product tend to produce in beginning another product run, increasing lead time (the time it takes from order placement to shipment). Long lead times may cause impatient customers to take their business elsewhere, a well-hidden cost. In rapidly advancing technologies, product may be obsolete even before it is shipped. Since new quality problems often remain hidden until discovered by a customer, a pipeline full of defective product could be a major liability.

Steps in Production Planning

The first step in planning of any type is to identify the goals you wish to achieve. The trade-off between idle machines and idle inventory will exist whether or not management cares to acknowledge it, and the schedule will enforce a given trade-off level, whether or not it is appropriate to the particular industry.

One company serving a seasonal market may decide to level their resource load by carrying more inventory. Another may decide to provide better service by carrying less inventory and more resources. Ideally, this should be a conscious decision rather than a random one. Once an inventory strategy is selected, the company should establish a procedure for quoting delivery dates that are in fact achievable. If customers know that the product cannot be delivered on the desired date, they may adjust their own schedules, or plan to order earlier next time. If they find out at the last minute, they may get upset and go elsewhere next time. If demand exceeds capacity, it may be necessary to consider other orders already promised when calculating a reasonable promise date.

The next several steps break down the production process into the required tasks and figure out when each one has to start. Sometimes one task can be accomplished with any of several different resources (equipment and/or workers), in which case the assignment of tasks to resources can happen at planning time or at execution time. If several tasks need the same resource during the same time period, the plan should provide workers with a means to determine priority.

These several steps have a traditional set of names. *Process planning (routing)* determines the sequence of operations needed to produce the product. *Loading* sets aside the necessary time on each machine or work station to process the desired quantity. *Schedul-*

ing establishes when each step of the work will be performed. *Dispatching* is the official authorization to do the work. In *flexible manufacturing systems* (discussed later) these conventional steps may occur automatically under computer control. Finally, *production control* is the system whereby deviations from the planned schedule are reported to the production planning and control office so that schedule adjustments can be made.

There is no such thing as a perfect plan, simply because the data on which the plan is based may change before the plan is executed. The time to process a given part is a statistical quantity that may vary unpredictably from time to time. Also, machines break down and parts get scrapped. Orders may be canceled or top-priority rush orders may be added. Sometimes, the data base itself is in error. On the positive side, there may be a *learning curve* (discussed earlier in this chapter) such that the worker learns how to do a task faster than the data base indicates, and the scheduling system should take advantage of such efficiencies.

Production Planning and Control Systems

Materials Requirements Planning (MRP). Materials requirements planning refers to a set of time-phased order-point techniques to support manufacturing schedules. MRP began development in the 1950s as the cost of computer calculation began to decrease. At its simplest, it provides a schedule for ordering raw material and parts and performing production operations to provide the products of production (end items) on time. MRP begins with a *master production schedule (MPS)* that identifies when end items must be available to meet customer or other commitments.

For example, assume a customer has been promised that one unit of product A will be shipped six weeks from now, a need that becomes part of the MPS. The next document needed is the *bill of materials (BOM)* for product A, which shows that it is produced from one unit of material M, one component C, and two parts P. In addition, we need to know the supplier lead times for each item we have to buy, and the sequence and duration of such production activities as machining, assembly, and testing that will take place in our plant. The relationships are illustrated in Figure 11-8.

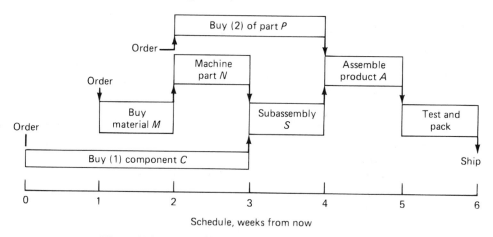

Figure 11-8. Schedule of production lead times for product A.

MRP converts this information into instructions for purchasing to place orders now for one unit of component *C*, one week from now for one unit of material *M*, and two weeks from now for two of part *P*. This will provide the suppliers the normal time required (three, one, and two weeks, respectively) to get them to us when we need them. Further, production planning is advised of the need to schedule the start of machining in two weeks, subassembly in three, assembly in four, and testing in five weeks. Similar requirements for all of the end items shown on the MPS are combined into comprehensive purchasing and production schedules, after checking the *inventory file* to identify material or subassemblies currently on hand or on order.

Manufacturing resource planning (MRP II). The original MRP acted as if each order could be scheduled independently of the others, assuming that enough capacity existed to assign simultaneous orders to different resources. However, the capacity of both equipment and skilled workers is usually limited. The past decade has seen the shift from the simple MRP to one incorporating machine capacity and personnel planning, and a trend toward total integrated manufacturing control systems. As this took place, the terminology *manufacturing resource planning* and the acronym MRP II have gradually replaced the earlier MRP. The *Tool and Manufacturing Engineers Handbook*[12] lists the following information as being provided in a modern MRP II computer system:

- Customer demand activity
- Production plans
- Production schedules and their execution
- Purchasing management
- Inventory management
- Product cost reporting
- Support of and financial applications of accounts receivable, accounts payable, general ledger, and payroll

Synchronized manufacturing. In 1979, an Israeli named Eliyahu Goldratt developed OPT, a proprietary capacity-sensitive scheduling software that was supposed to correct the deficiencies of MRP and MRP II. In the process of implementing that software, he discovered that many of the scheduling problems found in industry were the result of not properly recognizing the relationships between inventory and capacity. The concepts that he developed came to be known by the same name as his software (OPT) and are best described in his book, *The Goal*.[13] The same concepts are now advocated by many practitioners under the generic name *synchronized manufacturing,* and they can be used quite effectively in combination with MRP software packages, as described in *Regaining Control*.[14]

Just-in-time (JIT). JIT is a method involving very small raw material or in-process inventory quantities, small manufacturing lots, and frequent deliveries, such that a small batch of each component or subassembly is produced and delivered "just in time" to be used in the next production step. It was initially developed by the Toyota Motor Company and later was adopted by other Japanese companies. The Toyota system uses a series of cards called *kanban* (pronounced "kahn-bahn"), the Japanese for "a visible record or plate," to direct production. As described by Reda:

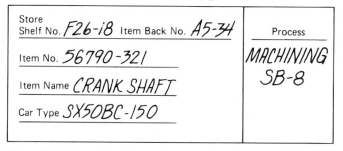

Withdrawal Kanban

Store Shelf No. *5E215* Item Back No. *A2-15*	Preceding Process
Item No. *35670807*	*FORGING B-2*
Item Name *DRIVE PINION*	
Car Type *SX50BC*	Subsequent Process

Box Capacity	Box Type	Issued No.	*MACHINING M-6*
20	*B*	*4/8*	

Production-ordering Kanban

Store Shelf No. *F26-18* Item Back No. *A5-34*	Process
Item No. *56790-321*	*MACHINING SB-8*
Item Name *CRANK SHAFT*	
Car Type *SX50BC-150*	

Figure 11-9. Examples of kanban cards.
(From Hussein M. Reda, "A Review of 'Kanban'—the Japanese 'Just-in-Time'
Production System," *Engineering Management International,* 4, 1987, p. 146.)

This card (kanban) is primarily used to signal the need to either deliver (withdrawal kanban) or produce (production-ordering kanban) more parts. A withdrawal kanban specifies the quantity required at succeeding processes (which are to be withdrawn from preceding processes), while a production-ordering kanban orders preceding processes to produce replacement parts.[15]

Figure 11-9 provides examples of kanban cards, and Figure 11-10 illustrates the mechanics of their use. The quantity ordered on each kanban is typically very small, so that defects and production problems in one location can cause the entire production line to shut down. Toyota reduced the incidence of such disruptions with a number of related innovations:

- Smooth production schedules for final assembly of the end item, with little month-to-month variation
- An incessant effort to eliminate the causes of defects
- Plant layout in flexible manufacturing cells such as those already described under "group technology" above
- Workers able and willing to work at different processes as demand requires

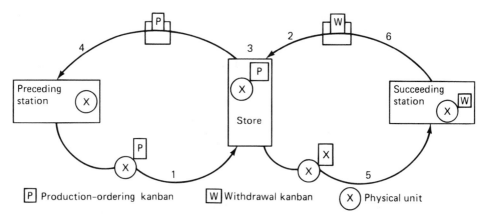

Figure 11-10. Mechanics of a simple kanban cycle. (1) Part produced at preceding station and P-kanban attached to it is sent to the store. (2) When the part is needed at a succeeding station, a W-kanban is sent to the store to withdraw the part. (3) At the store, the P-kanban is removed from the part and the W-kanban attached to it. The P-kanban is then collected in a "production-ordering" box. (4) At short time intervals, the P-kanban is then sent to the preceding station, constituting a production order. (5) The part with the W-kanban goes to the succeeding station to meet the demand. (6) The W-kanban is detached from the part and collected in a "withdrawal" box.
(From Hussein M. Reda, "A Review of 'Kanban'—the Japanese 'Just-in-Time' Production System," *Engineering Management International,* 4, 1987, p. 145.)

- Worker involvement in identifying and correcting problems as they occur through mechanisms such as the *quality circles* described in Chapter 12
- Reduction in the number of suppliers, offset by great emphasis on the quality and delivery schedules of those that remain

Parts coming from other plants are typically delivered one truckload at a time in the JIT system, and the production line often is fed directly out of the truck that just arrived. This is particularly feasible in Japan, where manufacturers tend to be located in the Tokyo area and to have long-term relationships with trusted suppliers also located in the same area. In the United States, with suppliers often thousands of miles away, some adaptations have to be made. To make JIT work better, General Motors has encouraged its major suppliers to build parts plants near GM assembly plants. To keep assembly plants constantly informed of the location of supplies en route to them, some American trucking firms now carry radios capable of relaying their status to the plant at any time via the *Geostar* satellite.[16] At the new (1992) General Motors Opel plant in Eisenach, Germany, final assembly schedules are transmitted electronically to the nearby Lear seating plant; car seats are built four hours before they are needed and trucked to the Opel plant.[17]

Model changeover. Production planning is especially vital in mass production where major changes are necessary between annual (or other periodic) models. In an example reported by Treece in *Business Week,*[18] General Motors Corporation shut down the production lines at their Oshawa, Ontario, plant November 19, 1993, to install new welding robots and other machinery to build the 1995 Chevy Lumina, and it was 87 days until the first production line started up on February 19, 1994—slowly. By April 1 it had built a total of only 288 cars, and did not expect to reach its full line speed of 60 cars an hour until August. The Toyota plant in Georgetown, KY, on the other hand, made the changeover to the 1992 Camry in 18 days, and the Honda plant in Marysville, OH, changed over to the 1994 Accord in just 3 days, and reached full speed in just six weeks. Treece explains:

> To start, Toyota and other Japanese auto companies make new-car designs as compatible as possible with existing equipment. To make that easier, they tend to use more flexible automation—such as welding robots that can be reprogrammed. And their factories have extra space beside the assembly line, so that new gear can be tested well before it is needed.
>
> At Marysville, Honda began debugging the 1994 model Accord's machinery a year before the production changeover. More was added over Christmas of 1992, and then "about every weekend until July, something was being put in place," says plant manager Steve Yoder. Even as the old line ran, Honda made prototype new models by shunting them onto the new equipment, as if in a railroad switching yard.

Weeks when plants are shut down or run below design speed are weeks when profits are not made, and American car companies have played catch-up. The Ford plant in Kansas City, MO, was shut down 60 days to launch the 1995 Ford Contour/Mercury Mystique, but it has been gradually installing more flexible equipment in the hope of launching the 1996 model in only three weeks (two of them the July company-wide vacation); Chrysler expects to achieve this in all plants by 1997, and General Motors to achieve a two-week changeover in all plants by the year 2000.

FLEXIBLE MANUFACTURING SYSTEMS

Need for Flexible Manufacturing Systems

A 1985 U.S. government report on the flexible manufacturing systems (FMS) industry begins by explaining:

> Automation in manufacturing in the past was only considered where large quantities (mass production) were required, such as in the automotive industry and in the household appliance industries. This level of production automation could be accomplished only by transfer machines and dedicated lines of machines and then only for production of a limited variety of different parts. However, on a worldwide basis, 75 percent of all metalworking manufacture takes place in small batch production, limiting the benefits of automation.[19]

Today, the demand for differentiated (tailored) products rather than mass-produced ones, and worldwide competition in almost every industry, join to create a compelling need for more efficient means of producing small batches of high-quality products. At the same time, the computer revolution in both hardware and software makes possible computer control of machining and other manufacturing operations to reduce the cost of setup for small batches. Robots and other computer-controlled devices for handling and transferring work between machines, automatic guided vehicles (AGVs) for movement of work and tools, automated storage systems, and the development of computer-integrated manufacturing techniques and software combine to make flexible manufacturing systems possible.

Some Definitions

The U.S. Department of Commerce recognizes four basic categories of flexible manufacturing technology: stand-alone machines, the flexible manufacturing cell, the flexible manufacturing system, and the fully automated factory. Each is described below:

> The *stand-alone machine* is typically a machining center or turning center with some method of automatic material handling, such as multiple pallets or chuck [tool] changing arrangements. These provisions permit the machine to operate unattended for extended periods (often a full eight hour shift), changing tools and work pieces under direction of the machine control. This computer control can also initiate and control features such as probing, inspection, tool monitoring, and adaptive control.
>
> The *flexible manufacturing cell (FMC)* normally incorporates more than one machine tool, together with pallet changing equipment such as an industrial robot to move work into the cell, between machines in the cell, and out of the cell.
>
> The *flexible manufacturing system (FMS)* includes at least three elements: a number of work stations, an automated material handling system, and system supervisory computer control. . . . Automatic tool changing, in-process inspection, parts washing, automated storage and retrieval systems (AS/RS), and other computer-aided manufacturing (CAM) technologies are often included in the FMS. Central computer control over real time routing, load balancing, and production scheduling distinguish FMS from FMC.
>
> The *fully automated factory (FAF)* or "factory of the future" represents the full development of all aspects of computer integrated manufacturing (CIM). In the FAF, all functions of the factory will be computer controlled, integrated, and to varying degrees self optimizing.[20]

Advantages of an FMS

Hartley provides a good description of the FMS installed by Yamakazi, a manufacturer of machining centers (themselves FMS components) at its main plant in Oguchi in Japan:

- It consists of 18 machining centers instead of the 36 needed previously.
- Previous employment was 106 direct and 80 indirect workers. The new FMS "is manned by one person in the computer room, one person in the tool room, and four people at the loading/unloading station. These people are needed on two shifts only, the third shift being unmanned. Therefore the manning is down from 190 to 12."
- "The components for a complete machine [used to take] three *months* to pass through the machine shops, four weeks in assembly, and one week for inspection and adjustment. Now, the time spent in the machine shop has been reduced to only four days—for an average process time of 24 hours—while assembly now takes two weeks [because workpieces are available when needed], and inspection one week. Thus, a machining centre can now be produced in under four weeks, whereas it previously took over four months."
- Capital costs (in British pounds) compared as follows: land and building was reduced from £1,900,000 to £1,000,000; machinery and equipment increased from £4,000,000 to £7,500,000; work in process [working capital] decreased from £2,500,000 to £100,000; total investment therefore only increased from £8,400,000 to £8,600,000.
- Although the total capital costs above were comparable, labor costs dropped from

over £2,000,000 to about £130,000, so that average annual profit (assuming three-shift operation) increased from £800,000 to £1,900,000.[21]

Ranky[22] states that FMS provides the following benefits if designed and used successfully:

- Greater productivity, which means a greater output and a lower unit cost, on 45 to 85% smaller floor space.
- Quality is improved because the product is more uniform and consistent.
- The intelligent, self-correcting systems (machines equipped with sensory feedback systems) increase the overall reliability of production.
- Parts can be randomly produced in batches of one or in reasonably high numbers and the lead time can be reduced by 50 to 75%.
- FMS is the only available manufacturing environment to date where the time spent on the machine tool can be as high as 90% and the time spent cutting can again be over 90%. Compare this with stand-alone NC (numerically controlled) machines, where the part, from stock to finished item, spends only 5% of its time on the machine tool, and where the actual productive work takes only 30% of this 5%.

DISCUSSION QUESTIONS

11-1. Why is a vigorous production capability so important to the United States in the "postindustrial society"?

11-2. What are some of the positions that engineers fill in a large manufacturing plant?

11-3. What subjects will be important in the education of the manufacturing engineer for the twenty-first century?

11-4. Discuss some of the factors that would be most important in selecting a site for (**a**) a portland cement plant, (**b**) a research "think tank," and (**c**) a shoe factory.

11-5. Distinguish between (**a**) product layout, (**b**) process layout, and (**c**) group technology.

11-6. (**a**) If it costs $2 per unit to store an item for one year, $40 setup cost every time you produce a lot, and you use 1000 units per year, how many lots of what size should be manufactured each year? (**b**) How would your answer change if the setup cost can be reduced to $10?

11-7. Setup for a stamping operation required a time-consuming fixture installation and testing that took 4 hours each time a different part was to be produced; typically, 12 hours production was made for inventory of a given stamping before the machine was stopped to permit setup for a new part. After careful process analysis, fixtures and transfer methods were revised to permit setup in 15 minutes. Discuss the implications for this change on (**a**) optimum batch size, (**b**) order frequency, and (**c**) machine and labor productivity.

11-8. A production plant with fixed costs of $300,000 produces a product with variable costs of $40 per unit and sells them at $100 each. What is the break-even quantity and cost? Illustrate with a break-even chart.

11-9. A machine tool salesperson offers the plant of question 11-8 equipment that would increase their fixed cost by $180,000 but reduce their variable cost from $40 to $25. Should the plant accept this suggestion if they can sell their entire plant capacity of 10,000 units per year at $100 each? Illustrate by modifying the break-even chart of question 11-8.

11-10. A plant is beginning production of a light-alloy product and finds that it takes 400 hours to produce the first item. How many hours should it take to produce each of the following: (**a**) the second item; (**b**) the eighth item; (**c**) the thirty-seventh item? (*Hint:* Refer to Table 11-1.)

11-11. The first two units of a product cost a total of $9000 to produce. If you believe an 80% learning curve applies, how much would you expect the *fourth* unit to cost?

11-12. Distinguish between MRP and MRP II.

11-13. Describe how the kanban is used in the just-in-time production system.

11-14. What are some of the reasons for using a flexible manufacturing system?

11-15. What is the relationship between flexible manufacturing systems and the group technology concept introduced earlier in the chapter?

NOTES

1. From a presentation by Andrew Grove, president, Intel Corporation, to the plenary session "A New Era in Manufacturing" at the American Society for Engineering Education annual conference June 21, 1988, in Portland, OR.

2. *Statistical Abstract of the United States 1994,* 114th ed. (Washington, DC: U.S. Bureau of the Census, 1994), Table 1329, pp. 823–826.

3. Stephen S. Cohen and John Zysman, "Manufacturing Innovation and American Industrial Competitiveness," *Science,* 39, March 4, 1988, p. 1114.

4. Ramchandran Jaikumar, "Postindustrial Manufacturing," *Harvard Business Review,* 64:6, November–December 1986, p. 70.

5. Claire LeBuffe, "Degrees in '93: Signs of Confidence?" *ASEE Prism,* February 1994, p. 29.

6. A. T. Kearney, Inc., *Countdown to the Future: The Manufacturing Engineer of the 21st Century* (Dearborn, MI: Society of Manufacturing Engineers, 1988).

7. John F. Welch Jr., "Competitiveness: The Real Stuff of American Engineering," *Gateway Engineer,* February 1990, p. 9.

8. Harold T. Amrine, John A. Ritchey, Colin L. Moodie, and Joseph F. Kmec, *Manufacturing Organization and Management,* 6th ed. (Englewood Cliffs, NJ: Prentice-Hall, Inc., 1993), pp. 254–255.

9. Richard Muther, informal presentation to the Department of Engineering Management, University of Missouri—Rolla, October 10, 1988.

10. National Research Council, Manufacturing Studies Board, *Toward a New Era in U.S. Manufacturing: Need for a National Vision* (Washington, DC: National Academy Press, 1986).

11. This section and the following one ("Steps in Production Planning") were adapted from a term paper by the late Susan Burgess, a doctoral student in engineering management at the University of Missouri—Rolla.

12. Raymond F. Veilleaux and Louis W. Petro, eds., *Tool and Manufacturing Engineers Handbook, Volume 5: Manufacturing Management,* 4th ed. (Dearborn, MI: Society of Manufacturing Engineers, 1988), pp. 2–17.

13. Eliyahu M. Goldratt and Jeff Cox, *The Goal: A Process of Ongoing Improvement,* rev. ed. (Croton-on-Hudson, NY: North River Press, Inc., 1986).

14. Susan Burgess and Mokshagundam L. Srikanth, *Regaining Control: Get Me to the Shipping Dock on Time* (New Haven, CT: Spectrum Publishing, 1989).

15. Hussein M. Reda, "A Review of 'Kanban'—the Japanese 'Just-in-Time' Production System," *Engineering Management International,* 4, 1987, pp. 145–146.

16. "Truckers Reporting In via Satellite," *St. Louis Post-Dispatch,* December 14, 1988, business section.

17. "GM's German Lessons," *Business Week,* December 20, 1993, p. 67.

18. James B. Treece, "Motown's Struggle to Shift on the Fly," *Business Week,* July 11, 1994, pp. 111–112.

19. U.S. Dept. of Commerce, International Trade Administration, *A Competitive Assessment of the U.S. Flexible Manufacturing Systems Industry* (Washington, DC: U.S. Government Printing Office, July 1985), p. vii.

20. Adapted from U.S. Dept. of Commerce, *A Competitive Assessment,* pp. 2–4.

21. John Hartley, *FMS at Work* (Kempston, Bedford, UK: IFS (Publications) Ltd., 1984), pp. 157–160.

22. Paul Ranky, *The Design and Operation of FMS* (Kempston, Bedford, UK: IFS (Publications) Ltd., 1983), p. 4.

CHAPTER 12

Managing Production Operations

PREVIEW

We begin this chapter by defining product quality and introducing the categories of quality costs. The statistics of variables and attributes are introduced and applied to control charts and inspection sampling. We discuss the contributions of Taguchi, Deming, quality circles, and Total Quality Management to the quality revolution. We next consider the nature and importance of productivity in international competition, followed by the methods of work measurement.

The last part of the chapter relates to some of the manufacturing functions of interest to engineers, beginning with maintenance and plant engineering. We also discuss the functions of human resources management, purchasing and materials management, and packaging engineering.

ASSURING PRODUCT QUALITY

Some Definitions

Quality has been described[1] as *fitness for use* or *customer satisfaction.* It may be divided into two categories. *Quality of design* measures the extent to which fitness for use is incorporated into the product design through the specification of proper materials, tolerances, and other precautions. Quality of design will vary to some extent with the intended market: one would not expect the same features in a stripped-down Ford Escort and a fully equipped Lincoln Continental. *Quality of conformance* (or *quality of production*) measures how well the quality specified in the design is realized in manufacture and delivered to the customer. One author adds a third "aspect of quality," *quality of use,* measuring how the product is applied or employed, and what that does to its properties.[2]

Importance of Better Quality

The March 1987 *Consumer Reports* summarized the relative frequency of repair reported by their readers for 120,000 television sets made under 20 brand names and purchased between 1981 and 1985. Brands named Toshiba, MGA (Mitsubishi), Sony, and Hitachi led the list with a "frequency of repair index" of from 7 to 9. American brands (Philco, Sylvania, and Magnavox) were at the bottom, with an index of from 21 to 27. As a result, the last American-owned television plant in the United States ceased production in 1995; other brands and plants have disappeared or are owned by Japanese companies. As we learned in Chapter 10, for example, in the early 1970s an American-built Motorola 19-inch television set averaged 1.5 to 1.8 defects per set; three years after sale to a Japanese company the same plant was producing 19-inch TVs (under the Quasar name) averaging 0.03 to 0.04 defect per set.[3]

Similarly, the April 1987 (Annual Auto) issue of *Consumer Reports* summarized problems reported by readers on 333,000 automobiles. All models by Honda, Mazda, Nissan, and Toyota showed much lower frequency of repair indexes than did almost all models built by American companies, and a majority of American owners of Japanese cars said they would purchase the same brand next time.

Seven years later, the April 1994 issue again summarized readers' experience on 486,000 1991–93 vehicles. The quality revolution waged by American automobile companies over the last decade has begun to show results. Perhaps most notable is the Saturn, which competes for quality with Japanese small cars; this was achieved at a cost of billions by General Motors by starting a division from scratch with a new plant at Spring Hill, TN, selecting workers through intensive psychological testing, sending many to Japan for training, and then running the division using Japanese management principles. Another General Motors initiative was their New United Motors joint venture with Toyota at a former GM plant in Fremont, CA; using carefully selected workers under Japanese leadership, the plant produces high-quality Toyota Corolla and (GM) Geo Prizm small cars. General Motors has rotated many of their supervisors through the Fremont plant, but when they returned to existing plants with historic adversary relationships between management and labor they found it difficult to implement changes; nonetheless, some Buick and Oldsmobile models have taken their place among compact and midsize cars with lower frequency of repair.

This quality revolution requires both the use of methods of quality assurance and more enlightened, participatory management. Fortunately, many American companies *are* making progress in these directions. As a result, American automobile companies are regaining market share from the Japanese. More generally, when 20,000 consumers in 20 countries were asked how they rated the overall quality of manufactured goods from 12 major exporting nations, 38.5% answered "excellent" or "good" for Japanese products, 36.0% for German, and 34.3% for U.S., versus only 21.9% for the next best (Britain) down to 5.8% for the 12th (Russia).[4] (And because of currency exchange rates and labor costs, more and more

Japanese and, more recently, German companies are building products in the United States.)

Quality Costs

An important step in getting management support for improving quality is documenting the total cost of poor quality and of quality control efforts. The American Society for Quality Control (ASQC) has established four categories of costs to help in this analysis:[5]

1. *Prevention costs* are those incurred in advance of manufacture to prevent failures, such as quality planning, training, data analysis and reporting, process control, and motivation programs.
2. *Appraisal costs* include the costs of inspection of incoming parts and materials (whether by your supplier or by you when you receive it), inspection and test of your product in process and as a finished product, and maintenance of test equipment.
3. *Internal failure costs* are those that would not appear if there were no defects in the product before shipment to the customer. They include scrap (labor and material spent on unrepairable items), rework (the cost of making defective items fit for use, including necessary retesting), downtime and yield losses caused by defects, and the cost of material review and disposition of defectives.
4. *External failure costs* are those caused by defects found after the customer receives the product. These include the costs of investigating and adjusting complaints, the costs of replacing defective product returned by the customer, price reductions ("allowances") offered to compensate for substandard products, and warranty charges. The total costs to your customer in downtime and other damage may be much higher, and these may drive your customer to seek a more trustworthy supplier.

When these total quality costs are added up, in company after company they total far more than management realized—often of the same magnitude as total company profit. Typically, the prevention costs are found to be a very small percentage of the total. When a concerted effort is instituted to develop a comprehensive quality program, to find the primary reason for failures, and to modify design, processes, and employee training and motivation to minimize failures, savings in failure costs are commonly many times the cost invested in prevention. Even appraisal costs are reduced, since top-quality product does not require the same intensive level of inspection. Figure 12-1 shows the relationship of these components of quality cost with quality level. This classic figure suggests that there is some optimum economic quality short of 100% conformance that should be striven for. Merino[6] points out this may be a valid conclusion where quality is "inspected in" through intensive inspection and test, but that the modern approach of continuous improvement of product design and of the processes used in manufacture makes it possible to approach very close to 100% conformance without excessive prevention and appraisal cost.

Statistics of Quality

Statistics consists of gathering, organizing, analysis, and use of data. The methods of *statistical quality control* were developed in the United States in the 1930s and '40s (largely

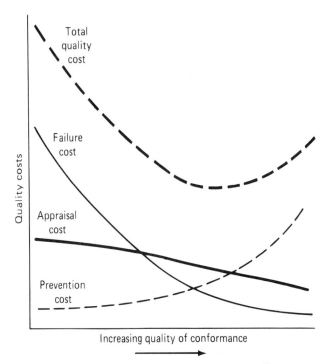

Figure 12-1. Effect of quality improvement on quality costs.

at Bell Laboratories). However, they received their most intensive application in Japan after World War II, as a result of visits of the American statistical quality control experts Deming and Juran at the invitation of General Douglas MacArthur's occupation forces. Only when the Japanese brought their quality and reliability to a level that threatened the American economy did American industry begin to pay attention. Unfortunately, American engineers and American business leaders are poorly prepared to respond to this problem. Gordon Geiger, in his presidential address to the Accrediting Board for Engineering and Technology (ABET, which accredits U.S. engineering curricula), highlights this:

> We can learn from the Japanese, who have applied the fundamentals of quality management to their processes and brought themselves from a state of complete ruin 40 years ago to [being], without question, the world's most powerful economic nation—a nation that today provides the financing for America's debt. Interestingly enough, the basis for much of their quality control is statistics and yet fewer than one percent of all engineering curricula in the United States require (and fewer than one percent of all engineering students take) a basic course in statistics prior to graduation with a degree in engineering.[7]

Lester Thurow, dean of MIT's Sloan School of Business, highlighted this weakness in discussing the problem faced by a Japanese firm when they built a plant in North Carolina. Although they were accustomed to using high school graduates for statistical quality control in Japan, they could not find a high school *or college* graduate able to do the job and thus had to hire someone with a graduate degree. He concludes by asking "How can you win in a technical era with mathematical illiterates?" [In the same *60 Minutes* television program,[8] Thurow stated that 70% of chief executive officers in Europe and Japan had a technological background, whereas only 30% did in the United States and (perhaps as a result) that production jobs are the "dumping ground" of U.S. industry, with lower pay and less promotion.] Fortunately, in some American companies managers from the top down are now being taught statistics as the basis for understanding and improving quality problems, and there is strong pressure within ABET to require training in statistics in all engineering curricula.

Statistical methods are used to evaluate some *quality characteristic,* such as the diameter of a hole, the weight of a package, or the tensile strength of a metal strip. Two types of statistical methods are used in quality control (Table 12-1). *Variables* methods involve *measuring* the quality characteristic (such as the hole diameter) on a sample of the item being controlled, then using a *continuous* probability distribution such as the *normal distribution* for analysis. *Attributes* methods involve *counting* as *defective* those items that do not fall within a stated specification, then using the fraction defective in a sample in *discrete* probability distributions such as the *binomial* or *Poisson* for analysis.

Each probability distribution is described by a "measure of central tendency" (average) and a "measure of dispersion" (spread). For example, consider a hole with a specified diameter x of 1.250 ± 0.010 inches (i.e., holes from 1.240 to 1.260 inches will meet specifications). Assume that this diameter was actually measured as 1.235, 1.245, 1.250, 1.256, and 1.259 inches in a sample of five items. Using the variables approach, the *mean* value (termed "x-bar") would be found as the sum of these n values (6.245 inches) divided by the number of measurements ($n = 5$), or 1.249 inches:

$$\bar{x} = \frac{1}{n} \sum_{j=1}^{n} x_j$$

Using the attributes approach in this same example, one would simply determine (perhaps using a "go/no go gauge") that only one of these five holes was defective (fell outside the specified range of 1.240 to 1.260). The sample would then be assigned a *fraction defective p* of $\frac{1}{5}$, or 0.20; the mean value "p-bar" would then be found by averaging the p values for a large number of samples.

The most common measure of dispersion from the mean value is the *sample standard deviation s,* which is the square root of the *sample variance V. V* is an unbiased estimator of the population variance and can be found as the sum of the squares of the deviations from the sample mean, divided by one less than the sample size:

$$V = \frac{1}{n-1} \sum_{j=1}^{n} (x_j - \bar{x})^2$$

TABLE 12-1   Some Statistical Methods Used in Quality Control

Class of Statistics	Action Involved	Probability Distributions	
		Type	Examples
Variables	Measuring	Continuous	Normal, exponential
Attributes	Counting	Discrete	Binomial, Poisson

In our example,

$$V = \frac{(1.235 - 1.249)^2 + (1.245 - 1.249)^2 + \cdots + (1.259 - 1.249)^2}{5 - 1}$$

$$= \frac{0.000,196 + 0.000,016 + 0.000,001 + 0.000,049 + 0.000,100}{4}$$

$$= 0.000,0905$$

and the sample standard deviation s (an estimate of the population standard deviation σ) is the square root of V, or 0.0095.

In the variables method, the *range R,* which is the difference between the highest and lowest values in a sample, is also a common measure of dispersion. In the example above, the sample range would be 1.235 to 1.259, or 0.024 inch. Ranges are easier to calculate but provide less information per sample than does the standard deviation.

Process Control Charts

In production operations it is important to ensure that a process is "in control," which means that it continues to produce items with unchanged quality characteristics. *Process control charts* (Figure 12-2) are used for this purpose. They consist of three parallel lines: a central line representing the mean value of a quality characteristic; an "upper control limit" (UCL), normally three standard deviations (3σ) above this mean value; and a "lower control limit" (LCL), normally three standard deviations below the mean. If the process stays in control, 99.73% of all future observations should fall between the UCL and the LCL, symmetrically dispersed about the mean. A great deal can be learned about such things as

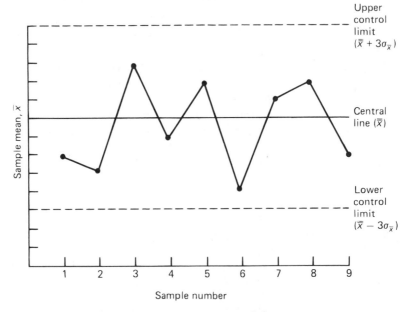

Figure 12-2. X-bar control chart.

raw material or operator changes, tool wear, or changes in machine settings by observing measurements that fall out of the control limits, bunch on one side of the central line, or follow some other nonrandom pattern.

For example, if you wish to control some measurable quality characteristic x using variables statistics, you may wish to maintain a control chart based on the mean (average) value $\bar{x}$ of measurements of random samples of five items. To do this, you would take samples of five items at regular periods until you had about 25 samples, use this information to calculate values for the central line (mean of these mean values, or $\bar{\bar{x}}$) and control limits, and construct an $\bar{x}$ chart similar to Figure 12-2. As you continue making the product, you continue to take samples of five items and enter their mean value on the control chart to assure that nothing has changed in your process or materials.

To control the variation (dispersion) about this mean value, you would maintain one of two types of charts: an "R-chart," measuring the difference between the highest and lowest value within each sample of five, or a "sigma-chart" (σ-chart), measuring the standard (root mean square) deviation of measurements from the mean value.

If you are using the attributes approach (counting but not measuring defects), you can control the level of these defects using either a p-chart (which measures sample *fraction defective* and is based on the binomial probability distribution) or a c-chart (which measures *defects per sample* and is based on the Poisson probability distribution). (A defective item is one that contains one or more defects.) In the attributes approach, control charts for standard deviation (σ_p or σ_c charts) are used to control variability.

Inspection and Sampling

Examining product to determine if it meets the specifications set for it, or *inspection,* is certainly the original method of quality control, and it is still the most common. Inspection may be performed on the raw materials and parts you receive from suppliers ("acceptance sampling"), on your finished product, or on your goods-in-process (before you invest the cost of the next production step in them). Examination of every item (100% inspection) may seem desirable, but it is often expensive unless it can be done automatically, and human inspectors cannot be expected to be continually vigilant and catch all defects. Most inspection is therefore done by sampling lots (batches) of product and accepting or rejecting the lot depending on the number of defectives in the sample. Sampling rules can be developed statistically for each situation, but it is more common to consult an established sampling table.

For example, suppose that you were using the common military standard MIL-STD-105E[9] for acceptance sampling by attributes (counting, not measuring) of a lot of 2000 items, and you considered 1.0% defective an *acceptable quality level* (AQL). If you looked up the normal "general inspection level II" for this lot size and AQL in MIL-STD-105D you would be directed to:

- Take a random sample of 125 items from the lot.
- Accept the lot if it contained no more than 3 defective items.
- Reject the lot if it contained 4 or more defective items, then either 100% inspect it (sort out the defectives) or return it to the producer as unsatisfactory.

Or, you might choose the *double-sampling* alternative:

- Take a sample of 80 items.
- Accept the lot with no more than 1 defective.
- Reject the lot with 4 or more defectives.
- With 2 or 3 defectives, take a second sample of 80 items and accept the lot if the total defectives in the two samples were 4 or less; otherwise, reject it.

In *multiple sampling,* from 1 to 7 sequential small (32-item) samples would be used in this same situation, with a more complex decision rule.

Sampling from a larger lot or to assure a higher quality (smaller AQL) would require a larger sample and a different decision rule; the reverse would permit a smaller sample. Sampling plans are also available on a variables basis, where you can take a smaller sample but use the actual measured value of your quality characteristic for lot acceptance; other plans are used for continuous rather than batch production.

Taguchi Methods

Specifications have traditionally been treated on an all-or-nothing basis—a measurement is either "in specification" and completely acceptable or "out of spec" and completely unacceptable. The first three (normal) distributions of a quality measurement shown in Figure 12-3 [(a) narrow spread, centered on the specification midpoint *m*, (b) narrow but off center, and (c) wider spread but centered] would therefore be almost equally acceptable. Even the "uniform distribution" in Figure 12-3d would be equivalent.

Gen'ichi Taguchi[10] believes instead that there is some "loss to society" whenever a quality characteristic deviates from its ideal value. In one common model used by Taguchi, that loss is assumed to be proportional to the square of the deviation from some target value *T*. In Figure 12-4, *T* is taken as the specification midpoint, and the dollar loss *L* varies with the actual value *y* as

$$L = k(y - T)^2$$

where *k* is a cost coefficient.

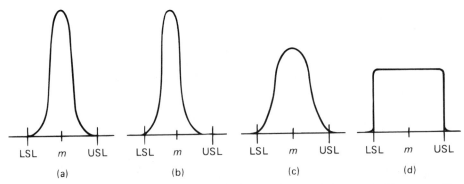

Figure 12-3. Four distributions of a quality characteristic in terms of the upper and lower specification limits (USL and LSL) and specification midpoint *m*.

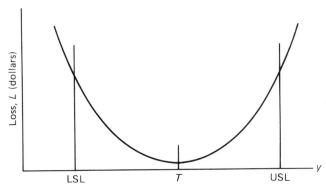

Figure 12-4. Illustration of the Taguchi loss factor.

Taguchi emphasizes the need for a continuous quality improvement program to reduce the variation of product performance characteristics about their target values. His methods include extensive experimentation in which product and process parameters are varied in a statistical matrix of tests. Results are then evaluated using ANOVA (analysis of variance) methods to identify the values that lead to least variation. These tests also show the parameters that cause most of the variation, leading to the most cost-effective design modifications and process improvements. Taguchi methods began to be employed in the U.S. automobile industry in late 1982, and they have since been widely adopted in that industry and are spreading to other U.S. industries.

Quality Circles

Production workers are the final determinants of quality, and their willing and informed involvement in the quality effort is essential. One approach to achieving this is through the institution of *quality circles.* In this technique, workers are gathered into groups of perhaps 10, which meet together, perhaps one hour a week, over an extended period. These quality circles (other names are used) are taught some basic methods of statistics and problem solving, and then they proceed to identify problems within their work area, develop alternatives, and formally propose a solution to management.

One example of the techniques a quality circle might be taught is *Pareto analysis,* named after an Italian economist. This analysis is based on the observation that, for many classes of items, most (perhaps 80%) of the total number of items will be from only a small set (perhaps 20%) of the possible types of items (Figure 12-5). For example, only 20% of juveniles might commit 80% of juvenile delinquencies, 20% of inventory items for airline operation might represent 80% of total investment in inventory, and in the present case, 80% of the total cost of failures might result from only 20% of the potential failure causes. By analyzing failure causes over a period of time, the quality circle or the quality engineer can identify the most important failure causes and concentrate improvement efforts on them.

The Japanese have more than 1 million such circles in operation at any one time, involving an average 10 people each, which have produced tens of millions of suggestions for

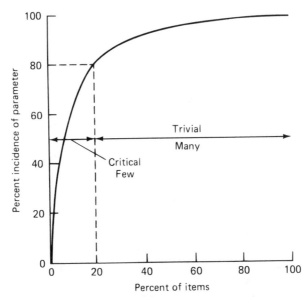

Figure 12-5. Pareto analysis.

improving products and production methods, and many American firms have also adopted this approach.

Deming's 14 Points

W. Edwards Deming, a statistician with the Bell System, was invited to Japan in 1950 to help their industrialists improve their reputation for poor-quality goods. He convinced them they could make their goods the highest quality in the world, and they proceeded to do so using his methods; they award the coveted Deming Prizes each year for greatest improvement of quality. Returning to the United States, he found American industry slower to respond, but when Japanese competition became a real threat to corporate survival his ideas gained increasing acceptance, and his name had become synonymous with quality by the time of his death in 1993. He is best known for the "14 points" developed in his 1982 book *Out of the Crisis,*[11] which are summarized here:

1. Create "constancy of purpose" that encourages everyone to cooperate in continually improving quality and meeting customer needs.
2. "Adopt the new philosophy" of defect prevention instead of the concept of "acceptable quality level" monitored by defect detection.
3. End dependence on mass inspection by building in quality to begin with.
4. End the practice of purchasing solely on price; develop long-term relationships with single suppliers based on product quality and trust.
5. Improve constantly and forever the system of production and service, to improve quality and productivity, and thus constantly decrease costs.
6. Institute modern methods of training in the organization's philosophy and goals as well as job performance.

7. Institute leadership, so that supervisors become coaches to help workers do a better job.
8. Eliminate fear, which impedes employee performance.
9. Break down barriers between departments.
10. Eliminate numerical goals and slogans—they don't work.
11. Eliminate work standards and numerical quotas, which emphasize quantity rather than quality.
12. Remove barriers that rob employees of their pride of workmanship.
13. Institute a vigorous program of education and training.
14. Structure the organization for quality.

Total Quality Management (TQM)

The quality revolution that has been sweeping industry for the last 15 years (and has begun making inroads in government and even higher education) has become known as *Total Quality Management (TQM)*. It incorporates the ideas of Deming, Taguchi, J. M. Juran, Armand Feigenbaum, and many others, and blends the techniques above with others (such as Quality Function Deployment,[12] which relates customer needs to product and process parameters using a sequence of diagrams called the "house of quality"). The exact implementation of TQM will vary with each industry and, indeed, with each organization pursuing it. Recognition for outstanding quality performance is provided in the United States by the coveted Malcom Baldrige National Quality Award, given each year to at most two organizations in each of three categories (manufacturing, service, and small business).

PRODUCTIVITY

Productivity Defined

Productivity can be defined as output produced per unit of resources applied. A simple measure might be units of production per labor hour or per labor dollar. For example, a primitive farmer in Africa may barely be able to feed his family by the sweat of his brow, while the large American family farm can feed many families; the American farmer is therefore considered much more productive. Among the reasons for this farm productivity is the application of resources other than human labor—much more land, much more equipment and fuel for it, the finest seed, and ample water and fertilizer. The American farm is therefore capital intensive, whereas the African farm is labor intensive.

Manufacturing Productivity—International Comparisons

For a quarter century after World War II, the United States was the industrial giant of the world; it enjoyed the highest productivity among the major industrial nations. As a result, American workers enjoyed a standard of living that was the envy of the world. As of 1960 our major international competitors had lower manufacturing productivity than the United States, but they achieved annual rates of productivity growth in the 1960–73 period much higher than that of the United States, and significantly higher for several countries through the 1980s (Table 12-2). Japan's progress was especially notable, with 10.5% annual im-

TABLE 12–2   Average Annual Percent Change in Manufacturing Output per Hour

Country	1960–1973	1979–1985	1985–1991	1991–1992
Canada	4.7	2.4	0.3	4.2
France	6.5	3.1	2.9	2.9
West Germany	5.7	2.1	1.8	0.5
Japan	10.5	3.9	4.6	−5.0
United Kingdom	4.3	4.4	4.0	4.9
United States	3.4	2.0	2.9	4.3

Source: U.S. Bureau of Labor Statistics.

provement 1960–73 and well above that of the United States through 1990. In the early 1990s, however, U.S. manufacturers have become more efficient while Europe and Japan were much slower to come out of economic recession yet were less willing than U.S. companies to reduce employment.

Table 12-3 illustrates the complexity of productivity comparisons for the United States and 10 industrial competitors (Japan, Canada, and eight European countries). Within each country, indexes are shown for 1970, 1980, 1985, 1990, and 1992, based on a 1982 basis of 100.0. Let's compare what happened in three countries—the United States, (West) Germany, and Japan—in the decade from 1982 to 1992.

Output per hour (part A of Table 12-3) increased just under 30% in the United States and Germany; hourly compensation (part B), however, increased 48% for the United States but 67% for Germany for the same period. As a result, unit labor costs in the national currency ($D = 100 \times B/A$) increased 14% for the United States, but 29% (in deutschemarks) for Germany. (Japanese labor costs increased 53% over the decade, more than in the United States, but productivity was up 43%, so that unit labor costs were only up 7% in yen.)

In international comparisons, however, exchange rates between currencies must be considered (part E of Table 12-3). To convert unit labor costs D_y in some year y to a common U.S. dollar basis F_y in that same year, we need to know both E_y and E_{1982}, the equivalent of a dollar in the national currency in year y and in 1982, the base year in which all labor cost indexes, whether in national currencies or dollars, are set at 100. We can then calculate an index for unit labor costs in U.S. dollars for any year y as $F_y = D_y \times E_{1982}/E_y$. For the Japanese unit labor cost in 1992, since the dollar would buy 249.1 yen in 1982 but only 126.8 in 1992, we can calculate:

$$F_{1992} = D_{1992} \times E_{1982}/E_{1992} = 106.9 \times 249.1/126.8 = 210.0$$

Thus, unit labor costs in U.S. dollars have increased 14% in the U.S. between 1982 and 1992, but, after adjustment for exchange rates, labor costs in our two biggest competitors, Japan and Germany, have more than doubled in the same period. There are a number of reasons for this. Real hourly compensation ($C = B$ divided by the consumer price index in each country) of manufacturing workers remained almost unchanged in the United States during the 1982–92 period, but increased 28% in Japan and 35% in Germany (the latter due to labor union strength). In 1991 U.S. manufacturing workers averaged 22.0 days off (vacation, holidays, and personal days) and worked 1904 hours; the German had 42.5 days off and, with a work week of less than 38 hours, worked only 1648 hours; comparable figures were 1755

TABLE 12-3 Selected Indexes of Manufacturing Activity—Selected Countries: 1970–1992 (1982 = 100)

Index	United States	Belgium	Canada	Denmark	France	Germany[1]	Italy	Japan	Netherlands	Sweden	United Kingdom
A. Output per hour:											
1970	(NA)	44.3	76.9	57.2	59.6	67.1	54.6	52.0	52.9	69.0	71.3
1980	92.9	87.5	99.9	98.0	90.5	98.5	95.5	92.1	93.9	96.3	90.3
1985	106.8	117.2	119.8	105.0	108.2	113.4	122.3	112.0	118.7	113.2	117.4
1990	122.2	133.9	120.6	107.5	127.2	125.5	141.1	144.5	130.1	124.9	146.3
1992	129.9	142.7	126.4	110.4	130.7	129.8	151.2	143.2	132.3	135.5	159.4
Average annual percent change:											
1979–85	2.0	6.1	2.4	2.1	3.1	2.1	5.0	3.9	4.2	3.0	4.4
1985–92	2.8	2.9	0.8	0.7	2.7	2.0	3.1	3.6	1.6	2.6	4.5
B. Compensation per hour national currency basis:[2]											
1970	(NA)	23.2	28.7	22.3	18.1	34.5	11.6	25.0	27.8	24.4	14.5
1980	83.3	86.3	78.6	83.4	72.7	89.2	70.2	89.0	88.5	84.5	77.9
1985	111.2	122.0	116.8	120.6	129.7	116.3	150.9	110.1	111.5	131.8	127.5
1990	134.7	147.8	146.9	155.1	159.3	148.0	213.3	138.3	123.3	197.5	187.5
1992	148.2	164.6	162.2	166.3	171.7	167.3	252.2	153.0	136.7	222.3	226.0
Average annual percent change:											
1979–85	6.9	7.8	8.7	8.1	12.8	5.9	16.7	4.7	4.8	9.6	12.1
1985–92	4.2	4.4	4.8	4.7	4.1	5.3	7.6	4.8	3.0	7.8	8.5
C. Real hourly compensation:[2 3]											
1970	(NA)	55.1	77.4	70.1	58.3	62.6	60.3	63.6	63.9	72.6	64.3
1980	97.5	101.0	97.9	102.6	92.2	99.9	96.9	96.0	100.0	103.0	94.9
1985	99.8	101.6	101.8	101.4	103.9	107.9	109.3	103.6	102.7	102.6	109.3
1990	99.4	110.9	102.9	107.5	109.8	128.6	117.3	121.7	109.2	114.4	120.6
1992	101.9	116.8	106.0	110.3	111.9	135.0	123.5	128.2	112.4	115.7	132.3
Average annual percent change:											
1979–85	0.1	0.8	0.7	-0.5	2.3	1.8	1.5	1.1	0.3	-0.5	2.9
1985–92	0.3	2.0	0.6	1.2	1.1	3.3	1.8	3.1	1.3	1.7	2.8

(cont'd)

TABLE 12–3 (cont'd)

Index	United States	Belgium	Canada	Denmark	France	Germany[1]	Italy	Japan	Netherlands	Sweden	United Kingdom
D. Unit labor costs, national currency:											
1970	(NA)	52.2	37.3	39.0	30.4	51.4	21.3	48.0	52.7	35.3	20.3
1980	89.6	98.7	78.7	85.1	80.3	90.6	73.5	96.7	94.2	87.7	86.3
1985	104.2	104.1	97.6	114.9	119.9	102.6	123.4	98.4	93.9	116.4	108.6
1990	110.3	110.4	121.8	144.2	125.3	117.9	151.2	95.7	94.7	158.1	128.2
1992	114.0	115.3	128.3	150.6	131.4	128.8	166.8	106.9	103.3	164.1	141.8
Average annual percent change:											
1979–85	4.9	1.6	6.1	5.9	9.4	3.8	11.1	0.7	0.6	6.4	7.4
1985–92	1.3	1.5	4.0	3.9	1.3	3.3	4.4	1.2	1.4	5.0	3.9
E. National currency per U.S. dollar											
1970	1.000	49.68	1.010	7.489	5.520	3.646	623.	357.6	3.597	5.170	0.417
1980	1.000	29.24	1.169	5.634	4.225	1.818	856.	226.6	1.989	4.231	0.430
1982 (basis year)	1.000	45.78	1.234	8.344	6.579	2.428	1354.	249.1	2.672	6.284	0.572
1985	1.000	59.34	1.366	10.600	8.980	2.942	1909.	238.5	3.318	8.603	0.771
1990	1.000	33.42	1.167	6.190	5.447	1.617	1198.	145.0	1.822	5.923	0.561
1992	1.000	32.15	1.208	6.037	5.294	1.562	1232.	126.8	1.759	5.826	0.566
1995 (May 1)	1.000	28.63	1.356	5.415	4.947	1.389	1665.	83.3	1.558	7.259	0.618
F. Unit labor costs, U.S. dollar basis:[4]											
1970	(NA)	48.2	44.1	43.4	36.2	34.2	46.0	33.4	38.9	42.8	27.9
1980	89.6	154.7	83.1	126.2	125.2	121.2	116.3	106.7	126.8	130.3	114.8
1985	104.2	80.3	88.2	90.4	87.8	84.6	87.5	102.7	75.6	85.0	80.6
1990	110.3	151.2	128.9	194.4	151.3	177.1	170.8	164.4	138.9	167.7	130.8
1992	114.0	164.2	131.1	208.2	163.3	200.3	183.3	210.0	157.0	176.9	143.3
Average annual percent change:											
1979–85	4.9	-9.7	3.4	-5.8	-3.4	-4.1	-3.3	-0.8	-7.5	-5.2	-1.0
1985–92	1.3	10.7	5.8	12.6	9.3	13.1	11.1	10.8	11.0	11.0	8.6

NA Not available; [1]Former West Germany; [2]Compensation includes, but real hourly compensation excludes, labor costs to employers that are not compensation to employees; [3]Compensation per hour divided by the index of consumer prices; [4]Indexes in national currency adjusted for changes in prevailing exchange rates.

Source: *Statistical Abstract of the United States 1994*, 114th ed., U.S. Bureau of the Census, Washington, DC, 1994, Tables 1390 and 1402, pp. 873 and 881.

hours for France (1990), 1780 for Italy, 1861 for Japan, and 1946 for Korea.[13] As a result, Germany and Japan are finding it more difficult to be competitive. By 1990 Japanese auto companies had already built 11 new assembly plants in the United States. Facing a 30% cost disadvantage against Japanese and U.S. competitors, Mercedes in 1993 sought a U.S. site to produce a luxury sport-utility vehicle, and they chose a coastal state (Alabama) because half of the production will be exported to other countries.[14]

The Manufacturing Studies Board summarizes their findings in a 1986 report *Toward a New Era in U.S. Manufacturing: The Need for a National Vision* with the conclusion:

> Manufacturing has already entered the early stages of revolutionary change caused by the convergence of three powerful forces:
>
> 1. The rapid spread of manufacturing capabilities worldwide has created intense competition on a global scale.
> 2. The emergence of advanced manufacturing technologies is dramatically changing both the products and processes of modern manufacturing.
> 3. There is growing evidence that changes in traditional management and labor practices, organizational structures, and decision-making criteria are needed to improve the effectiveness of manufacturing operations, provide new sources of competitiveness, and introduce new strategic opportunities.[15]

One of the greatest challenges to American industry in the last decade of the twentieth century is to recognize these forces and adapt to them. Today's American engineering student will be playing a major role in determining whether America can continue to meet this challenge as the next century begins.

WORK MEASUREMENT

by Dr. Bill Omurtag
Professor and Chairman
Engineering Management Department
University of Missouri–Rolla

Introduction

The question of what constitutes a fair day's work for a fair day's pay has occupied an important place in labor–management relations since time immemorial. This question was at the heart of Taylor's scientific management movement. Taylor correctly hypothesized that there must be one best way to do a task. He showed by experiment that a scientifically designed work method would significantly increase productivity of the worker, as measured in work output per unit time worked. This increase would then provide increased revenue and profits, which would allow higher pay to the productive workers, who could then afford a higher standard of living, thus fueling the engines of economic prosperity.

The basic idea behind the scientific management of work is the idea of specialization and systemization followed by the simplification and improvement that result from it. It is indeed the application of these principles in U.S. industry that largely accounts for the apparent superiority of its economy over those of other nations since the beginning of the twentieth century. This approach was particularly well suited for few products made in very large quantities for a mass market, as exemplified by Ford's Model T, which came in only one color (black) and was made using the assembly line.

Basic Time-Study Method

The basic steps in scientifically engineering manual work can be summarized as follows:

1. Clearly define the work cycle and establish a flow chart describing what is to be done at each step based on the actual work environment.
2. Select a typical employee who would normally do the work and establish rapport with the employee prior to observing the work and taking time measurements for each element in the work cycle.
3. Determine performance rating and additional allowance factors to adjust the directly observed times, and develop a work standard for each job.

This process would be applicable in measuring an existing job when a method change is being contemplated. A scientific study would provide a basis for determining the possible impact of the change on productivity, and would provide other valuable information for the decision makers involved. If the procedure described above is used in a formal way to lead to a negotiated agreement between labor and management, it produces a formal job standard for an already existing job using the time-study method.

If a new job or task is being contemplated, it is first broken down into a sequence of elemental motions. Standard times for each elemental motion, developed after extensive study, are then added to establish time standards for a job. The *methods-time measurement (MTM)* and similar predetermined elemental time systems can provide accurate time standards for many tasks without the disadvantages of performance rating encountered in time study. Since the elemental task times have been determined to represent a 100% work pace, only the additional work allowances need to be included in the calculation of a labor standard for a given job.

Work Sampling

In some cases the nature of work is not as easy to specialize, systematize, and simplify as is assembly-line work. In these cases an approach called *work sampling* may provide the best way to establish work performance standards. In work sampling, an operation is observed over a long period of time (from several hours to several weeks) using a statistically sound sampling procedure, to determine productive and idle work patterns and their relative frequencies. From this information plus the measured output of the observed operation, a performance standard is determined. The advantage of work sampling is that it can be used in less structured work environments, such as service and professional work, where time study may be ineffective. One of its disadvantages is that it cannot be used to estimate a job that does not already exist. Another major drawback is that the observer still has the burden of determining the performance rating of the observed employee to adjust the time standard.

Other Methods

A relatively new approach is known as *work physiology*. The main idea behind this approach is that the work done by an individual may somehow be accounted for by physiological factors such as oxygen consumption and other factors of metabolism. This approach does not appear to have many applications in industry at the present time; however, considerable research is being conducted on this as well as other human factors in many universities at the present.

The methods of measuring work described above perform well in measuring manual work, whose output can be counted or measured in like units consistently over time. This is what is commonly known as the direct labor or touch labor component of the cost of manufacturing a product. In modern-day manufacturing this cost component is around 5 to 20% of the total cost of a product, depending on the technology being employed by a given manufacturer. This is clearly a small amount compared with the indirect labor and other knowledge workers' cost contribution. How do we measure the indirect labor costs? Time study and work sampling or predetermined elemental time systems are not easily applied to measuring these types of costs. The fact is that only limited success has been achieved in measuring indirect or white-collar productivity as rigorously as the touch labor in industry. Future industrial engineers and labor leaders will have to find better methods of measuring white-collar work and future managers of people will have to have much better understanding of the ramifications of the question of what constitutes a fair day's work for a fair day's pay in increasingly service- and knowledge-based work environments.

Despite the recent deemphasis on work measurement in industry as a basis for compensation standards, the same tools are still used very effectively in determining more accurate cost estimates and production plans. The Japanese "miracles" in rapid die changing without interrupting production, and the quick setup and maintenance of expensive and versatile production machinery, could not have been accomplished without a careful application of the work design and measurement principles promulgated by Taylor and his followers. So it appears that the principles of scientific management are still at work for us if we only reexamine where they may be best suited and not insist on their use as labor reduction tools.

MAINTENANCE AND FACILITIES (PLANT) ENGINEERING

A wide variety of functions and activities required for the effective functioning of a manufacturing plant, but not concerned directly with production of the product, fall under an organization headed by the plant or facilities engineer or the maintenance manager or superintendent. In the first two subsections following, we discuss several types of maintenance and some aspects of maintenance management; in the third subsection, we outline briefly some of the other activities that fall under the umbrella of facilities or plant engineering.

Scope of Maintenance

The *Maintenance Engineering Handbook*[16] identifies the following primary functions of the maintenance (engineering) activity:

1. Maintenance of existing plant equipment
2. Maintenance of existing plant buildings and grounds
3. Equipment inspection and lubrication
4. Utilities generation and distribution
5. Alterations to existing equipment and buildings
6. New installations of equipment and buildings

Some of these functions, such as major alterations or additions to buildings and equipment, occur so irregularly that it is not economical to staff for such activity, and these are contracted out; even so, a plant engineer is normally appointed as project engineer to monitor progress of contractor activity to assure that changes will meet the needs of the plant. Maintenance of some items (elevators, computers, office equipment, rewinding of burned-out motors) is so specialized that it is normally contracted out. Some custodial activities, such as washing windows, care of grounds, and office janitorial service, may be contracted out if it is found to be more cost-effective. Contract services of these types may provide better methods and supervision of these ancillary activities than the plant affords, and they often have lower labor costs than those in the plant.

To give an idea of the variety of maintenance concerns in a plant of any size, I list the following topics, each of which is the subject of a separate chapter in the *Facilities and Plant Engineering Handbook:*[17]

- Roofing
- Flooring
- Refrigeration
- Air conditioning, heating, and ventilation
- Special-purpose rooms and their environment
- Electric-circuit protection
- Utilities
- Transportation equipment
- Materials handling systems
- Elevators
- Painting
- Corrosion protection
- Applied biology (insect, animal, and other pest control)
- Lubrication of machine tools

Types of Maintenance

The mainstream activities of maintaining plant equipment can be divided into corrective, preventive, and predictive maintenance; each is considered below. *Corrective maintenance* is simply repair work, made necessary when something breaks down or is found to be out of order. This is the activity that most of us think of when maintenance is mentioned. When equipment breaks down, especially machinery on which an integrated production line depends, the costs of lost production mount and pressure is on the maintenance team to get the equipment fixed and back into operation. Effective maintenance engineering requires thinking through the most likely types of breakdowns, assuring an adequate inventory of the most commonly needed or critical replacement parts, and providing spare capacity where breakdowns cannot reasonably be avoided.

Many mechanical systems *wear out:* their failure rates increase with time and the quality of performance falls off because bearings become loose, gears wear, O-rings and belts deteriorate, and grease hardens. These types of problems are reduced by periodic inspection, lubrication, and identification and replacement of worn parts. Efficient *preventive maintenance* requires documentation of all equipment to be included in the program and establishment of the most cost-effective schedule for inspection. Inspection checklists need to be established for each type of equipment, and inspectors trained to make simple repairs when problems are observed. Computers are useful to print out lists of inspection tasks that are due and maintain data on the time and material costs of inspection to support periodic analysis and revision of the preventive maintenance plan. To some extent preventive maintenance can be deferred or "scheduled around" more urgent corrective maintenance, but deferring it too long invites breakdowns and higher costs.

Predictive maintenance "is a preventive type of maintenance that involves the use of sensitive instruments (e.g., vibration analyzers, amplitude meters, audio gages, optical tooling, and pressure, temperature, and resistance gages) to predict trouble. . . . Critical equipment conditions can be measured periodically or on a continuous basis. This approach enables maintenance personnel to establish the imminence of need for overhaul."[18] Where diagnostic systems are built into equipment, production workers can observe warning signs during operation, catching incipient failures long before maintenance workers would see them.

Some Maintenance Management Considerations

Size of maintenance staff. Production supervisors naturally would like maintenance specialists of all types available immediately when a breakdown occurs, since the cost of idle maintenance time does not come out of their budget. When a plant is in full production and profits are high, it is easy to build the maintenance staff to a "comfortable" level, but when demand slows and costs are being trimmed, shortsighted managers will find maintenance an easy target for drastic cuts. Good management balances the cost of additional maintenance personnel against the probable costs of production loss and equipment damage if adequate maintenance is not maintained.

Work orders. To maintain control over maintenance costs, work is not ordinarily performed without a supporting work order signed by a foreman or supervisor. The work order states the problem and estimated the cost of repair, and provides space for workmen to document the time they spent on the problem and any materials or parts they used in solving it. Completed work orders provide data to analyze maintenance costs of each type of equipment, so that cost-saving decisions such as redesign or replacement can be made.

Work scheduling. In larger plants, maintenance scheduling is the responsibility of a separate unit of the maintenance organization; in smaller plants, this is done directly by the foreman. Schedules are only estimates, and may have to be changed if a breakdown emergency takes place.

Repair parts inventory. As industry becomes more automated, it has more complex operating equipment with more parts that can fail and require replacement. Like any other inventory, this can tie up large sums of money that might be put to more productive use, and good judgment and periodic review are required. Where equipment vendors will provide prompt repair service at an acceptable price, this eliminates the need for parts inventories as well as for specialized training, so that after-sale service is a real consideration in purchasing equipment.

Total Productive Maintenance (TPM)[19]

Total Productive Maintenance (TPM), a concept originated by the Japanese, is an integrated, top-down, system-oriented, life-cycle approach to maintenance, with the objective of maximizing productivity. Directed primarily to the commercial manufacturing environment, TPM:

1. Promotes the overall effectiveness and efficiency of equipment in the factory.
2. Establishes a complete preventive maintenance program for factory equipment based on life-cycle criteria.
3. Is implemented on a "team" basis involving various departments to include engineering, production operations, and maintenance.
4. Involves every employee in the company, from the top management to the workers on the shop floor. Even equipment operators are responsible for maintenance of the equipment they operate.
5. Is based on the promotion of preventive maintenance through "motivational management" (the establishment of autonomous small-group activities for the maintenance and support of equipment).

The objective of TPM is to eliminate equipment breakdowns, speed losses, minor stoppages, and so on. It promotes defect-free production, just-in-time (JIT) production, and automation. TPM includes continuous improvement in maintenance.

Other Facilities and Plant Engineering Functions

Some of the other activities that are often included in the responsibilities of the plant engineer, facilities engineer, or maintenance superintendent (often for lack of a better place to locate them) are:

1. Plant security (guards, fences, locks, theft control, emergency planning)
2. Fire protection (fuel and chemical storage, fire detection and extinguishment, loss prevention, and risk management)
3. Insurance administration
4. Salvage and waste disposal
5. Janitorial service
6. Pollution and noise abatement
7. Property accounting

OTHER MANUFACTURING FUNCTIONS

Human Resources (Personnel) Management

The many concerns for and about employees are centered in the personnel or industrial relations or (more recently) human resources department. A typical personnel department in a single-plant company employing several thousand persons might include the following sections:

1. Recruiting and employment (human resource planning, recruiting, interviewing, testing, transfers, and layoffs)
2. Equal employment opportunity (affirmative action, minority records and reports, complaint investigation)
3. Industrial relations (contract negotiations, contract administration, grievances, and arbitration)
4. Compensation (job analysis and evaluation, wage surveys, incentives and performance standards, managerial and professional compensation)
5. Education and training (orientation, skills training, management training, career planning, tuition assistance, organizational development)
6. Health and safety (industrial hygiene, safety engineering, first aid and medical, workers' compensation)
7. Employee benefits (insurance, pensions, profit sharing, food service, dependent day care, social programs)

Of special interest to the engineer is the safety engineering activity (under the health and safety section). As an example of the hazards of concern to the safety engineer and industrial hygienist, consider the following chapter titles from Hammer's *Occupational Safety Management and Engineering:*[20]

15. Acceleration, falls, falling objects, and other impacts
16. Mechanical injuries
17. Heat and temperature
18. Pressure vessels
19. Electrical hazards
20. Fires and fire suppression
21. Explosions and explosives
22. Hazards of toxic materials
23. Radiation
24. Vibration and noise

Safety personnel are involved in (1) identifying and analyzing hazards, (2) recommending protective devices and warnings, (3) providing safety training, (4) interpreting the Occupational Safety and Health Act (OSHA) and other codes and standards to management and other personnel, and (5) in workers' compensation insurance activity. In some of these areas, safety personnel share functions with fire prevention and other security personnel. They are also closely involved with plant insurance activities, since future plant fire, workers' compensation, medical, and liability insurance premiums will depend on the success of occupational safety and health programs.

Purchasing and Materials Management

Importance of purchasing. Purchasing is a vital contributor to producing a quality product at a profit. Half the value of the typical industrial product consists of materials and components purchased from other organizations. If a firm is making a 10% profit on its product, a dollar saved in more efficient purchasing has the same effect on profit as 10 dollars in added sales. Moreover, a quality defect in a supplier's component incorporated into the product has the same impact on your reputation as a mistake made internally. Zenz lists the following steps as being performed by purchasing in the large majority of organizations:

1. Recognition of need
2. Description of requirement
3. Selection of possible sources of supply
4. Determination of price and availability
5. Placement of the order
6. Follow-up and expediting of the order
7. Verification of the invoice
8. Processing of discrepancies and rejections
9. Closing of completed orders
10. Maintenance of records and files[21]

The engineer in purchasing. A survey of 12 purchasing manager associations showed about half of the respondents were college graduates; of these, 58% majored in business administration and 17% in engineering.[22] Certainly, an engineering education is of great value in the purchasing of highly technical components. Interestingly, one of the early articles written on purchasing was titled "The Engineer as a Purchasing Agent," and appeared in a 1908 engineering publication.[23]

Materials management. This is a more comprehensive organizational viewpoint in which all activities involved in bringing materials into and through the plant are combined under a materials manager. These activities commonly would include purchasing, inventory control, traffic and transportation, and receiving; they may include warehousing/stores and even production control. Purchasing is often responsible for make-or-buy analysis, value engineering/analysis (see Chapter 10), incoming inspection, and reclamation and salvage.

Even where these activities do not fall under a single manager, they must be performed in concert with each other. Engineering and purchasing personnel in particular must work closely together. Design engineers must be careful not to make specifications for purchased materials and components so restrictive that suppliers with less expensive but satisfactory products are ruled out; purchasing, on the other hand, must not make its decisions solely on price when a slightly more expensive choice may bring quality, reliability, delivery, or customer acceptance worth much more than the price differential.

Packaging Engineering

by Dr. Henry Sineath
PMMI Professor Emeritus of Engineering Management
University of Missouri–Rolla

Packaging is a $62 billion a year industry in the United States alone. It is a highly fragmented field involving materials, machines, and systems. Ninety-five percent of all parts manufactured in the United States are packaged. The output of one transformation process becomes the input to another and continues in this way until the packaging of the product for the final consumer. The challenge to the packaging engineer is to marry the product and the package, to transfer this combination throughout the distribution system of that product, and to deliver the packaged product at its destination at the same quality level that it left the manufacturing line.

Packaging engineering embraces a system of material and form, specification, machinery, and methods of unitizing secondary and tertiary packaging and a delivery system for the product. This system can be as simple as hand packaging or as complex as a completely computer-integrated system without human involvement. Throughout the system the package performs the functions of containment, protection, performance, and communication, and in many cases the package performance really involves the dispensing of the product from the package. Packaging engineering then proceeds beyond the manufacturing operation and is involved as a marketing tool; it provides excellent opportunities for cost reduction and product improvement throughout the distribution channel.

Successful introduction of modified or new products necessitates involvement of packaging engineering in the concept and design stages of the manufacturing operation. This permits the consideration of packaging in the manufacturing function before the product is ready for manufacturing. Consequently, the process engineering function requires a multidisciplinary approach to problem solving with particular need for highly developed coordination skills.

In summary, no matter how well one makes the product, one must deliver it to the user in a satisfying fashion. This requires that the product be protected throughout its distribution cycle to the point of use. It has been said that nothing happens until someone sells something. Corollary to this is that you probably will not sell it again unless it is packaged in a way that delivers the product in a satisfying fashion or condition for the final user.

DISCUSSION QUESTIONS

12-1. Distinguish between quality of design and quality of conformance.

12-2. Research and report on an American company or industry that (**a**) has had declining sales as a result of noncompetitive product quality, and/or (**b**) has grown stronger through a deliberate effort to improve quality.

12-3. The Ajax Corporation is proud of its quality control program. It employs two quality engineers at an average $35,000 salary for quality planning and engineering, test method development, and quality training. Every item is 100% inspected using six inspectors earning an average of

$20,000 a year. Nevertheless, in a typical year about 100 items averaging $200 in value must be scrapped, and another 200 require rework costing an average $75. Customer service costs estimated at $11,000 annually are generated by still more defects that slip through. Analyze the firm's quality cost situation and make recommendations.

12-4. (**a**) In your School of Engineering, determine which engineering curricula require a course in statistics, and estimate the proportion of engineers that graduate literate in statistics *or* (**b**) survey engineers and other employees in your company to estimate how many were trained in statistics in school or on the job.

12-5. Distinguish between the statistics of *attributes* and the statistics of *variables,* and comment on how they are applied in process control charts and in sampling.

12-6. Tomatoes are packaged in a can designed to hold a nominal 28 ounces of product. Five cans sampled randomly were found to contain 28.3, 27.3, 29.1, 28.5, and 27.8 ounces of product. (**a**) Calculate the mean value and range of sample data, and (**b**) estimate the variance and standard deviation of the sampled population.

12-7. An "*x*-bar" control chart is developed for recording the mean value of a quality characteristic using a sample size of three. The control chart has control limits (LCL and UCL) of 1.000 and 1.020 pounds respectively. If a new sample of three items has weights of 1.023, 0.999, and 1.025, what can we say about the lot (batch) it came from?

12-8. The involvement of production workers in quality circles seems to conflict with the concept of Frederick W. Taylor, founder of scientific management, that managers should define how work is to be done and workers simply perform as they are instructed and trained. Comment.

12-9. State the essence of Taguchi's teaching.

12-10. Examine Deming's 14 points and select three or four you consider most effective in assuring product quality. Explain your reasoning.

12-11. Choose an organization that claims to be using Total Quality Management. Discuss how they have implemented it and assess its effectiveness.

12-12. For most of the middle of this century U.S. factories were the most productive in the world. What has happened to change this?

12-13. Distinguish between the methods used in (**a**) time study of existing tasks, (**b**) time study of proposed new tasks, and (**c**) work sampling.

12-14. Comment on the application of the work measurement methods cited in question 12-13 to knowledge workers such as engineers.

12-15. Identify three types of maintenance, and distinguish between them.

12-16. Discuss how Total Productive Maintenance relates to other concepts developed in this chapter.

12-17. Where might engineering knowledge and skills be valuable in each of the following functions: (**a**) human resources management, (**b**) purchasing, and (**c**) other materials management activities?

12-18. Summarize the importance of packaging engineering to the typical manufacturing organization.

NOTES

1. J. M. Juran and Frank M. Gryna, *Quality Planning and Analysis,* 3rd ed. (New York: McGraw-Hill Book Company, 1993), p. 3.

2. W. Ernst Eder, "Total Quality Management—Defining Customers and Quality," *American Society for Engineering Education 1993 Annual Conference Proceedings,* p. 1388.

3. J. M. Juran, "Japanese and Western Quality: A Contrast," *Quality Progress,* December 1978, pp. 10–18.

4. "Up Front: Who Makes the Best Products?" *Business Week,* March 14, 1994, p. 8.

5. *Quality Costs—What and How* (Milwaukee, WI: American Society for Quality Control, 1971).

6. Donald N. Merino, "Optimizing the Cost of Quality Using Quality Economic Models," *American Society for Engineering Education 1991 Annual Conference Proceedings,* p. 94.

7. Gordon H. Geiger, "Reinventing Grinter's Wheel," unpublished presidential address to the Accrediting Board for Engineering and Technology annual meeting, November 12, 1987.

8. Lester Thurow, unpublished remarks on *60 Minutes* television program, February 7, 1988.

9. U.S. Dept. of Defense, *MIL-STD-105E, Military Standard: Sampling Procedures and Tables for Inspection by Attributes,* May 10, 1989.

10. See, for example, Gen'ichi Taguchi, Elsayed A. Elsayed, and Thomas C. Hsiang, *Quality Engineering in Production Systems* (New York: McGraw-Hill Book Company, 1989). For a very readable commentary, see Raghu N. Kackar, "Taguchi's Quality Philosophy: Analysis and Commentary," *Quality Progress,* December 1986, pp. 21–29.

11. W. Edwards Deming, *Out of the Crisis* (Cambridge, MA: The MIT Press, 1982).

12. See, for example, Yoji Akao, *Quality Function Deployment* (Cambridge, MA: Productivity Press, 1990).

13. "Maybe We All Deserve Raises," *Business Week,* August 2, 1993, p. 34.

14. "Why Mercedes Is Alabama Bound," *Business Week,* October 11, 1993, pp. 138–139.

15. National Research Council, Manufacturing Studies Board, *Toward a New Era in U.S. Manufacturing: The Need for a National Vision* (Washington, DC: National Academy Press, 1986), p. 71.

16. L. C. Morrow, ed., *Maintenance Engineering Handbook* (New York: McGraw-Hill Book Company, 1957), pp. 1–4.

17. Bernard T. Lewis and J. P. Marron, eds., *Facilities and Plant Engineering Handbook* (New York: McGraw-Hill Book Company, 1974).

18. Harold T. Amrine, John A. Ritchey, Colin L. Moodie, and Joseph F. Kmec, *Manufacturing Organization and Management,* 6th ed. (Englewood Cliffs, NJ: Prentice-Hall, Inc., 1993), p. 170.

19. Section taken from Benjamin S. Blanchard, Dinesh Verma, and Elmer L. Peterson, *Maintainability: A Key to Effective Serviceability and Maintenance Management* (New York: John Wiley & Sons, Inc., 1995), p. 17. For more information, see Seiichi Nakajima, *Total Productive Maintenance (TPM) Development Program* (Cambridge, MA: Productivity Press, Inc., 1989).

20. Willie Hammer, *Occupational Safety Management and Engineering,* 3rd ed. (Englewood Cliffs, NJ: Prentice-Hall, Inc., 1985).

21. Gary J. Zenz, *Purchasing and the Management of Materials,* 5th ed. (New York: John Wiley & Sons, Inc., 1981), p. 100.

22. Zenz, *Purchasing,* p. 89.

23. James M. Cremer, "The Engineer as a Purchasing Agent," *Cassier's Magazine,* August 1908, pp. 322–332.

CHAPTER 13

Engineers in Marketing and Service Activities

PREVIEW

Production does not end the engineer's involvement in the product life cycle. The more technical the product, the more that engineers are involved in marketing the product and supporting its use in the field. In this chapter, we analyze the nature of engineering involvement in the marketing of each of eight types of industrial products, and discuss the engineering involvement in after-sales service.

Many engineers work in industries producing a service rather than a physical product. We discuss the importance of the service-producing industries nationally, the ways they differ from manufacturing industries, the significance of engineers in their operation, and target a few specific industries (computer applications, public works, college teaching and research, and health services) for special consideration.

MARKETING AND THE ENGINEER

By Dr. Henry Metzner
Associate Professor of Engineering Management
University of Missouri–Rolla

Types of Marketing Relationships

Marketing is the interface between the firm and its customers. At its simplest, marketing is selling, the exchange of money for the output produced by the firm. In this sense, marketing is vital to any firm since it is the activity that produces revenues which sustain the enterprise. Most firms, although by no means all, require more than selling from the marketing activity.

Transaction-based relationships. Marketing is generally assigned the market-related tasks prior to sale, such as identifying the customer, studying the customer's needs, obtaining an opportunity to make an offer to the potential customer, and closing a deal. These types of activities are more common in transaction-oriented types of firms, where each purchase generally signals the end to the relationship between buyer and seller, at least in the short run. Attracting a customer to a shoe store and selling a pair of shoes is a common example. The individual customer may return for repeat business, but it is just as likely, or more likely, that the customer will turn to another store the next time that shoes are desired. An industrial example is that of the motor carrier. Each shipment is typically a completed transaction and does little to assure either party of continued custom. Although returned shoes and lost or damaged shipments lead to repeat contacts between the parties, these are the exception rather than the rule. These relationships are often characterized by cash transactions.

More sustained relationships. Where the contact between the parties to a transaction is more sustained, as in delivery of components over time from a single contract, or several concurrent contracts, the tasks of marketing may be more encompassing, including informing the customer of the status of the order, determining the disposition of damaged or below-standard deliveries, determining the user satisfaction with deliveries, and so on.

Highly committed relationships. Often the relationship between the customer and the firm is more involved than simply the transfer of standard goods or services. Where there is substantial and rapidly advancing technology as in the case of computers, numerically controlled machining systems, or medical diagnostic equipment, the customer's choice to do business with a particular supplier involves a long-term commitment to the technical idiosyncrasies of the supplier.

For example, computer operations are controlled by an operating system that is often unique to the specific producer of that computer. If the customer is to have access to new applications, they must be supplied in a format compatible with the operating system and with the physical hardware of the computer. The commitment between parties is long-term and substantial. The supplier must have an intimate understanding of the customer's environment, requirements, and planning in order to place priorities on its own product development.

Thus marketing and marketing functions are, in part, defined in the context of the length of relationship and commitment between the buyer and seller. For transaction-based relationships, marketing is simple; for long-term, highly committed relationships, marketing is complex and many-faceted.

Engineering Involvement in Marketing Industrial Products

The product or service produced also determines the complexity of the marketing function. The vast majority of products bought by the household are characterized by mass production to lower costs, by consumption as the major purpose of purchase, by uniform or near-uniform requirements (which implies high substitutability), and by purchases of small amounts. In contrast, many products in the industrial arena are unique, such as special production machines or tools. Many industrial products are bought to be resold either as a component of some larger product, or as an item from an assortment, as in the case of repair parts sold by wholesalers. Some items are incorporated in the production of a more complex product, such as sheet steel used for appliance cabinets or rivets used for fastening metal pieces together. These items are ordered in large quantities and delivered in a stream to match the production process, rather than in batches.

In all, there are eight types of industrial products: installations; accessories; raw materials; process materials; component parts; fabricated items; maintenance, repair, and operating items; and services. Of these, the first two (installations and accessories) are capital investments, depreciated (written off the books) over a period of three or more years. The next five are consumed or incorporated in the process of producing a product, and the last (services) is in a class of its own. The role of the engineer is different with each type, as discussed below and summarized in Table 13-1. *Installations* are large, custom constructions such as buildings, punch presses, ovens and kilns, automated assembly lines, and the like. They are expensive, often single purpose, designed for optimum performance, and typically complex. Installations are capital items and are typically depreciated (written off) over 5, 7, or 10 years for custom-designed production equipment and over $31\frac{1}{2}$ years for nonresidential buildings (U.S. Tax Reform Act of 1986).

Engineers are important to the selling process for installations. More often than not, installations are custom designed for the particular and special situation of the buyer. Therefore, engineers are drawn into the process early to suggest solutions for the buyer's situa-

TABLE 13–1   Engineering Involvement in Marketing Industrial Products

Type of Product	Description	Engineering Involvement
Installations	Large, durable custom constructions	Selling and performance of design service; cost estimation and construction supervision
Accessories	Shorter-lived capital goods (equipment)	Seller's engineers design for general customer
Raw materials	Extractive and agricultural products	Assessment of quality
Process materials	Goods that change form in production	Buyer's engineers establish specifications
Component parts	Catalog items that do not lose identity in production	Supplier's engineers design for general customer and introduce to user's engineers
Fabricated items	Custom-made items	Buyer's engineers design and specify; seller bids on manufacture
Maintenance, repair, and operating items	Consumed in process of production or use	Repair parts and methods specified by maker's engineers; users have little engineering involvement
Services	Involve only incidental product use	For engineering services, engineers sell as well as perform

tion and to price them. Often, this process is iterative, leading to changes in the initial demands of the buyer as practical solutions become more clearly defined. Sales of installations are often subject to a bidding process where several competing firms submit engineered solutions in competition. Price is not the only criterion.

For certain kinds of installations, in particular buildings, facilities such as airports and sanitation stations, and turnkey factories such as oil refineries, the design and construction may be separately bid and let. The design of such facilities is a service, and the sales of these services is particularly difficult since it is an act of faith on the part of the buyer. A long history of successful ventures and satisfied customers is the best indication of the designer's ability and, failing that, creative, imaginative client presentations designed to relieve risk perceptions are the most used sales tools. Personal friendships and political favoritism are also common means of reducing the risk in the eyes of the buyer.

The construction bid is typically price driven, since the end product will be the same irrespective of the contractor. The engineer plays a key role in the firm's survival. Initially, it is necessary to determine the price to be bid. The cost of materials and some subcontracting (such as cement work or electrical wiring), which are more or less the same for all competitors, form the base for bid. The labor and capital costs for deployment of capital equipment such as trucks, loaders, cranes, and the like are added to the base to produce the bid price. If the bid is successful but price fails to cover the actual expenses of the work, the contractor loses money. If the price bid is higher than that of competitors, the bid is lost. Accurate estimation is essential, and estimating ability is an invaluable engineering skill.

Accessories are capital items of lesser durability, mostly depreciable in five or seven years. These items include standard office machinery such as copiers and typewriters, small computers, desks and other furniture, and similar items. Also in this category are such items as automobiles, forklifts, electronic test equipment, and standard machine tools. These items are standard, modified by the selection of appropriate standard options, or customized by minor procedures such as painting or adding the customer's logo. Accessories are characterized by near substitutes and price competition, by standardization, by distribution through dealers, and by broad application in several industries.

Except in extreme cases, engineers are not involved in the marketing of accessories. Large-quantity orders—200 jet engine starters, for example—where special price concessions may be called for or a special production run is economical, may call for involvement of engineering. In some cases, engineers may become involved in the selection of appropriate accessories where the criteria for selection are obscure.

Raw materials are the crude products of extractive and agricultural enterprises. Ore, tobacco leaf, grain, coffee beans, tea leaves, cattle, fish catches, and quarry marble are raw materials. The major problem in raw materials is the complication that quality is not controllable and must be assessed, usually by the buyer, but often by an agency such as the Department of Agriculture. The quality of animal furs is dependent on the severity of the winter and availability of food; the quality of wine-making grapes is a function of weather. Price is a function of quality, demand, and supply, and it is usually set by auction. These products are characterized by limited, very specific and often colorful marketing institutions such as tobacco auctions.

Marketing is dominated by the institutions that are traditional or practical in the raw materials industries. Thus a farmer who wishes to "market" his cattle takes them to auction barns, where they are sold by bid. The miners of uranium ore truck their mine's output to crushers, where each truckload is assayed and paid according to the published daily price of uranium. Marketing is simple in concept; deliver and accept the day's auction price. Engineers may have a part in the assessment of raw materials or the determination of the value of a particular lot of used machinery, which is marketed in the same fashion.

Process materials are basic manufactured goods that change form after they reach the industrial customer. Thus metal bars or rods are machined into screws or drawn into wire, copper ingots are cast into electrodes for plating, plastics are molded, and so on. There may be many or few transfers between extraction and consumption, but most transactions occur in the processing of materials.

Process materials are used as feedstock in production processes and are highly interdependent with those processes. Therefore, it is very important that process materials meet specifications at each transfer. Reliable delivery is also a desirable attribute since it reduces the safety stock required by the customer to ensure continuous production. Price is clearly important since the volume of process materials fed into a factory can produce major savings on differences of a fraction of a cent between suppliers. Because of the specifications that determine the characteristics of process materials, they are usually considered commodities.

Because the specifications of the material to be procured for the production process are determined by the buyer, it is the buyer's engineers who are most active. Nonetheless, the seller's engineers are often consulted during the specification phase, particularly for substitutions or new materials that may make the entire process more cost-effective. Where the desired commodity is out of stock when ordered, the seller's engineers may also be active in determining appropriate substitutes given the specific circumstances of the buyer.

Component parts are catalog items, designed and produced by a supplier and offered for sale to a broad spectrum of customers, either directly or indirectly through distributors. Components do not lose their identity when incorporated into larger products. Examples of automotive components are alternators, carburetors, cotter pins, or leaf springs. Component items may be very simple or very complex.

Component parts are sold through sales forces, distributors, and agents. Since component parts are meant to be incorporated in the assembly process of original-equipment manufacturers (OEMs), reliability of delivery is an important aspect since it leads to reduced inventories. High quality can enhance the marketability of the final product and is therefore desirable. Price is important when there are several identical or highly substitutable items.

Engineers are used primarily to introduce new components. Components, being part of a larger assembly, must be designed into that assembly by the buyer. The seller's engineers aid the process of design by teaching designers how to use the new components, how they differ from current components, and why they should be incorporated. Particularly complex new devices can benefit from highly intense technical information transfer at introduction. Examples are new computer-on-a-chip devices, lightweight electric motors, and new sensors. Once established in the design, further contact between buyer and seller is related to marketing tasks such as price, delivery dates, payment schedules, and other nontechnical considerations.

Fabricated items are custom-made items. Also discrete items used in the assembly of larger products, fabricated items cannot be distinguished from components by inspection of the physical piece. The difference between fabricated items and components is the ownership of the product specification or design. It is the buyer who is responsible for the design of the fabricated item. Once specified, the buyer seeks out a manufacturer with the capability to produce the fabricated item to specification within the desired time frame. The buyer, in addition to actively replacing the marketing effort with a procurement effort, has assumed the risk inherent in the design itself, the risk of setting up the manufacture of the product, the risk of marketing the component, and the costs of design and tooling. The manufacturer of fabricated items is left with little marketing to do.

The maker of fabricated items needs to respond to requests for quotation on items placed for bid by prospective buyers, must prove capability and capacity, and must fulfill the contract. For the most part, such marketing is entirely passive. As a consequence, such firms concentrate on accurate bidding and efficient manufacture, and do little marketing. If any marketing is done, it is to make the capabilities and capacities of the firm known to the class of customers that make up its market. Typical of these firms are machine shops, plastics molders, and foundries.

Maintenance, repair, and operating items are consumed in the process of production. Operating items are supplies such as lubricants, cleaners, fuels, and the like. Repair items are parts required to return broken production machinery to operation. Maintenance items are those used in maintaining machinery, such as air filters, spark plugs, and the like, which wear and must be replaced periodically and predictably. Maintenance items and operating supplies, which include office stationery and supplies, are typically marketed through indirect channels because they tend to be purchased frequently in relatively small amounts and have relatively common characteristics, many competing sources, and competitive prices.

Repair parts are required to restore broken equipment to usability. Although sometimes identical to maintenance items, as in the case of a fouled spark plug, repair parts are typically components that fail rarely and randomly, making the storage of such items uneconomical for the using firm. Consequently, spares provisioning is done by the supplying firm as a service. This consolidates the spares requirements of several users and provides spare parts along with repair services for those firms that prefer to leave such infrequent maintenance to those who specialize in it. For critical equipment, spares and maintenance may be provided economically by the user, particularly where the lack of such equipment may cause expensive delays, where there is a large quantity of equivalent equipment, as in truck fleets, or where the level of required maintenance is so high that repair is economical, as in railroad rolling stock. Engineers are generally involved when substitutions are required for items that are not immediately available, or when spares packages and maintenance accessories such as test equipment are being specified for a particular customer.

Services are marketing offerings that are characterized by only incidental use of product, for example, the use of cleaners by a janitorial service, and by an inability to inventory. The range of services are as wide as the range of products, spanning the spectrum from utilities such as telephone and electricity to merger and acquisition advice. As is also true of product production, learning through repetition and specialization are critical to efficient and economical provision of services. Thus, hiring specialists in air-conditioning design or tax preparation can be more efficient than providing the same services internally. Corporate "downsizing" of the early 1990s has led to elimination of many staff specialists (as well as middle managers); their functions are provided in part by contract with specialized service firms. Increasingly the successful modern organization is concentrating on the "core competencies" critical to product success, and depending on long-term relationships with trusted suppliers of both goods and services for the rest.

The marketing of services is more difficult than the marketing of products. Even for standard services such as trash hauling or security, measures of performance and satisfaction are difficult to design, and it is difficult to compare different vendors on the same service. Because services are performed anew each time, as in the design of buildings, it is a leap of faith to presume that the *next* performance will be like the past ones, although the more standard the service, the more reasonable the assumption. Consequently, the marketing of services resembles the marketing of fabricated items; one sells capability and capacity. As in the case of fabricated items, and unlike the case of cataloged components, one cannot first test and then accept; one pays for performance and accepts the outcome. Also in parallel with fabricated items, the seller is less aware of the current requirements of the potential client or customer, who chooses the appropriate time to plan facilities expansion or to revise pension plans.

Engineers are often the providers of services such as design or testing. As such, it is engineers who do the selling of such services. In many engineering consulting firms, those in charge of client relations or market development are engineers. They monitor the environment, uncover those in a position to let bids, make and maintain contacts, propose work, and follow up. They also overview those in the consultancy who actually perform the services.

Systems integration is still another type of service. Systems are interconnected and interdependent sets of standard components configured to do a unique task for the buyer. These systems are assembled or integrated rather than produced. Typically, the components of the system are all produced by the same manufacturer; this is often the case with computer systems, which are composed of a central computer and different combinations of memory storage units such as tape or disk drives, output devices such as printers, and communications devices such as microwave transmitter-receivers. In some cases, particularly in the case of the automated factory floor, the equipment is not from a single source, but from many suppliers, and designed by a systems integrator to operate together to produce the desired function.

The systems integrator must be thoroughly familiar with the client's situation in order to tailor an effective system. Therefore, there is a great deal of interaction between the engineers involved on both the buyer's and seller's sides. The systems integrator may provide documentation, training, maintenance, repair, modernization, or other services as part of the package, and engineers would be part of the delivery system.

After-Sales Service

The majority of the tasks with which marketing deals are oriented toward the obtaining of revenue, that is, with making sales. Engineers are also involved with client-related tasks after the sales transaction, including delivery and payment, is completed. These client maintenance tasks include installation, warranty, field service, documentation, training, provisioning, providing repair facilities, providing retrofit, rebuild and overhaul, and supplying spares and supplies.

Many industrial and military purchases are very expensive, often millions of dollars per unit, and therefore must have a long service life to justify the expenditure. This implies that the units cannot be discarded when they malfunction, but must be repaired. Since these units are often also complex, it is common that not all repair will be within the technical capability of the user. It is equally likely that users requiring such expensive equipment will have some ability to service and maintain them in order to avoid dependence on outside agencies for routine procedures such as lubrication, calibration and adjustment, replacement of worn cutting tools, exchanging of dies, and the like. On the other hand, the expertise to undertake the replacement of broken parts or burned-out wiring or other such random and infrequent repairs is unlikely to be found at the using organization. Therefore, some split in the responsibility for such maintenance is likely between the buying and selling organizations.

To accommodate different levels of maintenance ability, firms typically form departments for servicing the continuing needs of their clients. These field service personnel are responsible for traveling to the site where equipment is located and making on-site repairs, where appropriate. The most common examples of these are telephone repair personnel and office copier technicians. Common are tasks such as installation, reconfiguring the customer's equipment through update or addition, repair and servicing, and some degree of customer demonstration and training.

Engineers are used in these functions as backup to crews of technicians where problems become severe or complicated by multiple, interactive failures of complex systems such as computers or switching networks. Engineers are also used when the systems are still in early development, since frequent upgrading and retrofit is necessary, documentation is slow to reflect the status of the customer's specific equipment, and malfunctions and their remedies are less familiar. When the systems are stable, and longer runs of production are completed without extensive and necessary retrofit of early models, maintenance is more predictable, and engineers are shifted to other responsibilities.

In the case of extensive systems where the clients' own operators and maintenance crews must be trained, engineers are responsible for these tasks as well as for installation, integration into existing systems, and operational checkout. It is sometimes necessary to design a spares complement to accompany the installation and subsequent maintenance of equipment, where the mean time between failures is short, the equipment is vital, and the maintenance action is not overly complicated. Spare bulbs for movie projectors, spare tires for automobiles, and replacement engines for jet aircraft are examples. The task of designing such kits is called *provisioning* and is influenced by the sophistication of the customer's ability to do maintenance, the number of similar units in service, the reliability of the spares themselves, and the importance of operational readiness of the equipment to the economic health of the customer.

For less sophisticated clients, or clients with either little investment or little reliance on the systems, such as buyers of test equipment or other accessories, the level of service may be more fundamental, and include periodic inspection, servicing, calibration, and maintenance. Naturally, the more fundamental the service, the larger the investment in field service will be.

In addition to those sustaining tasks done in the field at customers' locations, there is a support activity required in a centralized location. This support activity includes analysis of reported failures to determine the rate of production of spares, to note weakness that may require redesign of the system or some portion of it, to note and trace retrofit and other necessary configuration changes, to update documentation so that appropriate plans and schematics are available to the customer, to alter spares and repair parts inventories and provisioning to match equipment configurations, to schedule overhaul and rebuild activities, and to disseminate repair information to field crews to avoid redundant troubleshooting.

For large systems, continuing support may include the provision of maintenance beyond the capability of that possible on site. Examples are dry-dock repairs and maintenance of oceangoing vessels, rebuild of airliners, or overhaul of jet engines. For these purposes, special shops are outfitted and maintained, and engineers are required.

ENGINEERS IN SERVICE ORGANIZATIONS

Importance of Service-Producing Industries

America is said to have become a "postindustrial society." Although, as I have argued, the United States cannot afford to permit its manufacturing industry to be less than the best, it is abundantly true that manufacturing no longer employs most Americans. Between 1970 and 1993, while overall U.S. civilian employment increased just over 50% (from 78.7 million to 119.3 million), manufacturing employment actually decreased about 6% (from 20.7 million to 19.6 million), plummeting from 26.4% of total employment to 16.4% (see Table 13-2). Agricultural and mining employment combined dropped from 5.1 to 3.1% of employment, and construction stayed at about the same 6% level. The total goods-producing sector (above) dropped from ⅜ of civilian employment in 1970 to little more than ¼ in 1992. The service-producing sector, therefore, provides nearly ¾ of employment. Some of the most striking changes are discussed below:

- Professional and related service employment has more than doubled. Although hospital employment gained 77%, the big gainers were other health services and social services, which more than tripled, and legal services, which almost did.
- Business and repair services multiplied almost fivefold (387% gain); almost a million people were employed in computer and data processing, a category not worth a separate listing before.
- Entertainment and recreation employment almost tripled, reflecting changing lifestyles.

Characteristics of the Service Sector

The wide variety of service-producing industries makes identifying common characteristics hard, but a few traits are normally present. Most services are more intangible than are manufactured goods. They are usually performed in real time, often in the presence of the customer. Services can seldom be inventoried, so they must be performed on a schedule that fits the needs of the customer, a challenge in staffing anyone who has worked at a fast-food restaurant can understand.

Most professional or consulting services are customized, personalized, and labor in-

TABLE 13–2   U.S. Civilian Employment by Industry, 1970 and 1993 (Thousands)

Industry	1970			1993			Change
Agriculture	3,463			3,074			−11%
Mining	516			669			30%
Construction	4,818			7,220			50%
Manufacturing	20,746			19,557			−6%
Transportation, communication, and other public utilities	5,320			8,481			59%
Wholesale and retail trade	15,008			24,769			65%
Wholesale trade		2,672			4,606		72%
Retail trade		12,336			20,163		63%
Finance, insurance, real estate	3,945			7,962			102%
Services	20,385			41,817			105%
Business & repair services		1,403			6,838		387%
Advertising			147			284	93%
Serv. to dwellings & buildings			*			749	
Personnel supply services			*			698	
Computer and data processing			*			957	
Detective/protective serv.			*			475	
Automobile services			600			1,477	146%
Other business & repair serv.			656			2,198	
Personal services		4,276			4,443		4%
Private households			1,782			1,114	−37%
Hotels and lodging places			979			1,435	46%
Other personal services			1,515			1,894	25%
Entertainment and recreation		717			2,060		187%
Professional & related serv.		12,904			28,293		119%
Hospitals			2,843			5,032	77%
Nonhospital health services			1,628			5,521	239%
Elementary & secondary sch.			6,126			6,372 ⎫	47%
Colleges & universities			#			2,633 ⎭	
Social services			828			2,770	234%
Legal services			429			1,253	192%
Other professional services			1,050			4,712	349%
Public administration	4,476			5,756			29%
TOTAL EMPLOYED	78,677			119,306			52%

*Not available; # Included in elementary and secondary schools.

Source: *Statistical Abstract of the United States 1994*, 114th ed., U.S. Bureau of the Census, Washington, DC, 1994, Table 641, p. 412, taken from U.S. Bureau of Labor Statistics data.

tensive. Others, such as airline transportation or telephone or electrical service, are standardized and very capital intensive. Some have both aspects: a stockbroker in the "front office" needs to have a customized approach to the needs and interests of each client; in the "back room" the clerical functions of record keeping and stock transfer need to be carried out efficiently and accurately—more in the mass production philosophy of manufacturing.

Like manufacturing, service-producing industries prosper by providing value for the customer, but often in a more immediate and personalized manner. Peters and Waterman[1] identify Delta Airlines, Marriott (hotels), McDonald's (restaurants), Disney Productions, Wal-Mart (stores), and both Bechtel and Fluor (in project management) as service-producing industries that have achieved excellence in performance. Lovelock[2] has edited an excellent book on the marketing, operations, and human resources aspects of managing service industries. Tosh[3] has summarized findings from research on service activities; his findings emphasize the need to provide a quality of service that creates a satisfied customer:

1. The average service business today loses 10% of its volume [annually] due to poor or inattentive service.
2. 96% of unhappy customers never complain, but 90% never return, each one tells at least nine others, and 13% tell more than 20 others.
3. Each happy customer tells at least five others, some of whom will become your customers.
4. The best opportunity to increase sales and market share is through your present customer base, because it costs five times as much to attract a new customer as it does to maintain an existing one.
5. There is almost a perfect correlation between employees' perception of an organization's human resource policies and practices and the customer's perceptions of quality and service.
6. The best indication of an organization's long-term financial success is the customer's perception of the relative quality of service.

Service employment has the reputation of low pay, but this is true only in part. As of August 1993, the hourly U.S. wage rate in the private service sector stood at $10.20, or $11.38 when the lower-paying retail area was excluded, compared with $11.79 in manufacturing.[4]

Technical Employment in the Service Sector

Manufacturing employed almost half (47%) of engineers in 1992, about 36% of engineering and science technicians, and about 28% of computer analysts, engineers, and scientists, but only 15% of other scientists and 13% of computer programmers (see Table 13-3). Another 47% of engineering employment is in the service-producing industries. While engineers represented about 3.6% of manufacturing employees in 1992, they comprised only 0.7% of employees in service-producing industries. Table 13-3 compares the employment of engineers and other technical specialists by type of employer; data for engineers is also shown for the three principal engineering disciplines. (Slight differences in totals reflect source differences.) Service-producing engineering employment in 1992 could be summarized as follows:

- About 22% of engineers (302,000) worked in the service-producing industries lumped together as "services" in these tables. These include educational services, research services, consulting services (especially for civil engineers), and a wide range of other services, including hospitals and computing services.
- Another 14% of engineers (190,000) worked for government; about 39% of these were civil engineers (although only 13% of all engineers were CEs); many of these were in highway and other public works for state and local government;

TABLE 13-3 Civilian Employment of Engineers (by Discipline), Scientists, Engineering and Science Technicians, and Computer Programmers by Industry Class in 1992 (Thousands)

Industry	Total Employed	Engineers*					Computer Engineers, Scientists	Scientists (except computer)	Engineer'g & Science Technicians	Computer Programmers
		Total	Civil	Electrical	Mechanical	Other				
Mining, oil/gas extraction	631	22	1	1	2	18	4	15	19	3
Construction	4,471	26	11	6	4	5	1	0	26	1
Manufacturing	18,040	641	8	168	137	328	184	88	450	81
Goods Producing Employees	23,142	689	20	175	143	351	189	104	495	85
Transportation and utilities	5,709	75	6	34	5	30	27	6	74	24
Wholesale and retail trade	25,391	56	0	36	7	13	27	4	84	57
Finance, insurance, real estate	6,571	11	1	1	1	8	78	10	5	75
Services	28,422	302	66	79	52	105	230	229	400	246
Government	18,653	190	74	40	14	63	83	188	161	50
Service Producing Employees	84,746	634	147	190	79	218	445	438	724	452
Self-Employed	8,794	29	5	6	5	13	32	102	26	16
Total Civilian Employment@	121,093	1,354	173	370	227	584	666	654	1,253	555

*Except computer engineers; #Computer system analysts, engineers, and scientists

@Total row includes agriculture, forestry, fishing, and private households

Source: *Statistical Abstract of the United States 1994*, 114th ed., U.S. Bureau of the Census, Washington, DC, 1994, Tables 979 and (total column) 642, pp. 614 and 413, from U.S. Bureau of Labor Statistics data.

many of the rest were military or civilian employees of the Department of Defense.

- About 6% (75,000) were employed in the transportation, communication, and public utilities industries.
- The remainder (5%, or about 67,000) were in wholesale or retail trade or in FIRE (finance, insurance, and real estate). Note that only 3000 of the 11,000 in the latter category held C.E., E.E., or M.E. degrees; one example of employment in this area is the (often industrial) engineers working for commercial insurance firms who advise manufacturers of fire and safety hazards in their plants.

Some Specific Service Industry Examples

Computer applications. People call our times the "information age" because of the unparalleled ease and economy with which information can be stored and manipulated. The development of massive integrated circuits has made possible computers of immensely increased capability and decreased cost; they have permeated our entire lives. Figure 13-1 gives an idea of the impact of computer technology. Of the areas portrayed in that figure, just one (CAD/CAM) relates to manufacturing and a second (consumer appliance controls) to products of manufacture. The rest are applications of computers to services: finance, general business operation, health sciences (medical), communications, government, and transportation. Engineering and computer science employment in this area is of at least three types: work for the manufacturers of computers and components, for systems consultants to the users, and for the users (mostly service industries) themselves.

Government service. At the height of World War II there were about 116 jobs per thousand people in manufacturing, versus 40 in government (23 in state and local government and 17 in federal government). In July 1992, government passed manufacturing as an employer for the first time: about 70 jobs per thousand in manufacturing versus about 75 in government, of which 12 was federal and 63 state and local government employment per thousand people.[5] The largest share of the budgets of local governments (except perhaps for education) and the biggest share of the 13.8 million people employed by state and local governments fall in the area of public works. Running this domain in each local government is a city engineer or public works director, who is usually, except in smaller towns, a registered professional civil engineer; in larger cities, a number of his or her department heads and professional staff will also be engineers. The responsibility of the public works director will include all or part of the following:

- Streets, highways, and bridges: their specification, maintenance, lighting, traffic control systems, and snow removal. (Detail design and construction of new roads and bridges are more likely to be subcontracted to engineering consulting firms, who will be led by engineers, and construction firms, who may be.)
- Water purification and distribution; sewage retrieval and pickup; solid waste disposal.
- Parks, playgrounds, airport and/or cemetery operation.
- Zoning, building inspection, and code enforcement.
- Vehicle maintenance.

At the state level, the largest employment of engineers is in the state highway (or

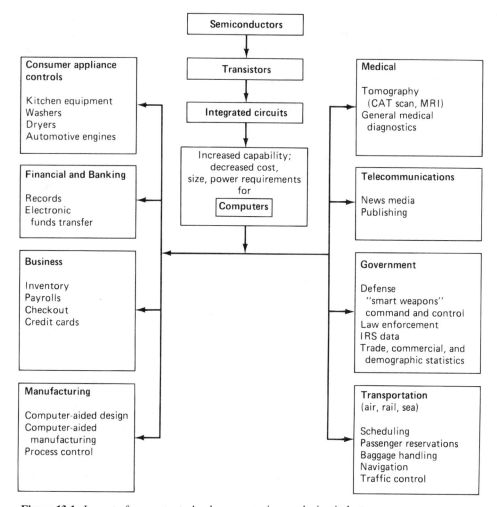

Figure 13-1. Impact of computer technology on service-producing industry.
(From National Academy of Engineering, *The Technological Dimensions of International Competitiveness,*
National Academy Press, Washington, DC, 1988, p. 14.)

transportation) department, but others are employed in energy, environmental, and various
regulatory functions. At the federal level,

> A majority of the 105,000 civilian engineers employed by the U.S. government is employed by
> the defense, intelligence, and space agencies, the Environmental Protection Agency, and the De-
> partments of Agriculture, Energy, and Transportation. However, the need for engineers extends
> across the entire federal government system.[6]

Starting salaries in government are often less than in private industry, but many engineers find they are offered much greater responsibility early in their career and earlier pro-

motion in government service. Government employees may have to learn to work within a bureaucratic system (true for some larger companies as well), but layoffs are rarer for engineers in government and vacations and other fringe benefits are attractive. The military services offer many opportunities for engineers, both officer and civilian, in development, maintenance, and operation of today's high-technology systems.

College teaching and research. Although these two careers are not necessarily related, they require the same basic preparation: graduate education. New assistant professors in engineering schools overwhelmingly are expected to have a "terminal degree" (usually Ph.D. or D.Sc.), and if they are teaching graduate subjects, they will normally be expected to do research and publish literature in their field—that is, to be a scholar. Engineers working in research or advanced design soon find they need more technological depth than a B.S. degree provides, and most of them will seek an M.S. degree; senior researchers, even in industry, may find a doctorate desirable as well.

Most engineers who are American citizens with a good undergraduate academic record can find financial support as a graduate teaching or research assistant to help earn graduate degrees: the M.S. degree in one to two years and the Ph.D., if desired, in another two to four years, depending on the individual, the school, and the amount of time spent providing teaching or research assistance for financial support. Although students who are American citizens often get first preference for financial support (and for faculty positions), the tenacity of students from other countries has been such that since about 1983, more engineering doctorates have been granted each year to graduate students holding temporary visas than to American citizens. In metropolitan areas, local universities often provide graduate courses in the evening that make it possible to earn a master's degree in three years or less, part time, and many engineers earn a master's degree in their discipline, in engineering management, or an M.B.A. in this way while working full time.

The position of professor of engineering can be a very satisfying one, with a great deal of personal freedom in the way time is spent and the subject areas pursued in teaching and research. Any engineer with an above-average academic capability can find opportunity in a faculty career. Members of minorities and women with credentials to teach engineering are especially rare and especially sought after.

Biomedical engineering and the health services. Table 13-2 shows that employment in health services increased dramatically from 1970 to 1993. One reason is the increasing age of the American people and the concomitant need for more hospital and nursing home care, but the reason more people are reaching such an age is the rapid expansion of medical knowledge and capability, and the technology that underlies it. Open-heart coronary bypass surgery, with its use of external blood circulation, a forest of monitors and indicators, and temporary use of a pacemaker, requires much more intensive use of highly skilled people than does the alternative of a decade or so previously—a massive, fatal heart attack. (If that resource had not been available to the author in May 1988, you might have been spared this book!)

The medical explosion has increased the need for medical specialists of all types: chemists, biochemists, and pharmacists. The increasing technical complexity has required the service of a new breed of engineer—the biomedical engineer. We will look at this specialist in some detail to provide one example of the increasing specialization of modern technology and of its practitioners; other, more numerous classes of engineer can be subdivided even further into specialties. Attinger[7] provides a definition and identifies the growing subdivisions of the field:

> Biomedical engineering can be broadly defined as the application of engineering concepts, methods, and techniques to biology and medicine. Because of the breadth of the field, several subspecialties have been emerging:
>
> Bioengineering is concerned with the quantitative analysis, both theoretical and experimental, of the structural and functional properties of the components of biological systems. . . .
>
> Medical engineering, or biomedical technology, deals with the design, development, application, and evaluation of the instrumentation, computers, materials, diagnostic and therapeutic devices, artificial organs and prostheses, and medical information systems for use in medical research and practice. . . .
>
> Clinical engineering uses engineering concepts and technology to improve health-care delivery in hospitals and clinics. . . .
>
> Health-care systems engineering deals with problems in the analysis of health-care concepts and health-care systems, such as [the] socioeconomic and psychosocial determinants of health. It is also concerned with the design and implementation of more efficient and less costly modes of health-care delivery. . . .
>
> Biochemical engineering, agrobioengineering, and genetic engineering are now emerging as new subspecialties in the rapidly developing field of biotechnology. . . .

The application of "engineering concepts and technology . . . in hospitals and clinics," whether labeled "clinical engineering" or not, can include use of:

- Electrical engineering, to reduce the estimated 1200 people electrocuted annually during routine diagnostic and therapeutic procedures.
- Chemical engineering, to purify water by ion exchange, distillation, electrodialysis, or reverse osmosis for use in delicate testing procedures.
- Civil engineering, for disposal of toxic wastes.
- Industrial engineering, for improving the efficiency of operation of hospital functions and departments (a function now performed by "management engineers" in many hospitals). One doctor reports[8] that "up to 40% of all hospital costs are related to the generation and storage of information," which offers opportunity for the management engineer.

U.S. Service Productivity

When we first think of productivity in the service-producing sector we tend to think first of one-on-one consultations with a doctor or lawyer—or perhaps a professor in a lecture hall—and conclude that achieving productivity increases in providing services to match those achieved in manufacturing is unlikely. However, the U.S. leads the world in service efficiency in a number of areas and, as a result, has a favorable balance of payments in services that offsets part of our negative balance of payments in manufactured goods. A 1992 study by McKinsey Global Institute offers some examples:[9]

- U.S. retailers are 4% more productive than German ones, 18% more than British,

31% more than French, and 56% more than Japanese. U.S. retailing is much less regulated, and stores such as Wal-Mart and Toys 'R' Us drive out less efficient stores and compel others to become more productive.

- Airline employees in Europe are 28% less productive than those in the United States, largely because of the imperative of deregulation in the United States.
- German bank employees are 32% and British 36% less productive than their U.S. counterparts, largely because of automated teller machines and back-office computer efficiency in the United States.
- Labor productivity among telephone companies is comparable in the United States, Japan, and France, but about 20% less in Germany and 40% less in Britain because of monopoly domination; factoring in capital investment as well as labor, the United States is far more efficient than Japan or Europe.

DISCUSSION QUESTIONS

13-1. How do the tasks of marketing vary with the type of relationship existing (transaction-based, more sustained, or highly committed)?

13-2. In what ways are the purchase of engineering services to design an installation and the purchase of accessories such as an automobile or office equipment fundamentally different?

13-3. In purchasing process materials, specifically steel sheet and rods, a buyer (commonly a business administration graduate) is offered material with slightly different specifications at a significant reduction in price. What should be his or her response?

13-4. Distinguish between component parts and fabricated items.

13-5. Although it is normally more efficient for the producer of repair parts to maintain an inventory of them at his factory, what alternatives might be considered when large users of the equipment are located in a remote region (such as the Middle East)?

13-6. For a high-technology product and producer with which you are familiar, describe the provisions for after-sales service that have been established.

13-7. In what ways are goods-producing and service-producing industries generally different?

13-8. Are the leading engineering disciplines—civil, electrical/electronic, mechanical, and (by difference) other—needed by the several types of service-producing industries in proportion to the numbers of such engineers? If not, identify some exceptions. (*Hint:* See Table 13-3.)

13-9. Provide an additional example of a change in the way a service industry does business as a result of wide, economical availability of computers.

13-10. Find an application of engineering concepts and techniques to the health services in the literature and write a summary describing it.

NOTES

1. Thomas J. Peters and Robert H. Waterman Jr., *In Search of Excellence: Lessons from America's Best-Run Companies* (New York: Harper & Row, Publishers, Inc., 1982).

2. Christopher H. Lovelock, *Managing Services: Marketing, Operations, and Human Resources* (Englewood Cliffs, NJ: Prentice-Hall, Inc., 1988).

3. John J. Tosh, presentation at the 1988 Human Resources Conference of the American Management Association.

4. "Business Outlook," *Business Week,* September 20, 1993, p. 28.

5. Bureau of Labor Statistics July 1992 data, graphed in *Government Executive,* October 1992, p. 8.

6. Margeret Opsata, "Working for the State and Uncle Sam," *Graduating Engineer,* March 1991, pp. 48–52.

7. E. O. Attinger, "Biomedical Engineering" field definition in *Peterson's Graduate Programs in Engineering and Applied Sciences 1988* (Princeton, NJ: Peterson's Guides, Inc., 1987), p. 285.

8. William M. Tierney, quoted in John Cary, "Physician, Reengineer Thyself," *Business Week,* June 14, 1993, p. 60.

9. David Wessel, "U.S. Workers Excel in Productivity Poll," *Wall Street Journal,* October 13, 1992, p. 2.

PART 4

Managing Projects

CHAPTER **14**

Project Planning and Acquisition

PREVIEW

In this chapter we first consider what makes a project a project. We then discuss the process of proposal anticipation and preparation that is essential to the life of the project-driven organization. The main body of the chapter is a description of the management tools used in project planning (and later control): the statement of work, milestone schedule, work breakdown structure, bar chart, network scheduling systems (PERT, CPM, and other variations), and resource allocation methods. Finally, we define and compare the several types of fixed price and cost reimbursement contracts.

CHARACTERISTICS OF A PROJECT

A project represents a collection of tasks aimed toward a single set of objectives, culminating in a definable end point and having a finite life span. A project is a one-of-a-kind activity, aimed at producing some product or outcome that has never existed before. (There have been earlier aircraft or oil refineries or office buildings, but none of them were exactly like the one being created by *this* project.) Responsibility for a project is normally assigned to a single individual, assisted by a close-knit project team. The term "program" is sometimes used interchangeably with "project," but more often a program is a more comprehensive undertaking, which may in turn consist of a number of projects.

Formal project management methods received their greatest impetus in U.S. aerospace programs and complex construction projects of the 1960s, and the methods have spread to many other complex, dynamic activities. Project management methods should be considered (1) where close interaction of a variety of technologies, divisions, or separate organizations is required; (2) when completion within a tight schedule and budget is necessary; and/or (3) for activities involving significant technical and/or economic risk to the organization. However, organizations that have never used formal project management before should specify it only when necessary, since the transition to it from traditional ways of doing business can be time consuming and stressful.

The three essential considerations in project management are (1) time (project schedule), (2) cost (in dollars and other resources), and (3) performance (the extent to which objectives are achieved). The successful project manager will attempt to keep these three in balance (Figure 14-1). Since achieving maximum performance is often possible only at the expense of cost and schedule, difficult trade-off decisions involving compromises are often necessary.

THE PROJECT PROPOSAL PROCESS

Every type of project should be preceded by a detailed description of what is to be accomplished, together with a proposal or estimate of the time and cost required. This process has been carried furthest by aerospace and other R&D organizations that depend on a continuing sequence of external project awards for their livelihood, and we will discuss the proposal process in that context.

Preproposal Effort

First, successful organizations of this type begin work long before an RFP (request for proposal) is received from a potential customer. The successful project-driven organization is continuously identifying new business opportunities—areas of technology or types of activity where attractive projects are likely to be funded. The firm estimates the resources and capabilities that will be needed to meet expected future needs of potential customers, compares them with the resources they have on hand, then proceeds to develop the needed technical

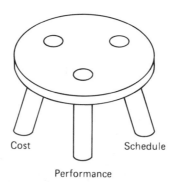

Figure 14-1. The "three-legged stool" of successful project management.

skills and acquire other needed resources (or at least identify sources for them) in advance.

Successful firms also maintain new business (marketing) groups that seek to identify specific customer needs as early as possible, well in advance of their issuing an RFP, so that the firm can make a preliminary bid decision (a decision whether to invest the resources it will take to prepare a proposal on a major project). Creating winning proposals is an expensive, time-consuming process that should only be begun on potential projects the firm believes fit their needs and for which they feel they have a reasonable possibility of capture; others should be declined to permit concentrating on the best opportunities. Assessment of the requirements of the project and the capabilities of the firm, compared with those of its competition, to meet these requirements should be part of this bid decision.

If the bid decision is favorable, successful firms will try to get an early start on developing a response to their best estimate of what the RFP will ask for. In many aerospace project opportunities, the firm that waits until they have an RFP in hand before evaluating it and starting on a proposal has little chance to capture the award. Figure 14-2 shows the relationship of the preliminary bid decision and other pre-RFP effort to later phases in the process of capturing new projects.

Proposal Preparation

By the time the RFP arrives, management often has appointed a proposal manager, who has prepared a budget for the proposal process and a letter ready for release calling on functional managers to provide members of the proposal team. The RFP is quickly examined to be sure it holds no surprises, and the tentative decision to prepare a bid is reconfirmed.

An RFP from the U.S. government typically includes a cover letter, a *statement of work* (which specifies the work to be performed), the required schedule, specification of the length and content desired in the proposal, and a stack of standard clauses (sometimes called "boilerplate") covering legal aspects of doing business with the federal government that may be several times as thick as the rest of the RFP. Worse than this, the RFP will "call out" specifications whose provisions then become a legally binding part of the contract. These specifications, in turn, call out other, "second tier" specifications that must be complied with. Dougherty[1] made the striking comparison between requirements in contracts used by the federal government and commercial airline customers for buying similar jet aircraft shown in Table 14-1. The bewildering mass of requirements incorporated in federal contracts deters potential contractors from dealing with the government and unnecessarily increases the cost of government contracts. Only time will tell whether the 1994 emphasis of Secretary of Defense Perry on reducing military specifications will significantly improve this situation.

A well-prepared "kickoff meeting" for the proposal team launches the proposal process. A representative of senior management may give a short pep talk on the importance of the project to the company and introduce the proposal manager, who will do much or all of the following:

- Give an overview of what the RFP asks for.
- Provide the best estimate from company intelligence as to what the customer *really* wants and the factors the customer will use in determining the contract winner.

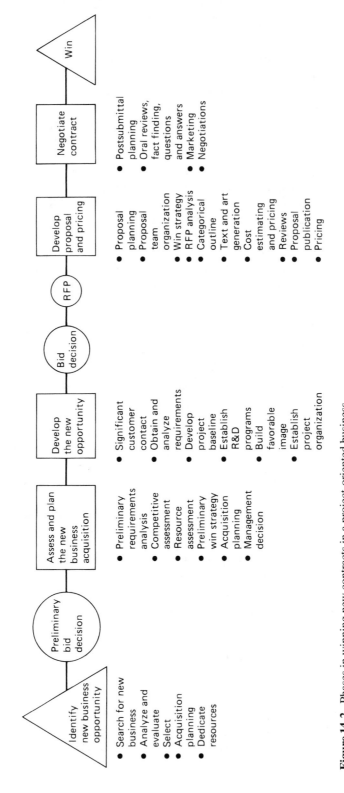

Figure 14-2. Phases in winning new contracts in a project-oriented business.
(From Hans J. Thamhain, *Engineering Project Management*, ° John Wiley & Sons, Inc., New York, 1984, p. 55. Reprinted by permission of John Wiley & Sons, Inc.)

TABLE 14–1   Comparison of Military and Commercial Aircraft Contract Requirements

	Customer	
Item	Government	Commercial
Specifications	210	9
Documents		
Original	500	10
Through second tier	11,000	50
Data submittals required	30,000	250

Source: Frank Dougherty, Assistant for Contract and Quality Matters, DOD Directorate of Industrial Productivity/Quality, presentation on "Acquisition Streamlining," to the American Society for Quality Control Midwest Conference, St. Louis, MO, October 9, 1986.

- Identify the organization, schedule, and labor-hour allocations for the proposal effort.
- Provide handouts giving, in as much detail as preparation time has permitted, management's concept of how the project might be carried out, and instructions to the project proposal team.

Proposal personnel are usually all experienced people, and so they can work rapidly with minimum guidance. They interact and prepare drafts of their parts of the proposal, which are typed and reproduced by a clerical "night shift" so they are ready for review in the morning. (The increasing capability of engineers to prepare their own drafts on word processors, and of desktop publication using laser printers, provides a distinct advantage in preparing proposals, which must undergo a sequence of reviews and revisions on a tight time schedule.) Figure 14-3 shows a typical organization chart for a major proposal team.

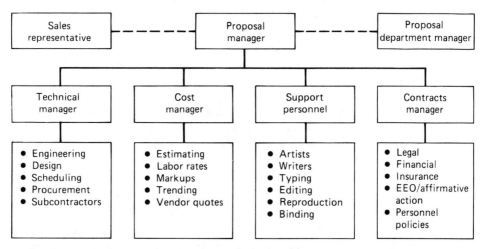

Figure 14-3. Typical organization chart for a major proposal.
(Adapted from Harold Kerzner, *Project Management: A Systems Approach to Planning, Scheduling, and Controlling,* 2nd ed., Van Nostrand Reinhold, Inc., New York, 1984, p. 823.)

Proposal Contents

The RFP will often specify separate management, technical, and cost proposals and their expected contents:

- The management proposal typically discusses the company, its organization, its relevant experience, its management methods and control systems, and describes the personnel proposed to lead the project.
- The technical proposal outlines the design concept proposed to meet the client's needs, with special emphasis on the approach planned to resolve the most difficult technical challenges posed by the project.
- The cost proposal not only includes a detailed price breakdown, but often also discusses aspects of inflation, contingencies, and contract change procedures.

The proposal package is critically reviewed by company senior management not involved in creation of the proposal, revised, printed, and delivered to the customer.

PROJECT PLANNING TOOLS

Since a project is a set of activities that has never been done exactly that way before, planning is extensive and critical. There are several planning tools that should be understood. These include (1) the statement of work, (2) the milestone schedule, (3) the work breakdown structure, (4) Gantt (bar) charts, (5) network scheduling systems (such as PERT or CPM), and (6) resource allocation methods.

Statement of Work

The project award will normally be accompanied by a statement of work (SOW) describing exactly what is to be provided in the project. The SOW will normally begin with the general scope of the work, then itemize the tasks to be performed, the contract end items (products) to be delivered, and the data and reports to be supplied. The SOW typically consists of a set of numbered paragraphs, often arranged in outline form and numbered in some decimal notation (e.g., paragraph 3.1.11) that can be cross-indexed with the *work breakdown structure* (discussed later in this section). Sometimes the SOW is that proposed by the contractor in the proposal bid package, and sometimes it has been created or modified by the customer. In any event, it is essential that customer and contractor come to a common understanding of exactly what each paragraph of the SOW means before work has progressed very far.

Milestone Schedule

Milestones are the key start and end dates for major project phases or activities. Examples from a typical aerospace project are project "go-ahead" (start), design reviews ending each design phase, 90% drawing release, start of each major test phase, delivery of first prototype and first production item, and the customer's required operational capability date (see Figure 14-4 for an example).

Such a schedule is essential for detailed planning, since reaching a major milestone

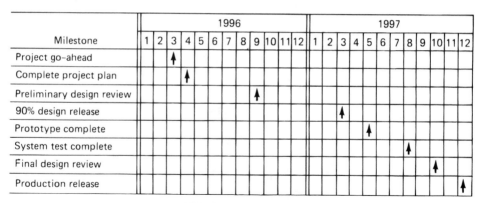

Figure 14-4. Typical milestone schedule.

point typically requires the coordinated effort of a great many people. For example, a major design review may require completion to a specified level of component or subsystem design by dozens of design groups (in subcontractors as well as your own departments), analyses of reliability, maintainability, producibility, safety, and other aspects of the design, and plans for testing, training operators, production tooling, and logistic support. In turn, accomplishing all of these analyses and reports will require "backing up" in time from each major milestone to hundreds of earlier supporting schedule points, at which specified information must be transferred between project entities to enable the receiving group to do their design or analysis, all in order to meet the major milestone.

Work Breakdown Structure

A work breakdown structure (WBS) is a product-oriented "family tree" of work effort that provides a level-by-level subdivision of the work to be performed in a contract. The WBS provides a common framework or outline that can be used to:

- Describe the total program/project effort
- Plan and schedule effort
- Estimate costs and budgets
- Support network schedule construction
- Assign responsibilities and authorize work
- Track time, cost, and performance

For example, Figure 14-5 illustrates a work breakdown structure for developing a jet engine. The top level is the entire project or program, which can be given a unique number or code, such as "XYZ," to distinguish it from other projects. The first item at the second level $(XYZ-1)^2$ is traditionally the "end item" to be delivered, in this case the jet engine itself. Other items at the second level might be the training of user maintenance and repair people (XYZ-2), creating the necessary ground support equipment (GSE) for starting and maintaining the engine (-3), system testing (-4), and the project management (-5) needed to integrate and manage all these activities. Each of these second-level items is divided further. For example, the engine is divided into the major subsystems of fan (XYZ-1.1), compressor (-1.2), and

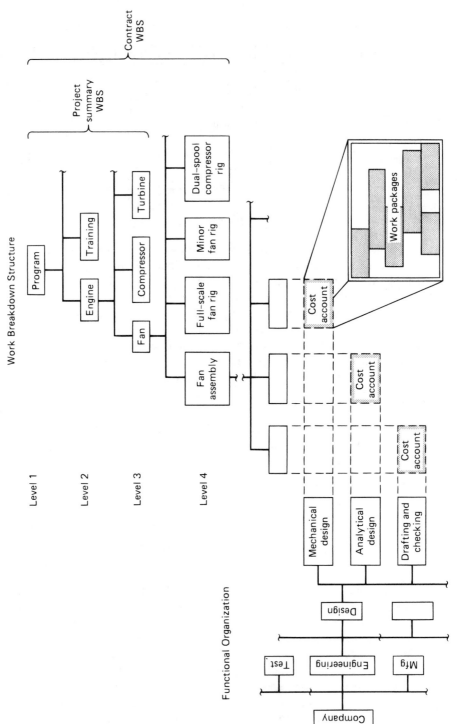

Figure 14-5. Integration of work breakdown structure and organization structure.

turbine (-1.3); the fan further into the fan assembly (XYZ-1.1.1), full-scale fan rig (-1.1.2), and so forth. Similarly, project management (XYZ-5) might be divided into project management per se (-5.1), configuration management (-5.2), and reliability engineering (-5.3).

The second dimension, shown on the lower left of the figure, is the functional organization, and this also can be coded. For example, manufacturing may be coded 1000 and engineering 2000, the latter might be further divided into 2100 for plant engineering, 2200 for design engineering, 2210 for mechanical design, 2220 for analytical design, and so forth. These two dimensions meet in a cost account such as XYZ-1.1.1-2210, which represents all the mechanical design (2210) performed on the fan assembly (-1.1.1). The cost account consists of one or more *work packages,* which form the ultimate unit by which work is controlled. The work package incorporates a set of tasks to be performed, a schedule, and a budget in labor-hours and other costs. The work package is the responsibility of one person (the "work package manager") from the organizational unit in which most of the work in the package will be done, and where possible it is structured to have a short duration and defined end point.

Every project activity that consumes resources is included in some work package, permitting progress on a particular end item of the work breakdown structure to be evaluated. For example, the project team member responsible for the XYZ engine compressor would receive reports on all work packages of the form XYZ-1.2.a-bbbb, where a is any (level 4) subdivision of –1.2 compressor effort, and bbbb is the account number of any organization with a compressor effort work package. Similarly, the functional manager responsible for mechanical design would be interested in all work packages ending in –2210 (such as –1.1.1-2210, –1.1.2-2210, –1.2.1-2210), for the XYZ project and all other projects as well. Beginning with the work package data base, a computer-based management information system can easily tailor reports in either the project or the organization dimension, and it can summarize information by adding work package data together to provide reports in less detail for any level of management or for the customer.

Gantt (Bar) Charts

Henry L. Gantt, one of the pioneers of the scientific management movement, is generally credited with initiating the concept of a class of charts in which the progress of some set or sequence of activities or resources in the vertical dimension is plotted against time in the horizontal dimension. The first reported application was in 1915, when Gantt was keeping track of the time between ordering and delivery of each lot of ammunition produced in the United States for Allied forces in World War I.

Gantt charts have found many other applications. In the job-shop or batch-production environment, Gantt charts are used to schedule the use of production machines, and elsewhere for the planning and control of work crews. In project management, it is tasks or activities (project performance) that must be charted against time (project schedule). Three things must be established in the project planning process before Gantt charts can be created:

1. The tasks (activities) needed to complete the project
2. The precedence relationships of the tasks (which tasks must be complete before other specified tasks can begin)
3. The expected duration of each task

TABLE 14–2 Information for Planning House Project

Task	Follows Task(s)	Weeks Duration	Task Description	Manning Level
A	Start	1.0	Clear site	3
B	Start	0.6	Obtain lumber and other basic materials	1
C	Start	2.0	Obtain other materials and components	1
D	B	2.0	Prefabricate wall panels	4
E	B	0.9	Prefabricate roof trusses	3
F	A, B	1.0	Form and pour footings and floor slab	3
G	D, F	0.3	Erect wall panels	4
H	E, G	0.2	Erect roof trusses	4
J	C, H	0.5	Complete roof	3
K	J	2.0	Finish interior	4
L	J	1.0	Finish exterior	2
M	L	0.4	Clean up site	1
N	K, M	0.2	Final inspection and approval	1

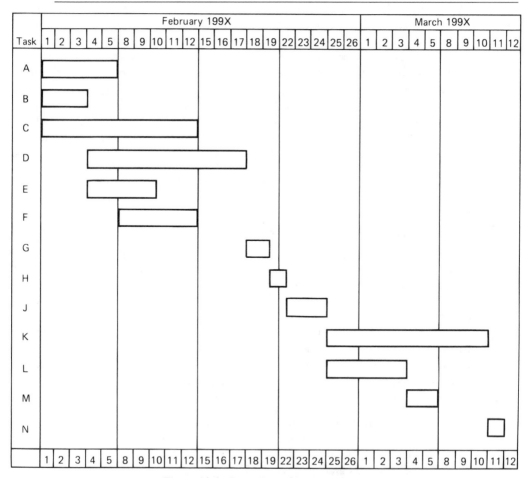

Figure 14-6. Gantt chart of house project.

Table 14-2 illustrates these three items (plus a "manning level" for later use) for a simple project: building a single-story residence on a concrete slab by first prefabricating wall panels (with electrical wiring and plumbing inside) and roof trusses. Time durations have been given in weeks assuming a five-day work week, so that an 8-hour day is 0.2 week. Given these durations and precedence relationships, a simple bar chart can be drawn in which each task is represented by a solid bar (Figure 14-6).

Bar charts can be elaborated with symbols and arrows above and below the bars to indicate significant points within the task or relationships to other tasks. A bar chart showing only the major subdivisions of project activity (composite tasks) is often known as a *master schedule.* Figure 14-7 illustrates such a schedule in "line-bubble" format for a minicomputer development project; symbols above the lines (bars) give the originally scheduled start and finish dates for each task; open symbols below each line show the latest reschedule, and solid symbols show that the date tasks were actually begun and completed.

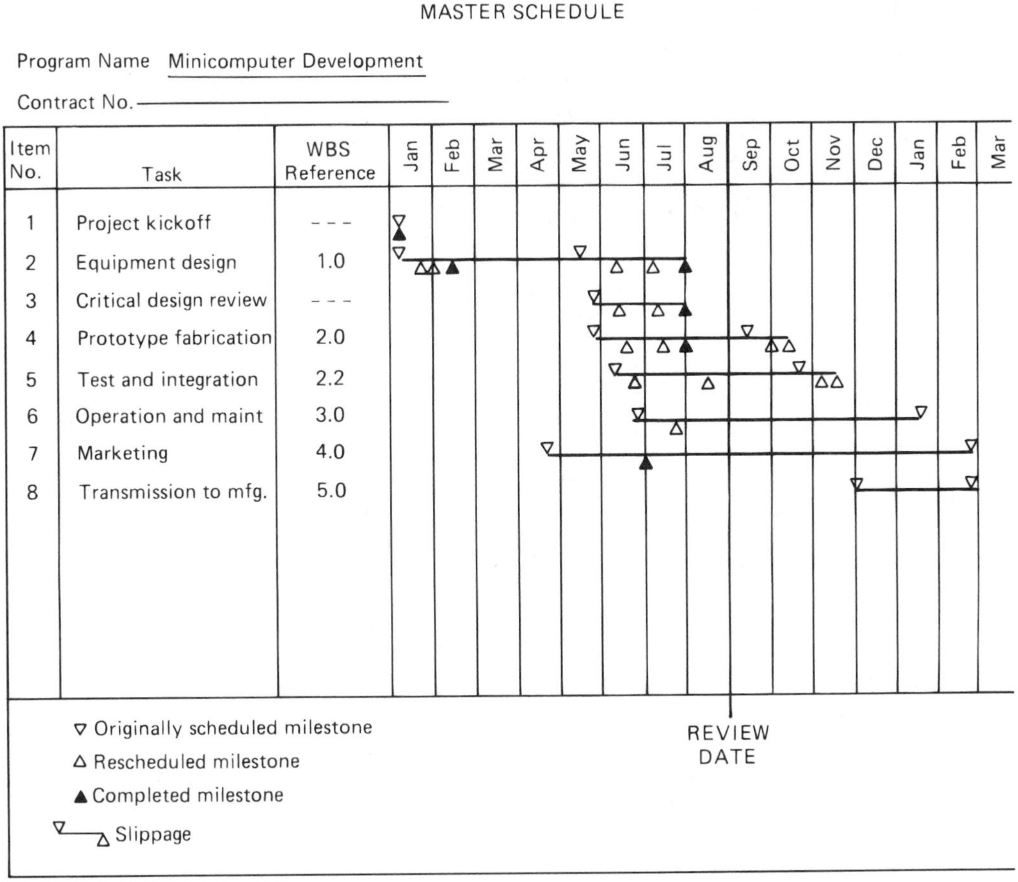

MASTER SCHEDULE

Program Name   Minicomputer Development

Contract No.

Item No.	Task	WBS Reference	Jan	Feb	Mar	Apr	May	Jun	Jul	Aug	Sep	Oct	Nov	Dec	Jan	Feb	Mar
1	Project kickoff	- - -															
2	Equipment design	1.0															
3	Critical design review	- - -															
4	Prototype fabrication	2.0															
5	Test and integration	2.2															
6	Operation and maint	3.0															
7	Marketing	4.0															
8	Transmission to mfg.	5.0															

▽ Originally scheduled milestone

△ Rescheduled milestone

▲ Completed milestone

▽△ Slippage

REVIEW DATE

Figure 14-7. Example of master schedule (in line-bubble format).

(From Harold Kerzner and Hans J. Thamhain, *Project Management Operating Guidelines: Directives, Procedures, and Forms,* Van Nostrand Reinhold Company, Inc., New York, 1986, p. 216.)

Gantt charts are easy to understand and use, and they provide a good tool for managing small projects without an excessive number of tasks. Bars can be filled in with a felt-tip pen to show the extent of task completion, making obvious the tasks that are behind or ahead of schedule. At one time Gantt charts were considered an inflexible tool, since a schedule change in some critical task might make it necessary to redraw many charts by hand. This disadvantage has been eliminated by project management software packages.

Figure 14-8 shows a simple Gantt chart as it would appear on a computer screen drawn by Harvard Project Manager 3.0,[3] an early personal computer software. Task A on Figure 14-8, for example, was planned (top solid bar) to begin October 7 and end before October 14, but it could be permitted to end as late as October 16 (thin extension of top bar). The task actually began October 9, but finished as originally scheduled (bottom solid bar for task A). If delays or changes in the project make replanning necessary, one simply changes a few parameters of computer input, and the computer completely replans the project and can reproduce a new set of Gantt charts and schedule reports at minimal cost.

Network Scheduling Systems

About 1958 two similar systems for network-based project scheduling were devised: the *program evaluation review technique (PERT)* was created by Booz, Allen, and Hamilton (management consultants) and Lockheed Aircraft Corporation for use in development of the Po-

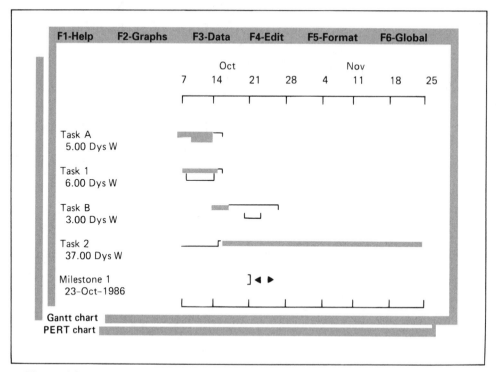

Figure 14-8. Gantt chart drawn by Harvard Project Manager 3.0.
(Copyright 1988 by Software Publishing Corporation.)

laris ballistic missile, and the *critical path method (CPM)* was developed by the DuPont Company for chemical plant construction. In the intervening years the features of each have been added to the other, but the terminology "PERT" is still used in aerospace and related industry, and "CPM" is preferred in the construction industry.

A network can be portrayed using either of two graphical techniques: the *arrow* or the *circle* diagram. Figure 14-9 is the arrow diagram for the house project, based on the same data (Table 14-2) used for the bar chart in Figure 14-6. The arrows represent *activities* or *tasks,* which have time durations and consume resources (dollar cost and use of people and equipment); the circles represent *events,* which indicate the start and/or end of one or more activities. An activity may be given its own symbol (such as A in Figure 14-9), or it can be designated by its predecessor and successor events (activity 1,2 or 1–2 instead of A). No activity may begin until all activities ending in its predecessor event have been completed. *Dummy activities,* shown by dashed arrows such as 3–2 in Figure 14-9, simply show a precedence relationship between events in an arrow diagram, and they consume neither time nor any other resource. For example, activity F (2–4) cannot start until both A and B (1–2 and 1–3) are complete, whereas activity D (3–4) or E (3–5) depends only on the completion of activity B (1–3). Durations (here in weeks) are shown below each arrow.

The *critical path* is the longest path through the network, calculated by a computer software algorithm (or, in this simple case, by hand). In our house problem the critical path, shown with heavier arrows, is B–D–G–H–J–K–N (1–3–4–5–6–7–9–10) and has a duration of 5.8 weeks. Activities not on the critical path allow a degree of scheduling flexibility (called *slack* or *float*) that the project manager can use in obtaining the best use of resources. For example, activity E (3–5) has a duration of 0.9 week, whereas the parallel path D–G (3–4–5) has a duration of 2.3 weeks; activity E therefore has a slack of 1.4 weeks, and its start may be delayed that much without affecting the ending date of the project. Similarly, activity C (1–6) has a duration of 2.0 weeks, and the parallel path 1–3–4–5–6 has a duration of 3.1 weeks; the start of activity C may therefore be delayed up to 1.1 weeks without penalty.

Exactly the same relationships can be shown on a *circle* diagram [also known as an *activity-on-node (AON)* diagram], in which activities are shown within the circles and the arrows simply show dependency relationships between activities. Figure 14-10 shows the AON equivalent for the house project. All paths must begin with the "start" symbol and terminate at the "end." No equivalent to the dummy activity of the arrow diagram is required. Computer software such as Harvard Project Manager 3.0 will print out AON diagrams (using rectangles rather than circles for nodes).

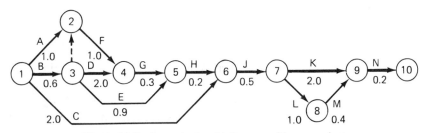

Figure 14-9. Arrow (network) diagram of house project.

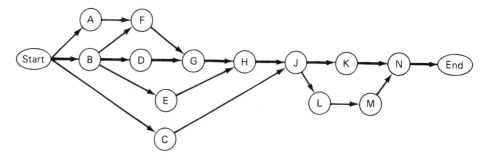

Figure 14-10. Circle, or activity-on-node, (network) diagram of house project.

PERT Treatment of Uncertainty

A special feature developed with PERT is the treatment of activity durations (and therefore total project duration) as variables rather than constants. To use this feature, estimators are asked to provide three estimates of the duration of any activity that might vary:

1. An *optimistic time (a)* that would only be improved upon once in 100 attempts
2. A *most likely time (m)* that would occur most often if the activity were repeated many times (statistically, the *mode*)
3. A *pessimistic time (b)* that would only be exceeded once in 100 attempts

The developers of PERT assumed that the probability distribution of possible durations of an activity fits a *beta distribution,* which need not be symmetrical (*m* need not be equidistant between *a* and *b*). The expected time (or mean value) t_e in the beta distribution can be approximated by

$$t_e = \frac{a + 4m + b}{6}$$

For example, if an activity were estimated to have an optimistic time of 10 weeks, a most likely time of 13 weeks, and a pessimistic time of 19 weeks, one would predict an expected (mean) time of

$$t_e = \frac{10 + 4(13) + 19}{6} = \frac{81}{6} = 13.5 \text{ weeks}$$

Assuming the optimistic (*a*) and pessimistic (*b*) estimates for duration of an activity are three standard deviations on either side of the mean t_e, the standard deviation σ for the activity becomes

$$\sigma = \frac{b - a}{6} = \frac{19 - 10}{6} = 1.5 \text{ weeks}$$

The expected length of the critical path T_e for the entire project is obtained simply by adding the expected times t_e for (only) those activities lying on the critical path. Standard deviations cannot be added in the same way—only variances (the squares of standard deviations) can. The standard deviation σ_T of the total project duration therefore becomes the *root mean square* of the standard deviations of activities lying along the critical path:

$$\sigma_T = \sqrt{\sum(\sigma^2)}$$

According to the *central limit theorem* of statistics, the probability distribution of the average or sum of a set of variables tends toward (approaches) the symmetrical *normal distribution,* even though the original variables fit other distributions. Knowing the expected time (mean) and standard deviation for the critical path permits us to draw a normal distribution fitting those two criteria.

For example, if the mean duration of the critical path T_e were calculated as 58.0 weeks and its standard deviation as 3.0 weeks, the critical path length would have the probability distribution shown in Figure 14-11. Then if you had a contract to complete the project in 61.0 weeks [which is (61.0 - 58.0)/3.0, or 1.0, standard deviation longer than the mean of 58.0 weeks], you would estimate an 84% probability (50% + 34%) of completing the project within that time. The probability of completion within a 52-week year, 2.0 standard deviations (6.0 weeks) less than the mean T_e, on the other hand, would only be (50.0 - 34.1 - 13.6 =) 2.3%.

PERT calculations normally consider only the longest (critical) path. If there is a second near-critical path with a duration close to the critical one, ignoring it may lead to an overly optimistic estimate of the probability of completion. This error can be eliminated by using the Monte Carlo (simulated sampling) method and averaging many simulated trials of the project, as discussed under "Simulation" in Chapter 4. The major errors in estimating project duration, however, lie in the accuracy with which the three estimates (or even one) can be provided for each activity, and the assumption that variations in individual activities are independent of one another. As a result, most projects using PERT use only one estimate of each task and avoid this calculation.

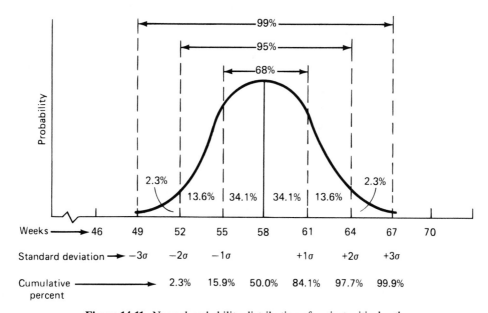

Figure 14-11. Normal probability distribution of project critical path.

Other Project Network Models

Quite a few variations and extensions of PERT and CPM have been developed to provide additional capability. Wiest[4] compares a "baker's dozen" of them. We describe a few of the best known:

Precedence diagramming method (PDM). PERT and CPM assume that a succeeding activity can begin when, and only when, all preceding activities have ended. There are many occasions when greater flexibility might be desired. For example, you might wish a concrete footing to cure 2 weeks before you began to build upon it. In CPM you would have to insert between pouring the concrete and the successor activity an additional activity "cure concrete" with a time duration of 2 weeks (but no cost); in PDM you connect the two with a "finish to start" (FS) delay constraint of 2 weeks. In building a mile of road you might estimate it would take 10 weeks to clear and grade the mile and a total of 15 weeks to surface it, but obviously the surfacing can begin at one end before the grading is complete at the other. In PDM this overlap is handled easily by incorporating "start to start" (SS) and "finish to finish" (FF) constraints.

GERT (graphical evaluation and review technique). PERT and CPM assume that when predecessor activities are complete, specific successors can begin with certainty. GERT (or VERT, a related technique) assumes that alternate paths are possible, each with some probability. For example, one project activity might be a component test. The test might have a 0.9 probability of success, with assembly into a system the next activity; it would also have a 0.1 probability of failure, with component modification and retesting, instead, the succeeding activities. Time durations for complex projects using GERT can be estimated by running them many times in Monte Carlo simulations.

PERT/LOB. PERT and CPM are designed for projects, which are one-of-a-kind activities. For production of a series of identical items, each following the same network logic, PERT has been combined with the "line of balance" (LOB) methodology developed for production planning.

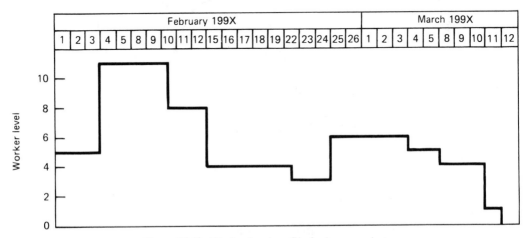

Figure 14-12. Worker-level profile for house project.

Resource Allocation

The bar chart schedule for our house project (Figure 14-6) was prepared by beginning each activity at the earliest possible time, but this may not lead to the best use of resources. In our data for the house project (Table 14-2) we included a "manning level" in the right-hand column. If we assume that all workers are "jacks of all trades" and therefore interchangeable, the personnel required in our "earliest possible" schedule will vary with time as shown in Figure 14-12. Eleven workers are needed, the last three for less than a week's work. They may not be available or may be expensive to import and train for that brief period.

Figure 14-13 provides a modified schedule that requires only eight workers, obtained by delaying the start of activity E by its slack (float) of 1.4 weeks. Modern project management software with resource allocation provisions can help in scheduling tasks that employ a variety of resources that may be in limited supply: qualified welders or other craftsmen, large cranes or other expensive equipment, or fixed delivery rates for common-use materials. The software proceeds through time by selecting among alternative activities that could begin at the same point, using a specified heuristic scheduling rule (such as first scheduling tasks with the least slack or with the earliest values of late finish time). The schedules that result are not always the shortest possible, but they are much better than would be obtained by a random selection among tasks.

We cannot complete our sample "house project" within the minimum 5.8 week time with fewer than eight workers. However, we may prefer a profile that does not reduce workers to three for a short period (February 22–24). This might be possible by reexamining our initial logic and perhaps concluding that some of the finishing work of activities K and L could begin before the roof (activity J) was complete. Another reason for deferring the start of activities not on the critical path, especially in projects of long duration, is to defer the expense they involve. In our project, for example, we can delay buying other materials and components (activity C) for a week, not only deferring the expense, but also reducing the storage space needed on the job site and the potential for theft and weather damage.

TYPES OF CONTRACTS

Contracts can be classified by the manner in which contract costs are borne. Two broad classes are fixed-price contracts and cost contracts; each has a number of variations. We discuss contract types in approximate order of decreasing contractor risk and increasing buyer risk.

Fixed-Price Contracts

Firm fixed-price contracts require the contractor to provide an agreed product and/or service for a specified price. The contractor realizes the total profit or loss and thus has maximum motivation to eliminate waste. The buyer must assure that the specifications define clearly what has been purchased for the fixed price.

Fixed-price with escalation contracts are firm fixed-price except that the contractor's price can be adjusted to incorporate any increases in specified labor and material rates. This

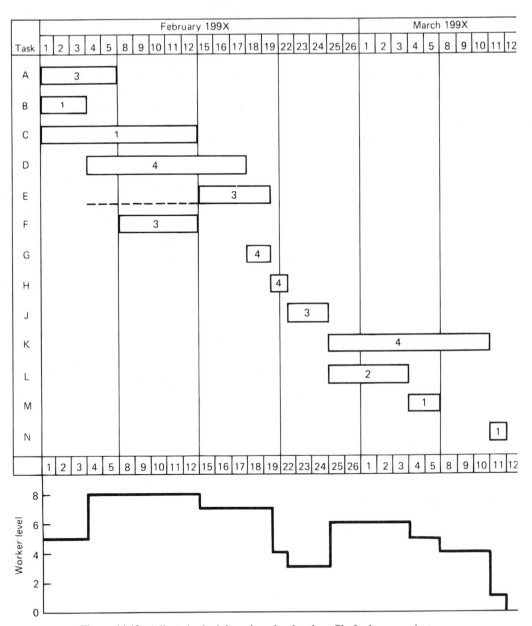

Figure 14-13. Adjusted schedule and worker-level profile for house project.

is especially useful in construction work where industry-wide union wage rates are scheduled for renegotiation during the life of the contract.

Fixed-price, redeterminable contracts can be adjusted later to reflect actual costs. The contractor therefore has less motivation to keep costs to a minimum before that adjustment is made.

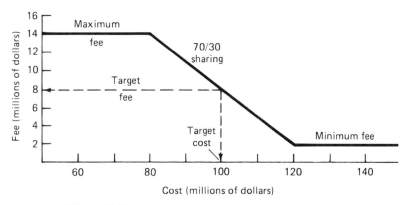

Figure 14-14. Example of cost plus incentive fee contract.

Fixed-price incentive contracts provide that the contractor and buyer share savings within a certain range, but establish a maximum cost for the buyer above which the *contractor* bears the total risk.

Cost Type Contracts

Cost plus incentive fee contracts establish an estimated target cost and target fee (profit). Within a specified range about the target, contractor and buyer share added costs or savings in an agreed ratio, but outside that range the *buyer* bears the total risk. In Figure 14-14, for example, the target (estimated) cost was $100 million, the target fee 8% or $8 million, and the customer/contractor sharing ratio 70:30 within a variation of ±20% of target cost. If the work actually cost $80 million or less, the contractor would gain 30% of the first $20 million saved, increasing his fee to $14 million on less work. If the cost overran to $150 million, the contractor would only have to share 30% of the first $20 million overrun, and would still receive a fee of $2 million regardless of cost.

Cost plus fixed-fee contracts require the buyer to pay all costs, plus an agreed-upon fee. In this example, if the initial estimate of costs were $100 million and the fee rate agreed were 8%, as above, the contractor would receive an $8 million fee for services regardless of cost overruns. This arrangement is not uncommon in R&D contracts, where the scope of work is difficult to estimate in advance. Obviously, the contractor's only motivation for reducing waste is the effect of costs on future contracts. Indeed, there is every motivation to load onto the contract new employees for training or improper overhead charges; both contractor and customer have a considerable accounting burden regarding allowable costs.

Time and materials contracts are common in repair, maintenance, and emergency situations, and they involve payment at agreed-upon rates (high enough to include profit) for hours worked, plus reimbursement of invoices for actual materials used. The only motivation for efficiency is the hope of future work; the advantage is its speed and economy for small jobs not worth detailed estimating.

A *letter contract* is a preliminary contract authorizing the contractor to proceed with specified work at customer cost in the interim until a formal contract is negotiated.

DISCUSSION QUESTIONS

14-1. Which of the following would be considered a project: (**a**) construction of a hydroelectric dam; (**b**) operation of a nuclear reactor; (**c**) development of an engine for the B-2 bomber through the first production prototype; (**d**) a production order for an additional twenty F-15 aircraft?

14-2. For a project-driven organization with which you are familiar, describe the approach used in advance of RFPs or bid requests to maintain competitiveness in being able to respond to them.

14-3. What would be some of the difficulties faced during the actual proposal preparation process, and how can they be minimized?

14-4. Prepare (**i**) a milestone schedule and (**ii**) a work breakdown structure for one of the following: (**a**) construction of a steel and concrete highway bridge, or (**b**) movement of the printing department of a daily newspaper from one building to another (assuming two identical presses, with one press continuously available for use).

14-5. Establish tasks, times, and precedence relationships for the project selected in question 14-4, and draw (**a**) a bar chart and (**b**) a network diagram schedule for it.

14-6. For the project outlined in the following table, prepare (**a**) a bar chart, (**b**) an arrow network diagram, and (**c**) a circle (AON) network diagram. (**d**) What and how long is the critical path?

Task	Follows Task(s)	Weeks Duration
A	Start	3.0
B	Start	7.0
C	A	2.0
D	B	7.0
E	B, C	5.0

14-7. For the project outlined in the following table, prepare (**a**) a bar chart, (**b**) an arrow network diagram, and (**c**) a circle (AON) network diagram. (**d**) What and how long is the critical path?

Task	Follows Task(s)	Weeks Duration	Manning Level
A	Start	6.0	3
B	Start	5.0	4
C	Start	5.0	3
D	A	3.0	2
E	A, B	6.0	5

14-8. For the project in question 14-7: (**a**) provide a worker-level profile assuming that all tasks begin as early as possible; (**b**) repeat the profile assuming that no more than nine people are available in any week and the manning level for a task cannot change; and (**c**) identify the project duration in each case.

14-9. Tasks X, Y, and Z must be completed in series to complete a project. The three time estimates (a, m, and b) for each task in days are X: 30, 45, and 66 days; Y: 24, 42, and 60 days; and Z: 26, 50, and 68 days. For each task, calculate (**a**) the expected time t_e and (**b**) the standard deviation σ. What is the (**c**) expected time T_e and (**d**) the standard deviation σ_T for the complete project?

14-10. If a project has an expected time of completion T_e of 45 weeks with a standard deviation σ_T of 7 weeks, what is the probability of completing it (**a**) within one year (52 weeks)? (**b**) Within 38 weeks?

14-11. Tasks *A, B,* and *C* must be completed in series to complete a project. The three time estimates (*a, m,* and *b*) for each task in days are *A:* 8, 11, and 14 days; *B:* 7, 10, and 19 days; and *C:* 10, 19, and 22 days. For each task, calculate (**a**) the expected time t_e and (**b**) the standard deviation σ. What is the (**c**) expected time T_e and (**d**) the standard deviation σ_T for the complete project? What is the probability of completing the project in (**e**) 40 weeks? (**f**) 46 weeks?

14-12. Describe what types of products or services each of the following contracts would be appropriate for, and give an example for each: (**a**) firm fixed-price; (**b**) cost plus fixed-fee; and (**c**) time and materials.

NOTES

1. Frank Dougherty, Assistant for Contract and Quality Matters, U.S. Dept. of Defense, Directorate of Industrial Productivity/Quality, presentation on "Acquisition Streamlining" to the American Society for Quality Control Midwest Conference, St. Louis, MO, October 9, 1986.

2. Often the first level (the project as a whole) is numbered 1.0, the second level 1.1, 1.2, 1.3, . . ., the third 1.1.1, 1.1.2, . . ., and so forth, so that *every* activity begins with the useless digit "1." for no apparent reason!

3. Copyright 1988 by Software Publishing Corporation.

4. Jerome D. Wiest, "Gene-Splicing PERT and CPM: The Engineering of Project Network Models," in Burton V. Dean, ed., *Project Management: Methods and Studies* (Amsterdam: Elsevier Science Publishers B.V., 1985), pp. 67–94.

CHAPTER 15

Project Organization, Leadership, and Control

PREVIEW

We begin this chapter by enumerating the elements needed in the project-driven organization and alternative organization designs. We pay special attention to the nature and functioning of the matrix management organization. Next, we look at the project manager: personal characteristics, career development, and the importance and content of a charter defining responsibilities and authority. Then, we consider methods of motivating effective project performance through team building and conflict management. Finally, we introduce key tools for monitoring and controlling project schedule and cost.

PROJECT ORGANIZATION

The Project-Driven Organization

Kerzner classifies organizations into two groups, based on whether or not their dynamics are primarily project driven. He distinguishes them as follows:

> In a project driven organization, such as construction or aerospace, all work is characterized through projects, with each project as a separate cost center having its own profit and loss statement. The total profit of the corporation is simply the summation of the profits on all projects. In a project-driven organization, everything centers around the projects.
>
> In the non-project-driven organization, such as low-technology manufacturing, profit and

loss are measured on vertical or functional lines. In this type of organization, projects exist merely to support the product lines or functional lines. Priority resources are assigned to the revenue-producing functional line activities rather than the projects.[1]

The legal forms and the patterns of departmentation (subdivision) for traditional (non-project-driven) organizational styles have been considered in Chapter 5. In this section, we take up organizational alternatives for the project-driven organization or division.

Elements of the Project-Driven Organization

Every project-driven organization needs four different categories or types of elements, and project organizations can be characterized by the number of these categories coming under the direct control of the individual project manager. These four categories are:

1. *The project office.* Every project needs a "unifying agent" of some type that bears primary responsibility for the project. In a small project, the project manager may serve this function alone. A larger project with any substantial design/development responsibility will usually have a *project engineer,* responsible for the technical integrity of the project and the cost and schedule of engineering activities. Another member of the project office is usually (by whatever name) the project administrator, responsible for project planning and control systems and documentation. Other functions frequently centered in the project office include design review and configuration/change control.

2. *Key functional support.* Support from certain key functional areas is central to project success, and activities in these areas often must be tailored to the specific needs of the projects. These key support areas may include:

   - Systems analysis, systems engineering, and integration
   - Product design and analysis
   - Quality assurance and reliability
   - Production planning
   - Product installation and test
   - Training, logistics planning, and field support

3. *Manufacturing and routine administration.* These activities are less likely to be under direct project control: manufacturing because it is usually too expensive to replicate for each project, and the others because they are service activities done essentially the same way for all projects. Type 3 activities commonly include:

   - Manufacturing
   - Accounting and finance
   - Purchasing and subcontracting, although project subcontract administration may fall under category 2
   - Personnel and industrial relations
   - Plant facilities and maintenance

4. *Future business.* Activities such as non-project-specific R&D and marketing are

necessary for the continuation of the project-driven organization. However, these activities are not properly part of specific projects, since projects by definition should have a definable end point.

Projectized Versus Functional Organizations

Conducting projects within the functional organization. A functional organization (such as that of Figure 15-1) is subdivided at the top into functional areas. When a project involving several of these areas is to be conducted, a project manager (or coordinator, or expeditor) is appointed to coordinate the activities of the various people working on the project. This person might be attached as staff to the general manager over all the functions involved, or may be the supervisor of the function most heavily involved in the project or a key subordinate of that supervisor, as shown in the figure. The designated project leader usually has no line authority over the bulk of project personnel—only over the immediate project office (category 1 of the four groups above) if one exists. He or she must therefore lead by persuasion and influence (the System II style introduced in Chapter 6). Nonetheless, projects are often conducted from within the functional organization—especially when they are small, short in duration, low in complexity, and schedule is not critical.

The projectized organization. In the fully projectized organization (Figure 15-2) the project or program manager is in direct control of all the elements needed to conduct that project (all of categories 1, 2, and 3). Such an organization is attractive for large, long-duration projects, especially those that are very complex, involve a number of different organizations, and require advancing the state of technology. For example, when the Space Division of North American Aviation (now part of Rockwell International Corporation) was

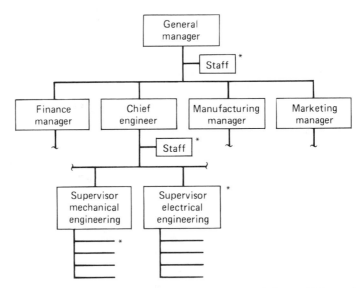

Figure 15-1. Typical functional organization [showing locations (*) from which a project manager or coordinator might lead a project].

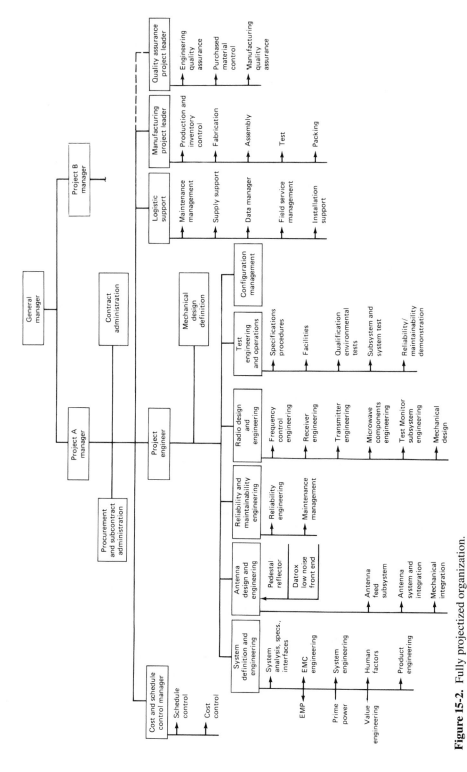

Figure 15-2. Fully projectized organization.
(Adapted from Russell D. Archibald, *Managing High-Technology Programs and Projects,* © John Wiley & Sons, Inc., New York, 1976, pp. 104–105. Reprinted by permission of John Wiley & Sons, Inc.)

awarded two multibillion dollar contracts (one for the Apollo command and service modules and the other for the second stage of the Saturn launch vehicle), management set up two separate program organizations in different locations. Each program was under a division vice president, and each had its own manufacturing plant and staff of specialists of all kinds. Little remained of the division outside of these two programs except the division president's staff and a vestigial new business operation. A variation of this design, the *partially projectized* organization, has the key activities (categories 1 and 2) directly under the project manager and the supplemental ones (category 3) remaining with the functional organization.

Comparing the two. The advantages and disadvantages of these two organization structures for project management can be compared as follows:

Functional Organization	**Projectized Organization**
Advantages	**Advantages**
Efficient use of technical personnel	Good project schedule and cost control
Career continuity and growth for technical personnel	Single point for customer contact
Good technology transfer between projects	Rapid reaction time possible
Good stability, security, and morale	Simpler project communication
	Training ground for general management
Disadvantages	**Disadvantages**
Weak customer interface	Uncertain technical direction
Weak project authority	Inefficient use of specialists
Poor horizontal communication	Insecurity regarding future job assignments
Discipline rather than program oriented	Poor crossfeed of technical information between projects
Slower work flow	

Matrix Management

A composite organization structure that combines many of the advantages of both functional and projectized management is the *matrix management* structure shown in Figure 15-3. In this system, the person assigned responsibility for a specific functional specialty on a specific project is accountable in two dimensions, reporting to both functional and project managers. This "two boss" reporting relationship defies the "unity of command" management principle articulated by Henri Fayol and, indeed, by most early management scholars. This leads to conflict unless the nature of these two reporting relationships is clearly understood.

Figure 15-4, which depicts a matrix structure simplified to show only one of many functional managers and one of several projects, helps explain the nature of these relationships and the benefits of the matrix organization. In the horizontal relationships, the project manager has control over the three key factors in project control: *what* has to be done (project tasks and performance), *when* it must be done (project schedule), and *how much* in the way of resources will be allocated (project budget). The project manager also provides a central focal point for customer contact, decisions on project changes, and project communication.

The functional manager, on the other hand, has primary responsibility for assignment of functional specialists, and he or she can therefore try to optimize the distribution of key specialists among the several project and nonproject assignments competing for them. The

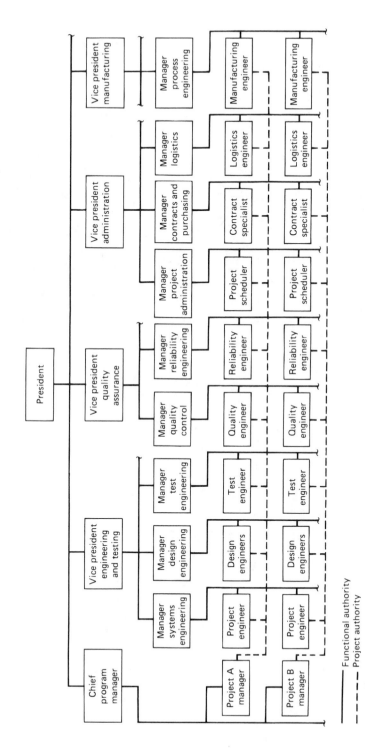

Figure 15-3. Typical matrix organization.

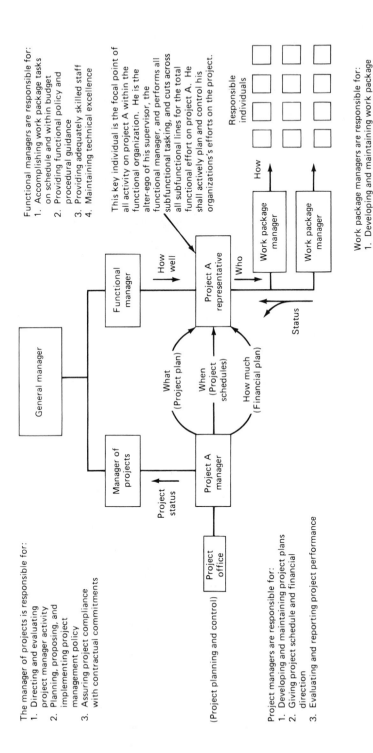

The manager of projects is responsible for:
1. Directing and evaluating project manager activity
2. Planning, proposing, and implementing project management policy
3. Assuring project compliance with contractual commitments

(Project planning and control)

Project managers are responsible for:
1. Developing and maintaining project plans
2. Giving project schedule and financial direction
3. Evaluating and reporting project performance

Functional managers are responsible for:
1. Accomplishing work package tasks on schedule and within budget
2. Providing functional policy and procedural guidance
3. Providing adequately skilled staff
4. Maintaining technical excellence

This key individual is the focal point of all activity on project A within the functional organization. He is the alter-ego of his supervisor, the functional manager, and performs all subfunctional tasking, and cuts across all subfunctional lines for the total functional effort on project A. He shall actively plan and control his organizations's efforts on the project.

Responsible individuals

Work package managers are responsible for:
1. Developing and maintaining work package plans for accomplishment
2. Establishing work package technical guidance
3. Establishing work package detailed schedule and operating budgets
4. Controlling and reporting work package performance

Figure 15-4. Matrix relationships.
(From David I. Cleland and William R. King, *Systems Analysis and Project Management*, 3rd ed., McGraw-Hill Book Company, New York, 1983, p. 353.)

functional manager also is responsible for assuring the quality of work done on projects or elsewhere in that specialty, and for selecting, evaluating, and rewarding work done in that specialty. The specialists, in turn, benefit from communication with their peers and from having a congenial "technical home" to return to when a project assignment is complete.

The functional specialists assigned to the matrix organization (or assigned temporarily to a projectized organization) develop an understanding of other functions that enhances their personal growth in the organization. They also develop contacts outside their narrow function that persist for some time, and research has shown a correlation between communication outside the specialized working group and the effectiveness of a technologist.[2]

Young[3] distinguishes between shifting and fixed matrix organizations. In the "shifting matrix," common in the aerospace industry, personnel are shifted between projects depending on the work load and project cycle, and the project team is disbanded at the end of the project. In the less common "fixed matrix" organization, the project manager is responsible for a successive series of projects, and functional personnel "are always assigned to the same project managers whatever the project."

The matrix organization has many applications aside from project management. Newman et al. generalize these applications:

> Matrix organization basically gives an operating manager two bosses. One boss deals with mobilizing resources, techniques of production, and other aspects of creating the product (or service). The other boss is concerned with creating a product that pleases the customer—the right performance characteristics, quality, delivery time, and so on. To avoid too much attention to either the input side or the output side, the two bosses negotiate do-able instructions for the operating managers.[4]

A classic application is a consumer products firm such as a soap manufacturer, where a *product manager* is assigned for each product line (e.g., home dishwasher detergents or industrial floor cleaners); the second dimension consists of the functional resources of R&D, production, sales, and advertising needed by all the product managers. Another example[5] is an advertising firm with one dimension of *account executives* responsible for satisfying a specific client, and the other the common resources such as "market research, copyrighting, art work, television production, media selection, and other functions."

Organization Structure and Project Success

Larson and Gobeli describe an extensive study sponsored by the Project Management Institute (PMI) in which the effectiveness of five project management structures were compared:

1. *Functional Organization:* The project is divided into segments and assigned to relevant functional areas and/or groups within functional areas. The project is coordinated by functional and upper levels of management.
2. *Functional Matrix:* A person is formally designated to oversee the project across different functional areas. This person has limited authority over functional people involved and serves primarily to plan and coordinate the project. The functional managers retain primary responsibility for their specific segments of the project.

3. *Balanced Matrix:* A person is assigned to oversee the project and interacts on an equal basis with functional managers. This person and the functional managers jointly direct workflow segments and approve technical and operational decisions.
4. *Project Matrix:* A manager is assigned to oversee the project and is responsible for the completion of the project. The functional manager's involvement is limited to assigning personnel as needed and providing advisory expertise.
5. *Project Team:* A manager is put in charge of a project team composed of a core group of personnel from several functional areas and/or groups, assigned on a full-time basis. The functional managers have no formal involvement.[6]

(Their two functional structures are versions of the "functional organization" described previously, the balanced and project matrix structures versions of the matrix structure, and the "project team" is a highly projectized organization.)

Larson and Gobeli mailed a questionnaire to PMI members, and they used data from 547 respondents who answered "a series of questions concerning a recently completed development project they were familiar with. . . . Respondents were simply asked to evaluate their project according to (a) meeting schedule, (b) controlling cost, (c) technical performance, and (d) overall performance with a response format of 'successful,' 'marginal,' and 'unsuccessful.'"[7] While the functional matrix (2) was perceived as more successful than the functional organization (1), the other three (3,4,5) were perceived as more successful than either of these in all four measures (a,b,c,d). Of the latter three (3,4,5), the project matrix (4) was judged best in controlling cost and, along with the project team (5), in meeting schedule.

THE PROJECT MANAGER

Characteristics of Effective Project Managers

Project managers need enthusiasm, stamina, and an appetite for hard work to withstand the special pressures of project management. Where possible, project managers should have seniority and position in the organization commensurate with that of the functional managers with whom they must negotiate. Whether they are project coordinators within a functional structure or project managers in a matrix structure, they will often find their formal authority incomplete, and they need a blend of technical, administrative, and interpersonal skills to provide effective leadership.

Technical skills. Many projects depend for their success on effective application of certain key technologies. The effective manager of such projects must understand the essentials of those technologies enough to evaluate whether the work done is of sufficient quality, even if he or she is not as expert as the specialists actually doing the work. Further, when an unfamiliar technology is involved in a problem on the project, the program manager must quickly be able to master the essential technology bearing on the problem from briefings by specialists, so that he or she can articulate the problem to client or general management and make effective decisions regarding resolution.

Administrative skills. Project managers must be experienced in planning, organizing, staffing, and control techniques as they apply to projects. In particular, they should understand the project planning techniques, such as the work breakdown structure, network systems, and others discussed in Chapter 14, design control methods such as design review and configuration/change control, and project cost control methods such as the "earned value" system discussed later in this chapter, especially as they are carried out in that particular company.

Interpersonal skills. Except in fully projectized organizations, project managers depend heavily on the work of others not under their line control. The ability to inspire, cajole, negotiate, and persuade others therefore becomes very important, and project managers need a good understanding of conflict resolution methods.

Developing Project Management Skills

Managers of large projects typically began in some specialty of engineering or business, learned project planning and control while applying their specialty in a project environment, and were assigned responsibility for a major project only after a series of project and functional assignments of increasing responsibility. However, engineers may find themselves assigned to small projects with little or no preparation. There are many short courses offered by universities, professional societies, and consultants, and many books available on the subject of project management, but none fully substitute for experience. Meetings and publications of the Project Management Institute, of engineering management divisions of the major engineering societies, and of the American Society for Engineering Management also help in acquiring project management skills.

The Manager's Charter

Of vital importance to the project manager is his or her *charter,* or scope of authority. It is highly desirable that the responsibilities and authority of the project manager be defined in writing in advance to clarify the interfaces between the project manager, functional managers, and others, and to reduce the potential for conflict and confusion. Following are some of the areas a project manager might like to see covered in such a charter:[8]

1. Specification of project priority relative to other activities
2. Designation as the primary contact with the customer
3. Authority to define the work to be performed by supporting departments in terms of cost, schedule, and performance
4. Control over the project budget, with signature authority on all work authorizations
5. Responsibility to schedule and hold design reviews, determine the agenda and representation, and establish responsibility for follow-up action
6. Responsibility for configuration and change control and for approving changes
7. Authority to constitute and chair the make-or-buy and source selection board
8. Responsibility for regular reporting to general management of project status and identification of any factors inhibiting project success
9. Participation in the merit review process for all personnel on loan to the project

Few project managers will be granted all the authorities suggested above, but the authority relationships with functional managers and among projects should be clarified where possible for more effective project performance.

MOTIVATING PROJECT PERFORMANCE

Team Building

Thamhain introduces his excellent chapter on "Team Building" by highlighting its importance to project/program management:

> Building the project team is one of the prime responsibilities of the project or program manager. Team building involves a whole spectrum of management skills required to identify, commit, and integrate various task groups from traditional functional organizations into a single program management system. This process has been known for centuries. However, it becomes more complex and requires more specialized management skills as bureaucratic hierarchies decline and horizontally oriented teams and work units evolve.[9]

The newly formed team begins with considerable lack of clarity about purposes, responsibilities, expectations, and a general lack of communication, commitment, and team spirit. Thamhain suggests specific measures for preventing such problems from developing in the first place:

1. The importance of the project to the organization, including its principal goals and objectives, should be [made] clear to all personnel who get involved with the project. . . .
2. Project leadership positions should be carefully defined and staffed at the beginning of the team formation stage. . . . The capabilities, interests, and commitments to the project should be assured before any of the lead personnel are signed up. One-on-one interviews are recommended for explaining the scope and project requirements, as well as the management philosophy, organizational structure, and rewards.
3. Members of the newly formed team should be closely located to facilitate communications and the development of a team spirit. Locating the project team in one office area is the ideal situation. However, this may be impractical, especially if team members share their time with assignments on other projects or the assignment is only for a short period of time. Regularly scheduled meetings are recommended as soon as the new project team is being formed. These meetings are particularly important where team members are geographically separated and do not see each other on a day-to-day basis.
4. All project assignments should be negotiated individually with each prospective team member. . . . The assignment interview should include a clear discussion of the specific task, the outcome, timing, responsibilities, reporting relation, potential rewards, and importance of the project to the company.
5. Management must define the basic team structure and operating concepts during the project formation stage. . . .

6. The project manager should involve at least all key personnel in the project definition and requirements analysis. . . .

7. The project manager should conduct team-building sessions throughout the project life cycle. An especially intense effort might be needed during the team formation stage. . . .

8. Project leaders should try to determine lack of team member commitment early in the life of the project and attempt to change possible negative views toward the project. . . . Finally, if a team member's professional interests lie elsewhere, the project leader should examine ways to satisfy part of the team member's interests or consider replacement.

9. It is critical for senior management to provide the proper environment for the project to function effectively. . . .

10. Project managers must understand the various barriers to team development and build a work environment conducive to the team's motivational needs. Specifically, management should watch out for the following barriers: (1) disinterested team members, (2) uninvolved management, (3) unclear goals and priorities, (4) funding uncertainty, (5) role conflict and power struggle, (6) incompetent project leadership, (7) lack of project charter, (8) insufficient planning and project definition, (9) poor communication, and (10) excessive conflict, especially personal conflict.

11. Project leaders should watch for changes in performance on an ongoing basis. If performance problems are observed, they should be dealt with quickly by the team [with the help of organizational development specialists if available].[10]

Managing Conflict

Sources of conflict. Conflict is inevitable in any organization, just as it is in any relationship. Conflict that is due to pettiness, lack of understanding of the other person or other group, or intolerance should be avoided. However, much conflict in organizations is natural, and stems from honest disagreement on priorities and the use of scarce resources. Modern emphasis on *conflict management* looks for ways to resolve such conflict positively to the overall benefit of the organization.

Thamhain[11] describes seven potential conflict sources, listed here in order of decreasing perceived intensity over the project life cycle:

1. *Conflict over schedules*
2. *Conflict over project priorities,* including conflict over the sequencing of activities and tasks to be undertaken
3. *Conflict over work-force resources,* especially in obtaining the desired quality and quantity of personnel from other functional and staff support areas
4. *Conflict over technical opinions and performance trade-offs*
5. *Conflict over administrative procedures* that define how the project will be managed, especially the project manager's reporting relationships, responsibilities, and authority
6. *Personality conflict*
7. *Conflict over cost* and the funds allocated to functional support groups

The relative importance of these seven conflict sources varies over the project life cycle, as illustrated in Figure 15-5. Thamhain reported the largest concern at project formation to be with administrative procedures (especially the charter establishing the relative authorities of project and functional managers); Posner reported[12] from research a decade later that this source of conflict has greatly diminished as project management organizations have been institutionalized. As the project progresses into the early and main project phases the overall level of conflict increases, first with staffing concerns and then with priorities, schedules, costs, and the technical issues that surface as subsystems are developed. Toward the end of the project the overall level of conflict is reduced, and the most likely forms of conflict relate to schedule slippages that delay project completion and (in Posner's later study) project costs.

Methods of conflict management. One obvious method of resolving conflict between two individuals is to appeal the matter up the chain of command to the level having authority over both individuals (the corporate equivalent of "I'll tell Daddy!"); and in some cases this will be necessary. However, higher executives simply do not have the time to solve everyone's squabbles, and they expect their subordinate managers and professionals to be mature enough to solve their own problems most of the time. Blake and Mouton[13] identify five methods for dealing with conflict:

1. *Withdrawal,* or retreat from actual or potential conflict
2. *Forcing* one's viewpoint at the potential expense of the other party
3. *Smoothing,* or emphasizing the points of agreement and deemphasizing areas of conflict
4. *Compromising* or negotiating, in which each party must give up something but each walks away partly satisfied
5. *Confronting or problem solving,* in which the parties focus on the issues, consider alternatives, and look for the best overall solution

Thamhain and Wilemon[14] examined the methods most and least favored by 100 effective engineering project managers. Confrontation (problem solving) was clearly preferred, followed by compromise and smoothing. Forcing, which leaves a residue of resentment that may backfire in the future, was the next-to-last choice, and withdrawal was considered least effective (except perhaps temporarily to allow tempers to cool).

Keys to Project Success

Baker, Murphy, and Fisher deduce the following definition of success from research conducted on 650 projects:

> If a project meets the technical performance specifications and/or mission to be performed, and if there is a high level of satisfaction concerning the project outcome among: key people in the parent organization [in which the project is carried out], key people in the client organization, key people on the project team, and key users or clientele of the project effort, the project is considered an overall success.[15]

Characteristics that strongly affect perceived failure of projects were found to include:

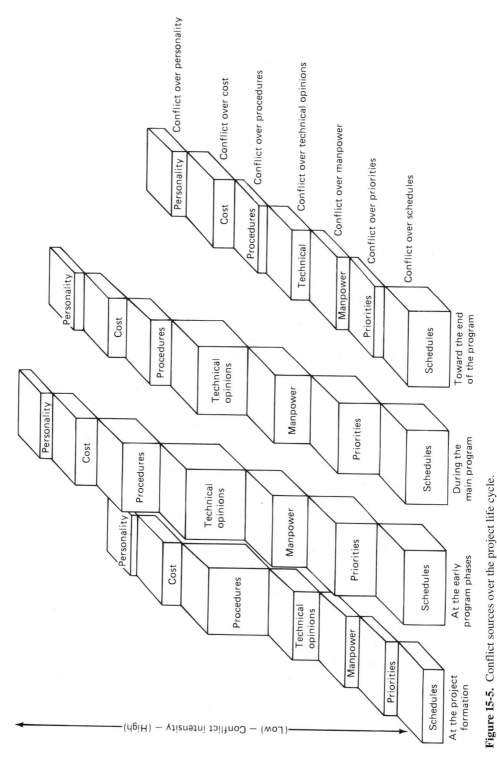

Figure 15-5. Conflict sources over the project life cycle.
(From Hans J. Thamhain, *Engineering Program Management*, © John Wiley & Sons, Inc., New York, 1984, p. 244. Reprinted by permission of John Wiley & Sons, Inc.)

319

- Inadequate project manager skills, influence, and authority
- Poor coordination and rapport with the client
- Poor coordination and rapport with the parent organization
- Lack of project team participation and team spirit
- Poor project control: inability to freeze design or close out the project, unrealistic schedules, inadequate change procedures, and/or inadequate status/progress reports
- Project of different type or more complex than handled previously and/or initially underfunded
- Poor relations with public officials or unfavorable public opinion

Characteristics associated with project success emphasized the *commitment* of the parent organization, project manager, and client to established schedules, budget, and technical performance. Frequent feedback from the parent organization and the client, adequate control procedures (especially change control), public support, and lack of excessive government red tape were also among these conditions deemed necessary, but not sufficient, for perceived success.

Baker et al. also made the following observations:

- Cost and schedule overruns were not among the characteristics significantly related to perceived failure, and meeting these targets was not significantly related to perceived success of past projects [but they may certainly affect the survival of managers *during* a project]. In the long run, what really matters is whether the parties associated with, and affected by, a project are satisfied.
- While judicious use of PERT/CPM systems contributes to better cost and schedule performance, their importance is far outweighed by the use of system management tools such as "work breakdown structures, life cycle planning, systems engineering, configuration management, and status reports."
- While each organizational structure discussed at the beginning of this chapter was associated with perceived success in certain situations, the projectized organization structure was most often so associated. This is consistent with the findings of Larson and Gobeli[16] discussed earlier in this chapter.
- The client and parent organizations need to agree on definite goals for the project and to develop close and supportive relationships with the project team, but then need to avoid meddling or interfering with the project team's decision-making process.
- Participative decision making within the project team is highly correlated with success, but public participation in projects affecting the public interest often delays and hampers projects and reduces the probability of success.
- The most important skills in the project manager are technical skills, followed by human and then administrative skills.
- "The research described in this paper [Baker et al.] supports the concept of a leader who is task oriented with a back-up social orientation for *most* project efforts. . . . a project manager may need to employ different leadership styles at different times during the project effort."[17]

Katz and Allen have found[18] that project performance reaches its highest level "when organizational influence is centered in the project manager and influence over technical details of the work is centered in the functional manager." In particular, they reported higher performance "when influence over salaries and promotions is perceived as balanced between project and functional managers" rather than being concentrated in the functional manager.

Customer (Client) Communications

The research of Baker et al. noted above makes clear the need for good coordination and rapport with the client (or customer), and this requires good communication. Cleland and Kerzner point out[19] that this communication takes place at three levels, as shown in Figure 15-6. Formal communications are normally between the project manager responsible for project performance in the contractor organization and the project manager or other person designated to monitor project performance in the customer organization. Usually, a separate contracts officer in one or both organizations will be involved when actual changes to the legal contract are discussed.

Often there is a senior management level of informal contact as well, providing checks and balances to assure that correct information has been transferred between contractor and customer and a channel to resolve any conflicts generated at the project manager level. This makes effective vertical communication essential as well, so that executives on each side are apprised of project progress and problems, and the executives themselves pass on to their subordinate project managers any insight or agreements resulting from higher-level communications.

A final, informal, level of communication in any large project is between employees of the client and contractor employee specialists. The customer will often demand this, feeling that essential information is filtered out in the summary that comes from the project manager. The danger of this channel is the temptation at the specialist level to propose and agree to "improvements" that change the scope of the project and require additional resources. En-

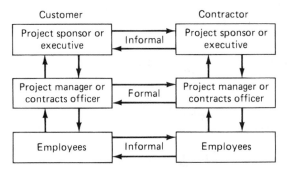

Figure 15-6. Customer (client) communications.
(From David I. Cleland and Harold Kerzner, *Engineering Team Management,* Van Nostrand Reinhold Company, Inc., New York, 1986, pp. 63–64.)

gineers and other specialists on both sides must learn that while they can agree that a change might be "nice to have," no work can be done on such a change until it has been negotiated through formal channels (usually using change control procedures such as those introduced in Chapter 10).

Early Learning Path

Balderston et al. emphasize, in their "Strategic Criteria for R&D Project Success,"[20] "that, in any development process, *the key management task is strategic creation of an early learning path.*" We discussed the phases of the engineering design process at the beginning of Chapter 10, emphasizing the need to reduce uncertainty in the early phases of the process, so that the later, more expensive phases are attempted only when as much early uncertainty as possible has been eliminated. Balderston et al. support this with their emphasis on the "early" learning path as shown in Figure 15-7a, in which project activities have been scheduled to gain maximum learning in the early period when cumulative project expenditure (Figure 15-7b) is still low.

These authors urge the project manager to focus on the specific technological approach used in each key area or subsystem:

- Where the *knowledge level* regarding the technological approach is low (uncertainty is high), the risk of project schedule and cost overruns due to the new approach is high. In this situation, they suggest adopting a contingency strategy by bringing along a more conventional, older strategy as a parallel development, with a system design that would accept either.
- Where the *analytic potential* of the approach is such that its performance cannot confidently be predicted by analysis, trial-and-error test and evaluation processes should be scheduled early.
- Where the *experience level* of the project staff with the technological approach is low, if the desired experience exists outside the organization its use should be considered and/or prototypes involving the proposed approach should be developed and tested early.

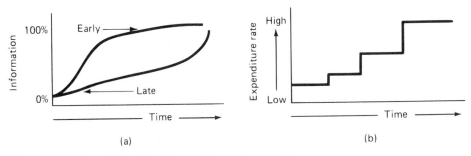

Figure 15-7. (a) Learning paths; (b) expenditure rate versus time.
(From Jack Balderston, Philip Birnbaum, Richard Goodman, and Michael Stahl, *Modern Management Techniques in Engineering and R&D*, Van Nostrand Reinhold Company, Inc., New York, 1984, pp. 135–149.)

CONTROLLING COST AND SCHEDULE

Project Control Systems

At the end of Part 2 we discussed the essential management function of control, which has been defined simply as "compelling events to conform to plans." There we found that the first step in control was to establish standards, and that this was done in the planning process. In Chapter 14 we introduced essential project planning methods such as the work breakdown structure, Gantt (bar) charts, and network scheduling systems such as PERT. Now we will see how these methods are used to help assure that project plans become reality.

Reducing Project Duration

Project managers often find during initial planning that the predicted project duration, found by summing up activity durations along the critical path, is too long to meet the required project completion date. Also, slippages along the critical path early in the project may predict a delay in project completion. This is an example of "feedforward control," since it provides early warning of future problems.

The first approach to reducing project duration to meet a desired completion date is to reexamine the logic used in sequencing activities. The project schedule may have been created by assuming that some activity X (grading a roadbed, for example) must be completed before activity Y (paving the road) can begin, and this sequence is later found to be on the critical path. The project manager may conclude that paving can begin as soon as half the roadbed is graded. Activities X and Y might be divided into smaller activities X_1, X_2, Y_1, and Y_2 such that Y_1 could begin as soon as X_1 was complete, and Y_2 as soon as both X_2 and Y_1 were complete, as shown in Figure 15-8a.

Precedence diagrams such as Figure 15-8b are an alternative tool of network planning, often used in construction, which can simplify the expression of a variety of precedence re-

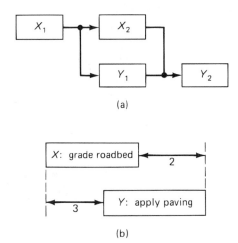

(a)

(b)

Figure 15-8. (a) Modified Gantt and (b) precedence diagrams.

lations between activities such as the roadbed/grading example just described. Figure 15-8b, for example, shows a start-to-start delay of three weeks and a finish-to-finish delay of two weeks between grading and paving.

In another variation, known as *fast tracking,* initial phases of manufacture or construction are begun before the design is complete for the remaining phases. The danger in this is that early work may have to be redone to accommodate unforeseen changes and unexpected results. Fast tracking should be attempted only when the advantages of time saving are compelling, and then only by experienced design and construction teams.

Crashing the Project

Another method of shortening a project is by reducing the duration of some activity along the critical path by applying more resources to it, or "crashing" it. The *normal time* T_n originally estimated for the activity is usually the time associated with the lowest cost (the *normal cost* C_n) to complete that activity. However, for many activities there will be some shorter duration (the *crash time* T_c) that can be achieved at some higher *crash cost* C_c) by using overtime, larger crews, more expensive equipment, or subcontractors. Each such activity along the critical path therefore has a *slope* defined as

$$\frac{C_c - C_n}{T_n - T_c}$$

(the negative of the normal algebraic slope definition) in terms of dollars per unit time reduction. The prudent project manager will add resources to "crash" the activities along the critical path having the lowest slope, as long as the cost of reduction is less than the benefits realized by a shorter project duration. This must be done with care, since reducing the duration of one path through the network often results in a parallel critical path having the same duration. When this occurs, an activity on each critical path must then be crashed at the same time to speed final project completion.

Consider the project shown in Figure 15-9. There are three paths through the network: A–C–E and B–D–E, with normal time durations of 18 weeks each, and the critical path

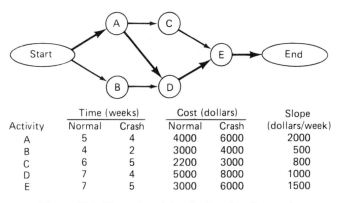

Activity	Time (weeks)		Cost (dollars)		Slope
	Normal	Crash	Normal	Crash	(dollars/week)
A	5	4	4000	6000	2000
B	4	2	3000	4000	500
C	6	5	2200	3000	800
D	7	4	5000	8000	1000
E	7	5	3000	6000	1500

Figure 15-9. Network and data for "crashing" example.

A–D–E, with a normal time duration of 19 weeks. The normal cost associated with completion of all activities in their normal time is $17,200.

In addition to normal cost and time, each activity has associated with it a higher cost C_c that would be required to complete the activity in a shorter time T_c. For example, activity B can be completed in four weeks for a normal cost of $3000, or in two weeks for an additional $1000. Assuming that costs are linear between these extremes, each activity then has a slope in dollars per week of reduction, as tabulated in the right-hand column in Figure 15-9 and shown graphically in Figure 15-10. The project manager will wish to crash the activity with the smallest slope *among those on the critical path.*

In our example, activity D would be reduced by one week at a cost of $1000, which is less than the cost of crashing A or E (and more than the $500 cost of crashing activity B or the $800 of C, but they would be of no help). This reduces critical path A–D–E to 18 weeks (and path B–D–E to 17 weeks). Note that path A–C–E remains at 18 weeks, and so it becomes a second critical path.

If we wish to shorten the project further, we must find a solution that shortens both paths A–C–E and A–D–E. Three solutions to save a second week are possible: to shorten activity A ($2000), to shorten both C and D ($800 + $1000, or $1800), or to shorten E ($1500). Naturally, we will choose the third solution, and we will complete the project in 17 weeks at a cost of $19,700 (17,200 + 1000 + 1500).

Earned Value System

In planning a project, activities are scheduled against time, and in budgeting the project, a cost for each activity is estimated. The *budgeted cost of work scheduled* (BCWS) can therefore be represented as a cumulative cost curve versus time. As the project progresses, labor, material, and other costs are carefully recorded. However, this *actual cost of work performed* (ACWP) cannot be compared effectively with the work scheduled to be accomplished by the same date, since some tasks will be behind schedule and some ahead of schedule. What is needed is a third measure, the *budgeted cost of work performed* (BCWP).

This "earned value" system works as follows: When a task (or separable part of a task) has been completed, the project is considered to have "earned the value" (BCWP) originally

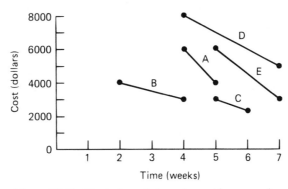

Figure 15-10. Illustration of slopes in crashing example.

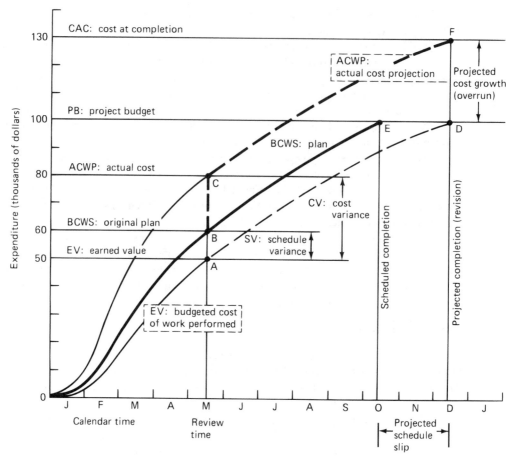

Figure 15-11. Illustration of earned-value system.
(From Hans J. Thamhain, *Engineering Program Management,* © John Wiley & Sons, Inc., New York, 1984, p. 136. Reprinted by permission of John Wiley & Sons, Inc.)

estimated (budgeted) for that task or segment. It then becomes possible to compare (as in Figure 15-11) the budgeted (estimated) and the actual costs of work performed to date to determine the *cost variance* CV (BCWP - ACWP) experienced thus far in the project.

Similarly, one can compare the value BCWP of work actually completed (performed) at some point in time with the value BCWS of work scheduled for completion by that time, to obtain the *schedule variance* SV (BCWP - BCWS). The analyst then projects the earned value (BCWP) curve to estimate the revised completion date, and projects the ACWP curve to that date to estimate the cost at completion (CAC). The U.S. Department of Defense requires contractors to demonstrate that their project control system will meet certain cost/schedule control system criteria (C/SCSC), in which a major requirement is ability to calculate cost and schedule variances in this way.

DISCUSSION QUESTIONS

15-1. The four categories or "elements of the project-driven organization" were written largely for projects in which systems were designed and manufactured. How would the categories and their contents differ for an architecture/engineering (A/E) firm whose projects involved the design of structures and management of their construction?

15-2. What types of projects are managed most effectively using the (fully or partially) projectized organization? What types are managed most effectively as part of the normal functional organization?

15-3. Discuss the ways in which the matrix management structure tends to reduce the problems of project management under (**a**) a fully projectized organization and (**b**) a normal functional organization. (**c**) What are the disadvantages of going to the matrix structure?

15-4. In the matrix management structure, the person responsible for the activity of a specific functional specialty on a specific project has two bosses. What considerations in a well-run matrix structure reduce the resulting potential for conflict?

15-5. In the text we discuss the technical, administrative, and interpersonal skills desirable in a project manager. What attributes do you believe are desirable in an engineering specialist working on a project in a matrix organization?

15-6. If you were a project manager, which three or four responsibility areas of the nine listed under "The Manager's Charter" (or others you think important) would you *most* want to have included in a charter granted you by management? State your reasons.

15-7. Discuss the precautions you might take as manager of a new project in building a project team to assure that the project does not get off "on the wrong foot."

15-8. Thamhain and Wilemon[21] tested the apparent preferences for each of the five modes of conflict resolution (compromise, confrontation, forcing, smoothing, or withdrawal) by measuring the extent of agreement of project managers with 15 aphorisms developed by Lawrence and Lorsch.[22] Which conflict resolution mode is suggested by each of these aphorisms: (**a**) "The arguments of the strongest always have the most weight"; (**b**) "Kill your enemies with kindness"; (**c**) "By digging and digging the truth is being discovered"; (**d**) "Better half a loaf than no bread"; and (**e**) "When two quarrel, he who keeps silence first is most praiseworthy"?

15-9. Define the five modes of conflict resolution (question 15-8) and, for at least three of them, describe a situation where each might be most effective.

15-10. Baker, Murphy, and Fisher conclude that project success is measured by the satisfaction regarding the technical performance of the product produced far more than achievement within planned cost and schedule. What reasons can you see for this?

15-11. Outline a strategy for effective communication between project personnel and the customer (client).

15-12. The text shows how to reduce the duration of the project shown in Figure 15-9 from 19 to 18 to 17 weeks at the lowest cost. Continue this process, showing the most economical way to reduce project duration from 17 weeks, week by week, to the minimum possible duration. What is the minimum project total cost at this minimum duration?

15-13. For the project outlined on the following table, (**a**) draw a network diagram (arrow or AON, as you prefer), and (**b**) identify the critical path and duration. (**c**) Identify the task(s) you would crash and the incremental cost to reduce project duration by (**i**) one week; (**ii**) a second week; (**iii**) a third week.

Task	Follows Task(s)	Duration, Weeks Normal	Crash	Cost, Dollars Normal	Crash
A	(start)	3	2	$500	$600
B	(start)	2	1	400	450
C	A, B	5	3	600	750
D	B	5	4	550	640
E	C, D	4	3	400	550

15-14. The following project carries a penalty cost of $200/day ($1400/week) for any delay in completion beyond 26 weeks. Any task can be accelerated by up to three weeks at a cost of $1000 per week reduction. Draw an arrow diagram, identify the critical path and duration, and determine which task(s), if any, you will crash, and by how much, to minimize project cost.

Task	Follows Task	Duration, Weeks
V	(start)	11
W	V	8
X	W	10
Y	V	20
Z	(start)	30

15-15. A project manager observes that she had budgeted expenditure of $450,000 by a specific date and had only spent $425,000 by then. What should she conclude?

NOTES

1. Harold Kerzner, *Project Management: A Systems Approach to Planning, Scheduling, and Controlling,* 4th ed. (New York: Van Nostrand Reinhold Company, Inc., 1994), p. 40.
2. Thomas J. Allen, *Managing the Flow of Technology: Technology Transfer and the Dissemination of Technological Information Within the R&D Organization* (Cambridge, MA: The MIT Press, 1977), pp. 110–113.
3. Edmund J. Young, "Project Organisation," in Dennis Lock, ed., *Project Management Handbook* (Aldershot, Hampshire, England: Gower Publishing Co. Ltd., 1987), p. 27.
4. William H. Newman, E. Kirby Warren, and Andrew R. McGill, *The Process of Management: Strategy, Action, Results,* 6th ed. (Englewood Cliffs, NJ: Prentice-Hall, Inc., 1987), p. 268.
5. Newman et al., *The Process of Management,* p. 270.
6. Erik W. Larson and David H. Gobeli, "Significance of Project Management Structure on Development Success," *IEEE Trans. on Engineering Management,* May 1989, p. 124.
7. Larson and Gobeli, "Significance of Project Management Structure," pp. 119–125.
8. Adapted from David Cleland and William R. King, *Systems Analysis and Project Management,* 3rd ed. (New York: McGraw-Hill Book Company, 1983), pp. 337–341, and elsewhere.
9. Hans J. Thamhain, *Engineering Program Management* (New York: John Wiley & Sons, Inc., copyright © 1984), p. 178. Reprinted by permission.
10. Thamhain, *Engineering Program Management,* pp. 191–194.
11. Thamhain, *Engineering Program Management,* pp. 238, 241.

12. Barry Z. Posner, "What's the Fighting About? Conflicts in Project Management," *IEEE Trans. on Engineering Management,* November 1986, pp. 207–211.

13. Robert R. Blake and Jane S. Mouton, *The Managerial Grid III: The Key to Leadership Excellence* (Houston, TX: Gulf Publishing Company, Book Division, 1985).

14. Hans Thamhain and David L. Wilemon, "The Effective Management of Conflict in Project-Oriented Work Environments," *Defense Management Journal,* July 1975, pp. 29–40.

15. Bruce N. Baker, David C. Murphy, and Dalmar Fisher, "Factors Affecting Project Success," in David I. Cleland and William R. King, eds., *Project Management Handbook* (New York: Van Nostrand Reinhold Company, Inc., 1983), Chapter 33, pp. 669–685. See also Murphy, Baker, and Fisher, *Determinants of Project Success,* NASA report NGR 22-03-028, NTIS Accession N-74-30392, September 15, 1974.

16. Larson and Gobeli, "Significance of Project Management Structure," pp. 119–125.

17. Baker et al., "Factors Affecting Project Success," pp. 677–678.

18. Ralph Katz and Thomas J. Allen, "Project Performance and the Locus of Influence in the R&D Matrix," *Academy of Management Journal,* 28:1, 1985, pp. 67–87.

19. David I. Cleland and Harold Kerzner, *Engineering Team Management* (New York: Van Nostrand Reinhold Company, Inc., 1986), pp. 63–64.

20. Jack Balderston, Philip Birnbaum, Richard Goodman, and Michael Stahl, *Modern Management Techniques in Engineering and R&D* (New York: Van Nostrand Reinhold Company, Inc., 1984), pp. 135–149.

21. Thamhain and Wilemon, "Effective Management," p. 30.

22. P. R. Lawrence and J. W. Lorsch, *Organization and Environment* (Boston: Harvard Business School, Division of Research, 1967).

PART 5

Managing Your Engineering Career

CHAPTER 16

Achieving Effectiveness as an Engineer

PREVIEW

In 1984 the American Association of Engineering Societies (AAES) conducted a major survey of engineers working in industry, resulting in 3357 usable responses: 69% from mailings to members of the major engineering societies, and 31% from visits to 22 high-tech companies. One class of questions asked "How satisfied are you with your engineering education?" Engineers responded they were "satisfied" or "very satisfied" with specific aspects of their engineering education as follows:[1]

- Technical education   88%
- Preparation for current career   66%
- Written communications skills   62%
- Oral communications skills   55%
- Human relations skills   52%
- Career guidance   24%

In the same survey, 55% of respondents expressed satisfaction with their career choice, but only 22% were satisfied with their career progress. Engineers spend four or more years in college being persuaded that if they solve the technical problems properly they will be rewarded (with high grades, invitations to honorary societies, and the like). Not surprisingly, when they go to work many engineers expect that all they have to do to succeed is to apply effectively the technical skills they have been taught to the engineering assignments given them. This is necessary, but not sufficient. Successful engineers learn to manage their careers with the same skill and care they apply to their technical assignments, and with a sufficient priority.

This chapter is concerned with some of the areas in which the basic engineering education is weak. First, we summarize the advice of some successful engineers, past and present, on "getting off to the right start" in early professional first assignments. Then we consider choices in career field and the sequence of stages in a career. Next, we discuss the need to communicate your ideas effectively, and some techniques for effective oral and written communication. Fourth is a discussion of how to stay technically competent in an age of exploding information. In the final section, we discuss the areas of professional society activity, registration, and certifications.

GETTING OFF TO THE RIGHT START

Background

In engineering education you have worked hard to survive a demanding curriculum and to build an academic record you can be proud of. The chances are that you landed your first job because the employer came to you: although less than 10% of college graduates are engineers, they receive about half of all on-campus job offers. Except in defense-related aerospace and electronics work, engineering employment has traditionally been reasonably secure, and most industries were less likely to lay off engineers than other workers.[2] The early 1990s, however, found "that jobs held by U.S. engineers are growing increasingly vulnerable [. . . due to] structural changes going on in the U.S. economy affecting the work environment confronting engineers, including corporate downsizing, reduced long-term research expenditures, defense cut-backs, automation, and globalization."[3] A "perverse consequence" of the increase in productivity of individual engineers using computer-aided design (CAD) is the elimination of the jobs of other engineers!

Even having a secure job, however, is not satisfying for most professionals. For a rewarding professional life, you must begin building yourself a personal reputation on which your future career success will depend. Several decades later you may come to reflect on the actions and decisions in your early career that made you successful, or that might have made you more successful; perhaps then you will be willing to share your hard-earned wisdom with young engineers who are following in your footsteps.

Fortunately, a number of successful engineers have done just that. One of the first was the famous French mining engineer and executive Henri Fayol, whose "Advice to Future Engineers" is as fresh and appropriate today as it was when he first published in 1916, more than 70 years ago:

> You are happy in the thought that you are going to be of use at last and you have the legitimate desire to win an honourable place by giving your service. The qualities which you will have to call into play are not precisely those which confer front rank at college. Thus health, the art of handling men, and bearing, which are not assessed in examinations, have a certain influence on an engineer's success. Circumstances, too, vary, so there is nothing surprising in the fact that the first class ["*A* students"] or even the highest of their year are not always those who do best. You

are not ready to take over the management of a business, even of a small one. College has given you no conceptions of management, nor of commerce, nor yet of accounting, which are requisite for a manager. Even if it had given you them, you would still be lacking in what is known as practical experience, and which is acquired only by contact with men and with things. . . .

Your future will rest much on your technical ability, but much more on your managerial ability. Even for a beginner knowledge of how to command, plan, organize, and control is the indispensable complement of technical knowledge. You will be judged not on what you know but on what you do and the engineer accomplishes but little without other people's assistance, even when he starts out. To learn how to handle men is a pressing necessity.[4]

A half-century ago W. J. King, discussing "What the Beginner Needs to Know at Once" in his *The Unwritten Laws of Engineering*[5] provides a framework that is still valid for the discussion following regarding your work, your boss, and your associates. I've woven into King's framework selected "keys to engineering and management success" by Bernard Sarchet,[6] founding president of the American Society for Engineering Management, as well as items from my own observation and reading.

Regarding Your Work

The new graduate makes his or her mark within the first few years in the organization. During this period he or she is tagged as a "comer" in the corporation (or a "front burner" as an Air Force officer or as a "tiger" somewhere else), or becomes labeled as "just one of the boys" (the feminine equivalent is obscure). The tiger will get the early promotions and the challenging assignments; the "late bloomer" will find it more difficult to be recognized and catch up later.

This makes it essential to give your best efforts to your early assignments, regardless of how trivial they may appear. Doing an exceptional job on a minor assignment is the best way to be recognized and assigned more important, more challenging, more satisfying work, since executives are continually searching for competent people to move up into more responsible positions. This applies to cooperative (co-op) work assignments of undergraduate engineering students in industry as well. Lee[7] found that higher-performance co-op students were those who "seemed to work harder and they also found their work assignments to be more challenging"; this work experience "also had an indirect effect on subsequent professional job performance" evaluated two years after graduation.

In the occasional situation where you have given a simple assignment your best for a reasonable period and have not been given anything more challenging, discuss your interest in getting more varied experience with your boss. If he or she cannot (or will not) do anything, ask permission to make an appointment with another manager who controls jobs you think you might like—or inquire of the personnel manager where there might be other opportunities in the organization; if this still doesn't work, you may need to find a more promising organization!

Don't wait for others—get things done! Just because you have asked a foreman, a vendor, or a colleague to provide something you need does not mean that it is going to happen in a timely fashion. Keep a "tickler" file and call (and call again if needed) to check on progress. Find another way to get it done or work two techniques in parallel if necessary.

Be understanding, but persistent, and learn to know the difference. The writer recalls a bleak Saturday night as a young Air Force project officer when a 150-pound object had to be loaded for shipment on an airplane immediately and there was no one around. A truck requested earlier appeared, but the airman who drove it apologized: "I'm sorry sir, but I'm a driver, and my commander says I'm not supposed to load!" A simple "I think I can lift *this* end" was enough to get the other half of the job done quickly.

Go the extra mile—and hour. Reputations are not made on a 40-hour week, and to be an effective professional you will at least have to do your professional reading largely on your own time; as you increase in responsibility you will also find that you need uninterrupted blocks of time that never seem to be available during the day for planning and thinking problems through. The fastest promotions generally go to those who put forth the extra effort and *meet deadlines*. This must be balanced against our other values—time spent raising our families, recreation to keep us whole and renewed, service to our community, and other investments of time that are important to us. Each of us must reach a balance (or make our compromises). These balances can be particularly difficult for married female professionals, unless they are fortunate enough to marry someone who truly *does* do his full share in home and family chores. Fortunately, success is measured in effectiveness, not just hours, as we will see under "Managing Your Time" in Chapter 18.

Look for visibility. You can do a good job every day, but you need to be seen to be recognized as a "rising star." Look for chances to make a presentation, to take leadership in a professional society chapter, to give a talk to the Rotary Club, or to organize the company United Fund drive. Sometimes a careful choice of car pool or lunch time and location or exercise site will put you in touch with established professionals who can give you more insight into the forces driving the organization. Learning the dividing line between making your capabilities visible and "brown nosing" takes maturity, but it is maturity that leads to greater responsibility.

Learn the corporate culture. Keep your eyes and ears open. Notice how successful engineers dress, and do likewise (but perhaps with just a touch more style); save your expressions of independence for important things. Notice how your more effective colleagues interact, and how they get things accomplished. If you cannot be comfortable and effective in your company's culture, perhaps you'd better go somewhere else!

Regarding Your Boss

Be as careful as you can in selecting your boss. King believes that "this is second in importance only to the selection of proper parents. . . . Long before the days of universities and textbooks, craftsmen in all the arts absorbed their skills by apprenticeship to master craftsmen." By observing the master engineer (or engineering manager) you can learn much more quickly the art of being an effective engineering professional. Then, says Sarchet, "*understand the boss* to the point that the decisions you make are the same ones he [or she] would make. You should strive to become his alter ego." Not only will you learn the art, but you will become so trusted and valuable that when this paragon is promoted, he or she will not want to tackle a bigger job without taking you along (often with a promotion) to help. (If, unfortunately, your first boss is not of this caliber, you still owe him or her your best while you are looking for a transfer that will increase the quality of your experience.)

Keep your boss informed. Ask yourself: "What does my boss need to know to do *his or her* job effectively?" In particular, never let your boss be caught by surprise. If something is going wrong or an assignment will be late, let the boss know. He or she may be able to reduce the consequences to both of you if he learns of it before *his* boss does. Again, if circumstances force you into a commitment to a higher executive or outsider, let your boss know the situation and to what you have committed as soon as feasible. Finally, if you are given a job to do, complete it or, if your initial effort convinces you it is not worth doing, tell your boss what you have found and get your boss's concurrence in dropping it; don't let your boss continue to think you are working on it. None of this implies that you should deluge your boss with unwanted trivia: more new professionals err by communicating too much than by communicating too little.

Make your boss's job easy. Your primary job is to help your boss carry out his or her responsibilities, so give top priority to whatever the boss wants done or ask the boss for guidance on conflicting priorities. Learn to do *completed staff work:* do not just come to your boss with a problem—state the problem, the alternatives you have considered, and your recommended action—if appropriate, even include any paperwork necessary for implementing your solution, so that if your boss agrees with you, all he or she has to do is sign.

Regarding Associates and Outsiders

Never invade the domain of any other division without permission of the manager in charge. King emphasizes that "this is a very common offense, which causes no end of trouble" and warns specifically not to offer a job to someone in another division (or even discuss it), not to commit the services of another group or make commitments on their time, and not to perform any function assigned to that group, without first discussing it.

If you are dissatisfied with the services of another section, make your complaint to the person most directly responsible for the problem, not his or her boss. You may be identifying a problem the person was not aware of (or was not aware it was so important to you). If you can do this tactfully, you will give the person a chance to initiate a correction before his or her boss learns of the problem; you often will get action with less friction and will build a trusting relationship with the other person, whom you may need to call on in the future.

Keep up the "old school ties" by staying in touch with past school friends, professors, old colleagues, and past bosses. Someday you may need help in finding a new job, getting a recommendation for graduate school, or some other venture. Also, you may need outside sources of information on people or other resources needed to solve a company problem; you will be measured not just by what you know, but by what you can find out when needed. *Networking* is the modern term for such a web of mutually supportive relationships.

In dealing with customers and other outsiders, remember that you represent your organization. Do not bad-mouth your organization or anyone in it—instead, put them in the best possible light (then take any problem back to the person who is responsible, so that it can be handled internally). Moreover, realize that the outsider is likely to regard you as being the technical, legal, and financial agent of your company, even if you have just recently joined it, and so be very careful of your commitments. In the Apollo program, young engineers from NASA and from contractor organizations met many times to discuss technical problems; considerable confusion (and some litigation) took place before engineers on both sides learned that, although they might come to an agreement as to what change *should* be made, they could not implement it until the contract representatives who had authority to commit each party negotiated the change.

CHARTING YOUR CAREER

Defining Career Success

Morrison and Vosburgh[8] point out that a career is defined differently for different people. For the "pure" professional the occupation is usually paramount, the link to the organization is tenuous, and the professional will go elsewhere if the organization does not provide the technical challenge, collegial relationships, and the values and rewards expected from the occupation; career and occupation become one. Other individuals become totally committed to an organization and transfer from design to project management to production to corporate staff within that organization; career and organization become one; large Japanese firms promote this orientation. Most engineering careers lie somewhere between these two extremes. In setting out to manage your own career you need to define what success in a career means to you, because this in turn influences what actions are appropriate.

Career Fields

There are a variety of broad fields of career endeavor for which your engineering education has given you a basic preparation. Some are entry-level opportunities for a new graduate, yet broad enough for a lifetime of challenge; others are best entered after you have gained some professional experience and, perhaps, graduate education. Leonard Smith identifies the following career fields:[9]

1. *Operational careers.* Many engineers begin their careers in operating areas such as manufacturing, purchasing, planning, customer service, and sales. Each of these areas will have applications of your engineering knowledge and skills. Operational assignments are very likely to lead into operational management and, if you prove successful, general management positions. Chapters 11, 12, and 13 have dealt with these areas.

2. *Research and design careers.* These careers include research in new engineering technology and the advanced design and preliminary development of sophisticated new systems. Engineers in this career field should plan to earn at least an M.S. degree in a technical specialty, and must exert a continuing effort to stay at the top of their technology, using some of the methods discussed later in this chapter. Chapters 9 and 10 and, from an advanced development point of view, Chapters 14 and 15 (project management) have attempted to give you some feeling for these careers.

3. *Engineering management careers.* As suggested in Chapter 1, about ⅔ of engineers in the past have spent about the last ⅔ of their careers in some level of management responsibility, although there is no guarantee these statistics will remain the same. The first half of Chapter 17 discusses the engineer's transition to manager.

4. *Entrepreneurial careers.* Quite a few engineers opt at some point in their career to form their own company, either alone or with selected colleagues. This company might be involved in research, design, manufacturing a product, or providing sales or service for a product. The risks of failure are greater, but so is the potential reward.

5. *Consulting careers.* These are careers in which you use your engineering knowledge and experience for the benefit of a variety of other organizations and individuals. For example, you might be asked to provide designs, give advice, solve problems, or provide expert testimony. To be successful, you will have to develop a substantial expertise and reputation in your selected area. You can combine this with an entrepreneurial career by forming your own consulting organization, but you will usually want to develop a specialty working for another organization first.

6. *Writing careers.* If you enjoy writing, you might find yourself writing for a technical magazine or journal, serving as a technical editor or abstractor, writing training or maintenance manuals, preparing sales literature for technical products, or writing a book like this one.

7. *Academic careers.* You may decide you want to teach others. With some courses in education you can fill a great need teaching math or science in our high schools, or with a master's degree you can teach these topics or preengineering subjects in junior colleges. For a career in university teaching in most engineering colleges, a doctorate is mandatory, but with a reasonable academic record you can usually get financial assistance in earning it. A career of university teaching can be combined effectively with research, consulting, and writing. Indeed, progress in research and technical publication are necessary for obtaining tenure at most engineering colleges.

8. *Other careers.* You always have the choice of leaving engineering and going into something entirely different. The training in problem solving you have received as an engineer can give you an advantage in thousands of other positions.

Career Stages

Super[10] proposed in 1957 the following five-step career sequence, together with fixed age periods that applied to them:

1. The *growth stage* (birth to age 14), from the first awareness of impending career decisions to the initial development of career aspirations, interests, and abilities.

2. *Exploration* (ages 15 to 24), involving making and trying out tentative choices, transition involving entering the labor market or advanced training, and trial in a beginning job.

3. *Establishment* (ages 24 to 44). "The first five years of this stage are considered to comprise a trial period in which one or two changes in the field of work would be made before a life work would be found or it became clear that life work would

be a series of unrelated jobs. The last fifteen years in this stage are classified as a stabilization period in which the individual acknowledges commitment to the life's work and to the organization by becoming socialized, progressing, and making a secure place in the field or organization."[11]

4. The *maintenance* stage (originally proposed from ages 45 to 65), in which the primary concern is to hold on to the place achieved in the world of work, and the person continues along established patterns.

5. The *disengagement* stage, in which physical and mental powers decline, and participation in the working world changes and then stops.

Super[12] and others later realized that these stages may occur at widely varying ages: establishment may be interrupted by a change in societal needs or a "midlife crisis," resulting in reentry into the exploration stage; disengagement may be deferred into the seventies or even eighties, or may be triggered by early retirement from industry in the early fifties (or the military in the forties), leading to exploration of a new career and a new cycle.

Dalton and Thompson[13] proposed a sequence of four career stages for professionals: *apprentice, colleague, mentor,* and *sponsor.* Morrison and Vosburgh summarize their findings:

1. The new professional serves as an apprentice and learns to be an effective subordinate who demonstrates willingness to do routine assignments, yet aggressively searches out new and more challenging tasks. By leaving this field too early, the individual does not learn from the experience of others. . . .

2. The young professional earns the way into the second, collegial stage by building a reputation as a technically competent individual. . . . The individual becomes less dependent and starts to contribute personal ideas about what to do in a given situation. Many professionals stay in this stage for the rest of their careers and have a reasonably successful career although their value to the organization dwindles over time. . . .

3. Movement into stage 3, mentor, takes place because the individual is able to take increased responsibility for influencing, guiding, directing, and developing other people. . . . The individual in this stage may serve with one or a combination of three roles: an informal mentor, a manager, or the "idea person" [gatekeeper]. . . . Eighty percent of those who make it to this stage were perceived by the organization to be above-average performers after age forty. [Mentors are busy professionals; the young engineer is well advised to seek out a potential mentor from among successful senior engineers or managers and *ask* if they would be willing to provide periodic coaching sessions; the National Society of Professional Engineers considers this process important enough that they provide a *Guidance Package for Mentors.*[14]]

4. Stage 4, sponsor, requires that the individual move up from influencing groups of individuals to affecting the direction of the organization or a major segment of it. . . . These people can play one or more of three roles: manager, internal entrepreneur, or idea innovator [the senior professional person in a field].[15]

COMMUNICATING YOUR IDEAS

Importance of Communication

Some definitions. Communication is the means by which information is made productive. The word stems from the Latin *communicare,* meaning "to impart" or "to make common." Weihrich and Koontz[16] define communication as "the transfer of information from the sender to the receiver, with the information being understood by the receiver." It is this *mutual* understanding that makes communication difficult.

Importance to the engineer. Engineering may be considered as a transformation process in which information is received, transformed in some way, and the results transmitted to others, as shown in Figure 16-1. The information *input* includes the statement of work, directives, methodologies learned in college or elsewhere, and company standards and practices. The "engineering" *transformation* involves a complex process of analysis and synthesis that requires substantial resources, time, and skill, but which is largely hidden to the outsider. The information *output* may be in the form of physical models, drawings, specifications, technical reports, and/or oral briefings.

Unless the information output is properly communicated, the meticulous engineering performed may be of little utility. Yet many engineers and scientists lose interest in a problem once they have solved it, and do not spend that extra effort in writing an effective report or documenting a computer program that would give their work real utility. Critics of engineering education have complained that "engineers can't communicate" for generations, with no resolution. At the November 1986 National Congress on Engineering Education, 71% of the engineers attending recommended that:

> Engineering curricula should include specific engineering courses which incorporate written, oral, graphical, and interpersonal communication skills, and these areas should be emphasized in the *evaluation* of student performance. It would also be useful to consider the development and implementation of English proficiency criteria as part of the admission and graduation criteria.[17]

Ability to communicate pays off for engineering undergraduates in co-op work assignments as well. Lee[18] reported that high performance co-op engineers were those developing more contacts—external contacts for co-op research engineers and contacts internal to the organization for those on design, analysis, and development assignments, and that "high performance co-op engineers acquired effective communication behaviors very rapidly."

Importance to the manager. Communication is the principal business of the manager, consuming an estimated 90% of his or her time. Mintzberg[19] estimates that managers spend

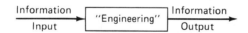

Figure 16-1. Engineering as a transformation process.

78% of their time just in oral communication (59% in scheduled and 10% in unscheduled meetings, 6% on the telephone, and 3% "managing by walking around" on tours); the remaining 22% of their time is deskwork, which consists in large measure of reading the communications of others and preparing written communications. Obviously, as engineers make the transition to manager, they must perfect their communication skills further if they are to be effective.

Modeling the Communication Process

Figure 16-2 diagrams the communications process concisely; let us discuss each step in turn:

- The thought of the sender must be *encoded* into English or some other language, a computer code, mathematical expression, or drawing with special consideration of the nature of the intended receiver.
- The code must then be *transmitted* using some selected *medium;* several are discussed in the next section.
- *Reception* of the message may be hindered because of distractions (*noise*) inhibiting the transmission or causing inattention in the reception.
- The message then must be *decoded,* which is effective only if sender and receiver both attach the same or similar meanings to the symbols used in the message.
- *Understanding* may be obstructed by prejudices, or by a desire not to hear or believe what is actually being said.
- *Feedback* that enables the sender to determine what the receiver actually understood of the message permits the correction of misunderstanding. Verbal feedback offers the same potential for misunderstanding as did the initial transmission, but face-to-face feedback is enhanced by nonverbal communication (discussed later).

Communication Methods Compared

Characteristics. Communications are transmitted in a variety of forms, each with its advantages and disadvantages. The nature of the most common methods is indicated in Table 16-1.

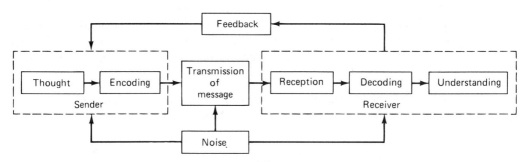

Figure 16-2. Communications process model.
(From Harold Koontz and Heinz Weihrich, *Management,* 9th ed., McGraw-Hill Book Company, Inc., New York, 1988, p. 463.)

TABLE 16-1 Characteristics of Common Communication Methods

Communication Method	Speed	Feedback	Record Kept?	Formality	Complexity	Cost
Informal conversation	Fast	High	No	Informal	Simple	Low
Telephone conversation	Fast	Medium	No	Informal	Simple	Low-medium
Formal oral presentation	Medium	High	Varies	Formal	Medium	Medium
Informal note	Medium	Low	Maybe	Informal	Simple	Low
Memo	Medium	Low	Yes	Informal	Low	Low-medium
Letter	Slow	Low	Yes	Formal	Medium	Medium
Formal report	Very slow	Low	Yes	Very formal	Complex	High

Source: Paul R. Timm, *Managerial Communication: A Finger on the Pulse*, 2nd ed., 1986, p. 59. Adapted by permission of Prentice-Hall, Inc., Englewood Cliffs, NJ.

Retention of information. Studies of learning and experience equate retention of information as follows:

We Tend to Remember:	When Our Involvement Is:
10% of what we read	Passive reading
20% of what we hear	Passive verbal receiving
30% of what we see	Passive visual receiving
50% of what we hear and see	Passive visual receiving
70% of what we say	Receiving and participating
90% of what we say and do	Being

Effectiveness. Timm[20] describes a communications experiment in which the effectiveness of four media were compared. In order of *decreasing effectiveness* they were:

- Oral plus written presentation
- Oral only
- Written only
- The *grapevine*

Timm's "oral plus written presentation" is strengthened further by (1) effective diagrams, illustrations, or demonstration; (2) feedback involving participation or "repeat back" presentation by the listener; and (3) where full comprehension is essential, simulation or on-the-job practice.

The *grapevine* of an organization is its informal communication system. It is a natural and inevitable occurrence in every organization that formal communication will be supplemented by informal transmission of information and rumors (both true and false) from employee to employee. The general pattern of a grapevine, shown in Figure 16-3, is not unlike a nuclear reaction. One person (fissioning atom) transmits a fact or rumor to several people (emits several neutrons). One or more (seldom all) of the recipients will pass the rumor on, often inaccurately, to one or several other people. When a rumor concerns something of great importance to most employees, it can spread at explosive speed (exceeding critical mass, as it were). Effective managers learn to monitor the grapevine (or persuade a secretary or other employee to do so), then make sure that factual information is published by bul-

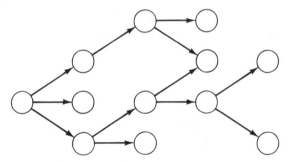

Figure 16-3. Grapevine.

letin board, newsletter, or by "seeding" the grapevine itself to moderate transmission of misinformation.

Other Factors in Effective Communication

Active listening. The art of effective listening is as important as effective communication. Listen positively and attentively, allowing the speaker to make his or her point. Analyze the speaker's attitude and frame of mind. Is the person an optimist or pessimist? Generally reliable or unpredictable? Try to reach beyond the speaker's words to his or her meaning. When in doubt, rephrase the speaker's words with a "Do I understand that . . . ?" Take notes of essential points unless that inhibits the communication. Finally, consider the speaker's nonverbal language (discussed below) as well.

Nonverbal communication. Albert Mehrabian[21] divides the relative influence of the verbal, vocal, and facial aspects of oral communication as follows:

- 7% Verbal (words used)
- 38% Vocal (pitch, stress, tone, length, and frequency of pauses)
- 55% Facial (expression, eye contact)

Obviously, the effective manager must learn, just as does the professional actor, that the way one makes a presentation is of paramount importance. This must include, in addition to the presentation from the neck up just described, the importance of *body language* (your posture, gestures, and body movement).

Communication Tools of Special Importance to the Engineer

The written report. The results of engineering studies are often documented in formal written reports, and the usefulness of the study is determined by whether the report is (1) read and (2) understood. A few important considerations for effective report writing include:

- Be sure that you have reserved sufficient time at the end of an assignment to report on it effectively.
- Begin your report with an *executive summary* of one or a very few pages, whose content answers the question: If the busy reader is only going to look at these few pages, what does that reader need to know?

- Consider putting your conclusions and recommendations at the front, followed by essential discussion, with peripheral material relegated to appendixes.
- Outline your report carefully, then write to your outline, and finally, take the time to review your work for clarity. Spelling and grammar will be judged as an indication of the value of the writer and the report. Use a word processor with a "spell-check" provision and a built-in thesaurus to improve your writing. If your grammar or proofreading skills are below standard, take positive action to improve them, and ask someone to review your work in the meantime.

The oral briefing. A briefing is an oral presentation of analyzed and synthesized information, presented to a person or group of people who have a need or desire for knowledge but who do not have the time to become thoroughly familiar with all the details of a subject. It may be extemporaneous, memorized, or paraphrased from notes. It often involves audiovisual aids such as viewgraphs, slides, flip charts, films, television, models, or samples, and it may be supplemented by a written report that most listeners will probably not take the time to read. Usually there will be an opportunity during or after the presentation for listeners to ask questions or make comments.

The essence of effective oral briefing is preparation and practice. Preparation includes steps of defining the objectives you wish to achieve, identifying your audience, outlining what you plan to cover, filling in the details, and preparing effective supporting materials. Practicing identifies weak areas in your presentation, helps ensure that the briefing flows smoothly, and assures that your presentation will fit within the allotted time. Effective briefing skills are essential for your success, for the busy executives who will make the final decision on whatever you are proposing often can be approached in no other way (and this may be their principal opportunity for them to assess your capability for greater responsibility).

Visual aids. Visual aids for the oral presentation commonly use either overhead projectors or 35-mm slide projectors (and increasingly, television tape). Viewgraphs (transparencies) for overhead projection can be prepared quickly on an office copier and are suitable for audiences of up to about 200 people in a semilit room; carefully prepared 35-mm slides look more professional and are suitable for larger audiences, but involve more cost and lead time. Dieter[22] provides some guidelines for using visual aids of either type:

1. Limit slides or viewgraphs to *no more than* one per minute.
2. Each slide should contain one idea.
3. Slides that present more than three curves on a single graph or 20 words or numbers on a slide are too complicated.
4. The first slide should give the title of your talk and names and affiliations of the authors.
5. The second slide should give a brief outline of your talk.
6. The last slide should summarize the message you delivered.
7. If you need to use a slide more than once, insert a second copy in the proper sequence.
8. If you have material to discuss between slides, insert a blank slide to avoid distraction.

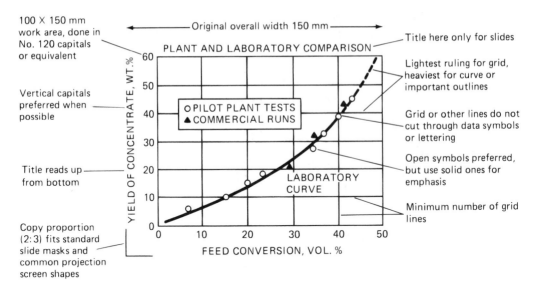

Figure 16-4. Effective graphical presentation.
(From *ANSI Y15.1M-1979*, Ameri can Society of Mechanical Engineers, New York, 1979.)

Material for a viewgraph should fit in a rectangle 6 inches high by 9 inches wide and should be clearly legible (unmagnified) at a distance of 6 feet; a 35-mm slide should be legible at a distance of 12 inches. Graphs should be prepared with a minimum of detail, as shown in Figure 16-4.

STAYING TECHNICALLY COMPETENT

The Threat of Obsolescence

The knowledge explosion. Well over 300 years ago, when the first scientific journals were begun, it was possible for a single scientist to be aware through correspondence of most of the scientific discoveries being made in the Western world. Once begun, scientific journals began multiplying like rabbits, doubling in number about every 15 years. Putka[23] estimated in 1987 that U.S. scientific journals then totaled about 5000. (These include journals dealing with very narrow disciplines such as the *International Review for the Sociology of Sport,* the *Journal of Molluscan Studies,* and the *Fibonacci Quarterly,* which deals only with applications of the Fibonacci number series.) Dieter[24] estimated in 1983 that current world output "amounts to two million technical papers a day, or a daily output that would fill seven sets of the *Encyclopaedia Brittanica.*"

Kaufman[25] reports that when the number of journals reached the critical number of 300 in the mid-nineteenth century, abstract journals began to appear to provide access to the literature. By the mid-twentieth century, abstract journals themselves reached the critical 300 level. Computer abstracting and computer searches of massive national data bases are cur-

TABLE 16–2   Sources of Sources of Technical Information

Indexes (and Computer-Access Name, If On-Line)

Applied Science and Technology Index
Science Abstracts (INSPEC): Series A, Physics; Series B, Electrical and Electronic; Series C, Computers
Engineering Index (COMPENDEX)
Chemical Abstracts (CA SEARCH)
Science Citation Index (SCISEARCH)
British Technology Index
Proceedings in Print (conference proceedings)
Index to Scientific and Technical Proceedings
Environment Abstracts (ENVIRONLINE)
Pollution Abstracts (POLLUTION ABSTRACTS)
Energy Index (ENERGYLINE)
Fuel and Energy Abstracts
Energy Information Abstracts (ENERGYLINE)

Literature (Subject) Guides

Guide to the Literature of Engineering, Mathematics and the Physical Sciences (Baltimore, MD: Johns Hopkins University Press)
Guide to Basic Information Sources in Engineering (New York: Wiley)
Scientific and Technical Information Sources (Cambridge, MA: MIT Press)
Science and Engineering Literature: A Guide to Reference Sources (Littleton, CO: Libraries Unlimited)

Journal Lists

1. *The Standard Periodical Directory* (New York: Oxbridge Communications)
2. *Ulrich's International Periodicals Directory* (New York: R. R. Bowker)
3. *Business Publications Rates and Data* (popularly called the "Standard Rate and Data Book"). Published monthly by Standard Rate and Data Service, Skokie, IL. Any relatively recent issue will list trade journals by field.

Source: Samuel E. Gluck, "Sources of Information for Engineering Management," in *Handbook of Engineering Management*, John E. Ullmann, ed., copyright © John Wiley & Sons, Inc., New York, 1986, p. 209. Reprinted by permission.

rent attempts to master the exponentially increasing flood of knowledge in science and technology, but today no single technologist can know, or even locate, all the relevant information even in a fairly narrow specialty. The engineering researcher (i.e., any engineer who needs to find some information) needs to begin first with a knowledge of the appropriate *sources of sources.* Gluck[26] has provided a compilation of these, of which Table 16-2 reproduces just the first portion.

Obsolescence defined. Obsolescence has been defined as "the process of passing out of use or usefulness" or even "the process of being replaced by something newer or better" (which, to the midcareer engineer, could be a new graduate equipped with the most modern education). Shannon amplifies this:

Persons are obsolescent technically if, when compared to other members of their profession, they are not familiar with, or are otherwise unfitted to apply, the knowledge, methods, and techniques that are generally considered important by members of their profession.[27]

Organizational obsolescence. Thompson and Dalton[28] believe that organizational obsolescence is a greater culprit, and they suggest three areas in which managers can make improvements and thus avoid having an obsolete organization: reward technical contribution, reduce barriers to movement, and focus on careers. However, individual engineers cannot and should not depend on their employer to combat obsolescence; they must take personal responsibility for their own career progress.

Obsolescence and Personal Characteristics

Kaufman[29] identifies a series of characteristics that reduce the onset of obsolescence:

1. Good capacity for knowledge acquisition, based on strong cognitive and intellectual abilities
2. Self-motivation to stay up to date
3. Strong interests in abstract ideas, leading to involvement in work in which they originate, develop, and test ideas (rather than greater interests in people and economics)
4. High achievement needs
5. The amount of energy applied at work ("usually evidenced by the willingness to work hard for long hours")
6. A good self-concept, based on feelings of competence, self-esteem, and confidence
7. Adaptability to change
8. At least a moderate propensity for taking risks

Most of these characteristics can be encouraged by the organization by challenging, but achievable, assignments and rewards for accomplishing them. Conversely, as the engineer moves into "jobs that emphasize the development and utilization of personnel or the estimation of costs and the control of expenses"—that is, into managerial work—it becomes increasingly difficult to stay at the forefront of technical knowledge, and few engineers return to advanced technical assignments after remaining in managerial work for any long period.

Methods of Reducing Obsolescence

Mastering the technical literature. If, as has been repeatedly stated, technical knowledge doubles in quantity every 10 years, the working engineer or scientist who quits learning upon leaving college will only know half what he or she needs to know 10 years later, and a quarter 20 years later. Technical managers have an even greater problem, since they need to stay generally knowledgeable about a wide range of technology, including areas they did not touch in school. Kaufman reports[30] that "among first-level industrial managers, more than three out of four feel it is important to read and use the library to obtain information, yet approximately half of them seldom or never engage in these activities." Indeed, "the average time engineers and scientists spend on keeping up to date with the professional literature is only *one hour a week*," a small fraction of the time required to stay competent.

TABLE 16–3 Sources of Messages Resulting in Technical Ideas Considered During the Course of 19 Projects

Channel	Seventeen Technological Projects		Two Scientific Research Projects	
	Number of Messages Produced	Percentage of Total	Number of Messages Produced	Percentage of Total
Literature	53	8	18	51
Vendors	101	14	0	0
Customer	132	19	0	0
Other sources external to the laboratory	67	9	5	14
Laboratory technical staff	44	6	1	3
Company research programs	37	5	1	3
Analysis and experimentation	216	31	3	9
Previous personal experience	56	8	7	20

Source: Thomas J. Allen, *Managing the Flow of Technology: Technology Transfer and the Dissemination of Technological Information within the R&D Organization*, © The MIT Press, Cambridge, MA, 1977, p. 45.

Dieter[31] reports "that engineers as professionals make much less use of the published technical literature than do scientists and other professionals." He believes this is largely because "all major indexing systems index by subjects," which is ideal for the research specialist but "is a severe handicap for the design engineer, who is involved in synthesis." He recommends that the engineer

> develop your own working files of technical information that is important to your work. A good way to do so is to put articles and information you want to have as ready reference into large three-ring binders with page dividers marking off different areas of interest. Material of less current interest may be stored in folders in filing cabinets. . . . Be selective. Adopt the policy of discarding outdated material when you replace it with newer information.[32]

Allen[33] points out that technological and scientific projects rely on different kinds of information sources. He found (see Table 16-3) that 17 technological projects received a third of their ideas from analysis and experimentation and another third from customer and vendor suggestions, whereas two scientific research projects obtained over half their ideas from the formal technical literature.

Continuing education. One of the major conclusions of the National Academy of Engineering in its 1988 report *The Technological Dimensions of International Competitiveness* highlights the national importance of continuing technical education:

> The long-range need for a capable and adaptive work force requires that continuing education become an integral part of the career development of each individual, particularly of every scientist and engineer. Industry, government, professional societies, and educational institutions share the responsibility for creating a [continuing education] system that will be of high quality and will encourage the employee and the employer to invest in obtaining skills of future value both to the individual and to industry.[34]

Certainly the need for lifelong learning is nothing new. Hsun Tzu (298–238 B.C.) stressed that:

> Learning continues until death and only then does it cease. . . . the objective of learning must never for an instant be given up. To pursue it is to be a man, to give it up is to become a beast.[35]

Many engineers try to keep themselves current by taking a series of educational courses; these may be formal graduate-level university courses, whether leading to a graduate degree or not, or in-house courses offered by their employer. Younger engineers tend to prefer formal graduate courses leading to a master's degree where a suitable program is available. Kaufman[36] reports that "there is consistent evidence that professionals who have taken graduate courses are perceived as less obsolescent by themselves, their supervisors, and their colleagues," that they are less likely to be laid off by their employers, and that professionals with a graduate degree are unemployed a shorter period if they do lose their jobs. Kaufman compares graduate and in-house noncredit courses:

> A study of 2,500 technical professionals in six organizations found that not only were those with graduate training at the M.S. level better performers than those who had only a B.S. but their high level of performance was maintained ten years after the B.S. holders began to decline in performance. It would appear that a heavy dosage of graduate courses can push obsolescence back by ten years. . . .
>
> Many organizations offer in-house noncredit courses for their professionals on a regular basis. The courses are offered primarily to supplement university-sponsored courses. They are usually more directly applicable to the work of the organization and are less demanding than are graduate courses, since they typically do not involve grades, examinations, or even homework. In-house courses are especially appealing to older professionals, who feel more comfortable and less threatened when they are learning material relevant to their jobs in a relatively noncompetitive manner. It might be expected that such courses would be ideal for helping the older professionals keep up to date . . . but the courses do not seem to help . . . enrolling in a large number of in-house courses was found to be unrelated to subsequent performance.[37]

Graduate coursework in engineering disciplines (or in engineering management) is normally available only in large metropolitan areas, where a critical mass of students exists. Engineers in midwestern towns and cities too small to justify graduate programs with "live" instructors "felt that either closed-circuit TV or TV tape was moderately effective. They were ambivalent about programmed learning, audio cassettes, and correspondence courses, and they disliked telephone conferences."[38] Fortunately, the advance of satellite technology has made graduate professional education by television a reality, and two institutions have been established to facilitate such education on a national scale: the Association for Media-Based Continuing Education for Engineers (AMCEE) and the National Technological University (NTU).

On-the-job activity. The most important vehicle to reduce obsolescence is personal growth on the job itself. In a survey of 290 professionals, Margulies and Raia[39] reported that 42% saw on-the-job problem solving and 20% saw colleague interaction as most important for professional growth, with 16% citing publishing and independent reading and only 14% citing formal coursework as most important. More recently, Farr reported:

Another group, one hundred engineers in several organizations, indicated that the best aids to updating their technical knowledge and skills were immersion in state-of-the-art technology in their work and having free time available to work on new ideas. The primary inhibitors to updating were non-challenging assignments and lots of nontechnical work.[40]

Supervisors of professionals play an important role in providing challenging work assignments to the engineer and adequate technician and clerical assistance to perform routine activities; if the supervisor does not provide such growth opportunities, the engineer must find them, even if in another organization. The employer can assist professional development in a number of other ways: by providing time and a supportive atmosphere for technical reading, self-learning, and preparation of technical papers; by providing a truly professional information acquisition and distribution system; and by a professional approach to selecting the kinds of continuing education programs to support and assessing results obtained from them.

PROFESSIONAL ACTIVITY

Professional Societies

Types and purpose of technical societies. There are a bewildering variety of professional societies seeking the membership and support of the engineer. The *1987 Directory of Engineering Societies and Related Organizations*[41] lists 266 "national engineering organizations" in the United States and Canada, as well as 111 foreign and international and 128 state, local, and regional engineering organizations. Weinert[42] divides the U.S. organizations roughly into four major groupings:

1. Those focused on established or emerging engineering disciplines. These include the five "founder societies," all headquartered in the United Engineering Center, 345 East 47th Street, New York. Here, we list them with the year of founding and 1986 total membership, including student members:

   - American Society of Civil Engineers (ASCE, 104,000, founded 1852)
   - American Institute of Mining, Metallurgical, and Petroleum Engineers (AIME, founded 1871, and now divided into four member societies)
   - American Society of Mechanical Engineers (ASME, 115,000, founded 1880)
   - Institute of Electrical and Electronic Engineers (IEEE, 261,000, founded 1884 as the AIEE, now international)
   - American Institute of Chemical Engineers (AIChE, 62,000, founded 1908)

2. Those focused on a broad occupational field, such as the Society of Automotive Engineers (SAE), the Society of American Military Engineers (SAME), or the American Institute of Plant Engineers (AIPE).

3. Weinert's "fastest growing group are those focused on a specific technology, group of technologies, or one of the specific materials' or forces of nature always referred to in classic definitions of engineering." These include such groups as the Ameri-

can Society of Heating, Refrigerating, and Air-Conditioning Engineers (ASHRAE), the Society of Plastics Engineers (SPE), the Society of Manufacturing Engineers (SME), and the American Society for Quality Control (ASQC). (In universities these technologies are usually covered, if at all, as subdivisions of major engineering disciplines, such as mechanical, chemical, or industrial engineering.)

4. His "final group is composed of those associations and societies formed either by individual engineers or by groups of societies to accomplish a specific purpose." This includes the Accrediting Board for Engineering and Technology (ABET), formed by the "founder societies" and other disciplinary societies to accredit engineering college programs; the National Society for Professional Engineers (NSPE), with its interest in professional registration, engineering ethics, and public policy; and the National Council of Engineering Examiners (NCEE), formed to coordinate the state licensing process. Although most large disciplinary societies have divisions or committees on engineering education and on management or engineering management, the American Society for Engineering Education (ASEE) provides an interdisciplinary forum for engineering education as an entity, and the American Society for Engineering Management (ASEM) an interdisciplinary forum in the management of technological activities. Finally, the American Association of Engineering Societies (AAES) is set up as an "umbrella organization" to represent the engineering profession as a whole, although some major societies do not participate in it and it is only partially successful.

Reasons for getting involved. Engineers owe it to their profession and to themselves to belong to at least one professional society and to support it with their dues and their effort. The larger disciplinary engineering societies maintain (through their activity in ABET) the quality of engineers entering the field. Societies of all four groupings provide a range of professional publications that you will need to keep professionally current in your field(s) of interest, hold annual (and often regional and local) conferences where you can keep abreast of new developments and share problems and solutions with leaders in your field of interest, and often sponsor educational programs important to your development. Many societies will have local sections in your city (or the metropolitan area nearest you). Local sections often meet for lunch or dinner monthly, giving you an opportunity to meet others in the area who share common problems and listen to speakers on topics of interest. Most local sections are short of volunteers for section activity, and service as a local section committee chair or officer can provide an early and satisfying opportunity to demonstrate your capacity for professional leadership.

Many technical professionals find it important to maintain an active membership in a number of societies because of their varied interests and responsibilities, and find that their memberships change as they progress through their career. Most engineers will first join the society of their undergraduate discipline (ASCE, ASME, IEEE), often as a student, and may continue in that society for life, as long as they continue to view themselves as belonging to that discipline. Current memberships often include a society focused on the industry they work in, and one or more on the function they are currently practicing. As professor emeritus of engineering management, for example, the writer currently maintains memberships in ASEE, ASEM, and IEEE/EMS, and has been for varying periods active in APWA, ASQC, IIE, NSPE, SAME, SRE, SSS, and TIMS; earlier, as a chemical and aerospace engineer, he was active in AIChE, ACS, ARS, and AIAA. Industrial employers are often willing to underwrite dues payments in appropriate societies for their employees, easing the burden of multiple memberships.

Technical papers and publications. Professions depend for their progress on the willingness of individuals to share their discoveries and observations with others. You can do this by offering to write a "paper" for presentation at a regional or national technical meeting; often you will find your employer willing to pay your expenses to attend a meeting at which you are speaking. Or you can offer an article to a professional magazine or journal. For a university professor, such publication (especially in a "refereed journal," where articles must first be reviewed by several of his or her peers) is usually necessary for promotion or even survival. As a result, professional society journals are often inundated with theoretical articles from faculty authors, but they are eager to receive more applied articles from practitioners. If you are working in industry and government, you may not have the same compelling reason to offer papers to meetings or publications to professional magazines, but you should look for opportunities to do so as a part of your personal growth and as a service to your profession. The writer is pleased that his three children, all engineers, have each begun their contribution to the professional literature.

Accreditation, Registration, and Certification

Three related topics are discussed in this section. *Accreditation* is a voluntary process where a designated agency grants recognition to an educational program that meets certain minimum standards. Engineering *registration* in the United States is granted by the several states to individuals meeting specified criteria. Professional *certification* of engineers is a voluntary process regulated by certain professional societies.

Accreditation. The Accrediting Board for Engineering and Technology (ABET) has the responsibility for accrediting those U.S. engineering curricula that apply for such consideration. ABET is controlled by a board of directors consisting of members designated by the major engineering professional societies, and is funded by those societies and the universities seeking accreditation. ABET has an Engineering Accreditation Commission (EAC) and a Technology Accreditation Commission (TAC) responsible for establishing accreditation criteria, visiting universities, and recommending accrediting action to the ABET Board in engineering and engineering technology, respectively.

The ABET visiting team includes a visitor for each curriculum being examined, chosen from a list of practicing engineers and engineering educators established by the engineering society responsible for that discipline. Visitors analyze curriculum content, faculty, facilities, funding, student preparation, and other factors against ABET guidelines. For example, an undergraduate engineering curriculum must be shown to include 16 semester hours (or equivalent) each of mathematics, science, engineering design, and humanities (and/or social science), and 32 semester hours of engineering science. This specifies about three years of content in the typical four-year program, not including "skill" courses such as English composition, which ABET also expects to be taught. Universities may choose to accredit at the five-year (master's) level rather than at the four-year level, but this is much less common. Accreditation visitors are reimbursed for their travel expenses but not for their time. The engineer who volunteers for such activity provides a service that is the hallmark of being a professional.

Professional engineering registration. Engineers who are eligible should seriously consider becoming registered with their state as a Professional Engineer. This is usually a two-step process. The first step, leading to the Engineer In Training (EIT) designation, usually requires an approved engineering degree (which in some states can only be from a curriculum accredited by ABET as just described) and successful completion of an eight-hour Fundamentals of Engineering (FE) test on topics such as calculus, physics, statics, thermodynamics, electrical circuits, and engineering economy (which may be taken while still an undergraduate). Not all topics are needed to pass the test, and a capable engineering student should be prepared for the examination by the end of his or her junior year. Between 80 and 85% of students and graduates of ABET-accredited engineering curricula pass the FE test, compared with 55% of those from nonaccredited engineering programs, 35% of those from accredited and 25% from nonaccredited engineering technology programs, and 35% of others (physical and computer scientists and engineers with little formal training).[43] The EIT provides an additional credential for the young engineer, and engineering students are well advised to sit for the engineering fundamentals examination while the content is still fresh in their minds (and before it changes), rather than waiting until their career path might take them in a direction requiring registration.

After graduation and about four years of acceptable engineering practice, the candidate can then sit for another eight-hour test on the principles and practices of engineering; if successful, he or she will be registered as a Professional Engineer and can then append the designation PE following his or her name. Between ½ and ⅔ of those attempting the PE test pass it. Although initial registration is by the state giving the examination, other states will usually extend reciprocity and offer registration to engineers already registered with a state whose requirements are at least as rigid.

Requirements for the EIT and PE differ from state to state. Some states permit graduates with approved science, less-than-approved engineering, or (in a few cases) engineering technology degrees to sit for the EIT and PE, commonly after a longer experience requirement; a number of states have a "long-established practice" (LEP) category requiring 12 to 20 years of responsible engineering practice and tests but without a degree specification. Lunch[44] has provided a state-by-state analysis of registration requirements.

Corporations engaging in the practice of engineering legally do so only under the direction of individuals in the firm who are registered professional engineers, and in some local government activities such as public works an engineer cannot progress very far without becoming registered. Civil engineers are most likely to need registration for career success, but others may, too. Although the majority of engineers in industry are not required to be registered, there is continual pressure to increase the kinds of positions for which society demands registration. About 18% (378,000 in 1992) of U.S. engineers are registered, including 44% of civil, 23% of mechanical, 9% of electrical, and 8% of chemical engineers.[45]

Certification. A number of professions in America have programs to examine individuals and grant those especially knowledgeable in their field recognition, often characterized by a designation they can place after their name. Common examples are the Certified Professional Accountant (CPA) and Certified Life Underwriter (CLU). A number of engineering professional societies, especially those whose body of knowledge does not commonly lead to an undergraduate degree, have chosen to do this as well. Some examples of certifications, designations, and the granting professional society are:

- Certified Quality Engineer (CQE) or Technician (CQT) or Certified Reliability Engineer (CRE) by the American Society for Quality Control
- Certified Manufacturing Engineer (CMfgE) or Technician (CMfgT) by the Society of Manufacturing Engineers
- Certified Professional Logistician (CPL) by the Society of Logistics Engineers
- Certified Data Processor (CDP) or Computer Programmer (CCP) or Systems Professional (CSP) by the Institute for Certification of Computer Professionals (which is sponsored and governed by nine computer- and systems-related societies)
- Certified Plant Engineer (CPE) by the American Institute of Plant Engineers

A typical engineering certification requires a combination of up to 10 years of education and experience, and passing of a day-long examination offered twice a year; several societies offer an associate certification to recognize earlier levels of preparation. Certification is not restricted to members of the sponsoring society, but nonmembers commonly pay a fee differential not greater than annual society dues. The sponsoring society may offer short courses and study guides to help prepare for examination, and may require recertification every three to five years at a lesser fee to demonstrate continuing competence.

A National Conference on Engineering Specialty Certification, attended by members of 24 engineering societies, was held April 12–13, 1988, in Atlanta. Attendees defeated a minority proposal that professional registration be prerequisite to certification and a proposal to proceed immediately with establishment of an umbrella organization, but instead they simply recommended that participating societies study the pros and cons that certification would hold for *their* members.

There is no certification program in the United States for engineering managers. In England, however, the Management Division of the Institution of Mechanical Engineers (IMechE) launched a Diploma in Engineering Management in 1982; members of other (royally) chartered engineering institutions as well as the IMechE are now eligible. The diploma consists of three parts: (1) a six-hour essay test of areas such as organization, managing people, finance, and legislation; (2) a written examination calling for detailed study of an option chosen from a list including innovation, engineering operations, and commercial operations; and (3) a final portion concerned with corporate business policy "by examination on a set case study or by report on a management project arranged between the candidate and his or her employer, and approved by the Institution."[46]

DISCUSSION QUESTIONS

16-1. How should Fayol's 1916 "Advice to Future Engineers" be modified to better suit the world 75 years later?

16-2. Suggest some ways in which newly graduated engineers can "make their mark" in early assignments on the first job.

16-3. The text urges you to choose your boss carefully. How can you try to accomplish this practically?

16-4. Your job exists because your boss has more work than one person can do. What does this tell you of your priorities?

16-5. If someone in an outside group declines to provide the support services you need, despite your most tactful requests and explanations, what do you do next?

16-6. For each of the eight "career fields" described under "Charting Your Career," tabulate the skills, interests, and special preparation that might assist in starting such a career.

16-7. How do the "career stages" of Super and/or those of Dalton and Thompson fit your concept of your own career?

16-8. Pick a specific engineering activity or project and identify the information input and output involved (see Figure 16-1).

16-9. Pick an engineering job and outline the communication requirements involved in it.

16-10. You must meet with your staff of 20 engineers and technicians to introduce them to a demanding new program they will all be involved in. (**a**) Discuss how you will prepare for this oral presentation. (**b**) What precautions will you take during this talk? (**c**) Why might you prepare a written communication in addition to the oral presentation?

16-11. Take a fresh look at a report you have written some time in the past. How might you have made it more useful to the busy reader?

16-12. Define and use the words in each of the following frequently confused pairs: (**a**) moral/morale; (**b**) personal/personnel; (**c**) principal/principle; (**d**) affect/effect; (**e**) capital/capitol.

16-13. Prepare three or four frames suitable for photocopying as viewgraphs or slides for use in an oral presentation on making effective oral presentations (or another assigned topic).

16-14. Discuss how the engineer who moves into managerial work can nonetheless stay up to date in technology.

16-15. As the new vice president for R&D in a high-technology firm, you are concerned about the increasing technical obsolescence of your professional staff. Discuss the actions you might take and the programs you might support to reduce this problem.

16-16. What are the regional and national sources of continuing education in your specialty? (Include courses offered through your employer if applicable.)

16-17. What are the significant primary and supporting professional societies in your area of professional interest, and how do they fit into Weinert's four groups?

16-18. Discuss the extent to which professional registration and/or certification is useful in your broad field of professional interest.

NOTES

1. *Toward the More Effective Utilization of Engineers: The National Engineering Utilization Survey* (Washington, DC: American Association of Engineering Societies, 1986), pp. 31–32, 116.

2. *Occupational Outlook Handbook,* Bulletin 2300, 1988–89 ed. (Washington, DC: Bureau of Labor Statistics, April 1988), p. 53.

3. *At the Crossroads: Crisis and Opportunity for American Engineers in the 1990s* (Washington, DC: American Association of Engineering Societies, 1994), reported in "Study Sees Structural Changes in Engineers' Work Environment," *Engineering Times,* March 1994, p. 3.

4. Henri Fayol, *Administration Industrielle et Générale,* Constance Storrs, trans. (London: Sir Isaac Pitman & Sons Ltd., 1949), pp. 89–90.

5. W. J. King, *The Unwritten Laws of Engineering* (New York: American Society of Mechanical Engineers, 1944).

6. John M. Amos and Bernard R. Sarchet, *Management for Engineers* (Englewood Cliffs, NJ: Prentice-Hall, Inc., 1981), pp. 342–344.

7. Denis M. S. Lee, "Job Challenge, Work Effort, and Job Performance of Young Engineers: A Causal Analysis," *IEEE Trans. on Engineering Management,* August 1992, pp. 214–226.

8. Robert F. Morrison and Richard M. Vosburgh, *Career Development for Engineers and Scientists: Organizational Programs and Individual Choices* (New York: Van Nostrand Reinhold Company, Inc., 1987), p. 169.

9. Leonard J. Smith, "Keeping Your Career on a Success Trajectory," in John E. Ullmann, ed., *Handbook of Engineering Management* (New York: copyright © John Wiley & Sons, Inc., 1986), pp. 729–736. Reprinted by permission.

10. Donald E. Super, *The Psychology of Careers* (New York: Harper & Row, Publishers, Inc., 1957).

11. Morrison and Vosburgh, *Career Development,* p. 178.

12. Donald E. Super, "A Life-Span, Life-Space Approach to Career Development," *Journal of Vocational Behavior,* 16, 1980, pp. 282–298.

13. G. W. Dalton and P. H. Thompson, *Novations: Strategies for Career Management* (Glenview, IL: Scott, Foresman and Company, 1986).

14. Allan Chapple, "Mentoring: Molding Managers for the Next Generation," *Engineering Times,* August 1992, p. 9.

15. Morrison and Vosburgh, *Career Development,* p. 180.

16. Heinz Weihrich and Harold Koontz, *Management: A Global Perspective,* 10th ed. (New York: McGraw-Hill, Inc., 1993), p. 573.

17. *Proceedings, National Congress on Engineering Education,* Washington, DC, November 20–22, 1986 (New York: Accrediting Board for Engineering and Technology, 1987), p. B-8.

18. Denis M. S. Lee, "Engineering Performance in the Era of Integrated Computer Design and Manufacturing: Some Basic Considerations for Engineering Education," Working Paper WPS-119-88, College of Management Science, University of Lowell, Lowell, MA, 1988, pp. 8–9.

19. Henry Mintzberg, *The Nature of Managerial Work* (New York: Harper & Row, Publishers, Inc., 1973), p. 39.

20. Paul Timm, *Managerial Communication: A Finger on the Pulse* (Englewood Cliffs, NJ: Prentice-Hall, Inc., 1980).

21. Albert Mehrabian, "Communication Without Words," *Psychology Today,* September 1968, pp. 53–55.

22. George E. Dieter, *Engineering Design: A Materials and Processing Approach* (New York: McGraw-Hill Book Company, 1983), p. 539.

23. Gary Putka, "Tracking Tuition: Why College Fees Are Rising So Sharply," *Wall Street Journal,* December 11, 1987, p. 25.

24. Dieter, *Engineering Design,* p. 495.

25. H. G. Kaufman, *Obsolescence and Professional Career Development* (New York: AMACOM, A Division of American Management Associations, Inc., 1974), pp. 10–11.

26. Samuel E. Gluck, "Sources of Information for Engineering Management," in Ullmann, *Handbook of Engineering Management,* pp. 207–227.
27. Robert E. Shannon, *Engineering Management* (New York: John Wiley & Sons, Inc., 1980), p. 126.
28. P. H. Thompson and G. W. Dalton, "Are R&D Organizations Obsolete?" *Harvard Business Review,* 54, November 1976, pp. 105–116.
29. Kaufman, *Obsolescence,* pp. 51–66.
30. Kaufman, *Obsolescence,* p. 15.
31. Dieter, *Engineering Design,* p. 497.
32. Dieter, *Engineering Design,* p. 496.
33. Thomas J. Allen, *Managing the Flow of Technology: Technology Transfer and Dissemination of Technological Information Within the R&D Organization* (Cambridge, MA: The MIT Press, 1977), p. 45.
34. *The Technological Dimensions of Engineering Competitiveness* (Washington, DC: National Academy of Engineering, Office of Administration, Finance, and Public Awareness, 1988), p. 11.
35. Edmund Young, "Management Thought for Today from the Ancient Chinese," *Management Bulletin,* October 1980, p. 31.
36. Kaufman, *Obsolescence,* p. 139.
37. Kaufman, *Obsolescence,* pp. 140–144.
38. John M. Amos and Daniel L. Babcock, "Continuing Education in a Rural Midwestern Area," *Proceedings of the 1980 Frontiers in Education Conference* (New York: Institute of Electrical and Electronics Engineers, 1980).
39. N. Margulies and A. Raia, "Scientists, Engineers, and Technological Obsolescence," *California Management Review,* 10:2, 1967, pp. 43–48.
40. J. L. Farr et al., "The Measurement of Organizational Factors Affecting the Technical Updating of Engineers," presentation to the annual meeting of the Academy of Management, Atlanta, GA, August 1979, as reported in Morrison and Vosburgh, *Career Development,* p. 26.
41. Gordon Davis, ed., *1987 Directory of Engineering Societies and Related Organizations* (Washington, DC: American Association of Engineering Societies, 1987).
42. Donald W. Weinert, "The Structure of Engineering," in Ullmann, *Handbook of Engineering Management* (New York: copyright © John Wiley & Sons, Inc., 1986), pp. 28–30. Reprinted by permission.
43. D. G. Sunar, *How to Become a Professional Engineer,* 4th ed. (Belmont, CA: Professional Publications, 1992), p. 20.
44. Milton F. Lunch, "Professional Practice and Organization of Engineering Firms: The Regulatory Framework," in Ullmann, *Handbook of Engineering Management,* pp. 63–105.
45. Paul Taylor, *Engineering Times,* January 1993.
46. Daniel L. Babcock, "Certification and Engineering Management," *1986 Annual Conference Proceedings* (Washington, DC: American Society for Engineering Education, June 1986).

CHAPTER 17

Managerial and International Opportunities for Engineers

PREVIEW

The first of the two major subjects of this chapter is the relation of management to the engineer's career, and the need for engineers in top management. We discuss the reasons technologists give for moving into management, and we consider the effectiveness of parallel career ladders in technology and management. Then we discuss methods of preparing for management responsibilities, including formal and informal courses and job experience.

The second chapter topic is international management. First considered is the importance of international trade to the United States, and then the nature of multinational organizations. We outline Japanese management styles and their utility for American companies. We discuss the significance of the new European common market, the breakup of the Communist bloc, and the North American Free Trade Area. Next, we outline problems and differences found in managing activities in third-world countries, and discuss Middle Eastern countries as an unusual example.

MANAGEMENT AND THE ENGINEER

Relation to the Engineering Career

As we learned in Chapter 1, management responsibilities are part of the normal career progression of the engineer. In the 1969 data of Table 17-1 (published in 1973), we see that two-

TABLE 17–1   Supervisory Responsibilities of Engineers as a Function of Age (Percent)

Age of Engineers	No Supervisory Responsibility	Indirect or Staff Supervision	Supervision of Team, Unit, Project, or Section	Management of Major Division, Program, or Company
25–30	38	25	31	7
30–35	22	21	41	16
35–40	16	18	41	25
40–45	12	16	36	37
45–50	12	15	32	40
50–55	13	16	29	42
55–60	14	17	27	43
60–65	17	16	25	42

Source: *Engineering Manpower Bulletin 25*, Engineers' Joint Council, New York, September 1973.

thirds of engineers typically were supervising at least a team, unit, project, or section by the time they reached their late thirties, and continued to do so for the remaining two-thirds of their careers. By midcareer, 40% of engineers were managing a major division, program, or company. (Perhaps the only large groups of engineers who pursued successful careers without the majority progressing into management were basic researchers and university professors—which helps explain the lack of preparation for this part of the engineering career in most undergraduate engineering programs!)

Although I haven't seen more recent data on engineers in management as comprehensive as this 1969 data, there seemed no reason to believe it had changed very much through about 1990 (either in the United States or in other countries such as Australia when I visited there in late 1991). In Japan, 95% of engineers traditionally switch to management in their late thirties or their forties, although recently some Japanese companies have been trying "to enhance the status of researchers so they will continue researching."[1] In the early 1990s, however, American corporations have undergone an organizational revolution (whether called downsizing, rightsizing, restructuring, or reengineering) that has led to massive reduction of middle management levels and positions and loss of staff specialists positions in favor of empowered teams, incorporating some specialist expertise, that are making decisions formerly reserved for managers. In a survey released in 1993, the percentage of male engineers "who feel they are part of management" (a somewhat different question) was 28% through age 29, 49% from 30 to 39, 55% from 40 to 49, and 64% age 50 or more; the comparable figures for women engineers were only 13% through age 29, 40% from 30 to 39, and 51% age 40 or more. Obviously, the engineer going into industry in the late 1990s can expect a lower probability of being "in management" by a particular age, but he or she can also expect to need a broader understanding of "management" concerns such as marketing and financing than was needed of nonmanagers in earlier hierarchical organizations.

The probability and speed of transition into management positions and responsibilities will also depend on which of the "career fields" discussed in Chapter 16 the engineer chooses to pursue, and when he or she makes the transition to it. Careers beginning in operational areas such as manufacturing, customer service, and sales involve business considerations from the beginning, and in the past they have commonly led to operations management positions. Careers in research or design (especially in advanced technology), in college teaching, or in technical writing can involve professional activity of increasing responsibility for much or all of a career without formally becoming a manager. Entrepreneurial and most consulting careers involve managerial concerns (if not title) almost from the beginning. In government or military service, on the other hand, one's power, prestige, and pay still seem to depend on the number of organizational levels below you, and here promotion into a managerial position still seems to come earlier, but often to involve less decision-making authority at a given level.

Need for Engineers in Top Management

Lawrence Grayson, recent president of the American Society for Engineering Education, reported in 1989 that engineers were not achieving the top levels in American corporations as frequently as their counterparts in other countries:

> Engineers must be prepared for leadership—leadership in technical, corporate, and national affairs. More and more problems facing this country have strong technical components. Yet, engineers are not attaining the appropriate leadership positions and therefore have not been able to make the decisions that the nation requires.
>
> In France, most of the leaders of business and government have graduated from the elite Grand Écoles. The approximately 175 schools concentrate primarily on teaching engineering and technology. In West Germany, a majority of the corporate leaders are alumni of the technical universities, whose graduate engineers have completed a period in industry and a thesis on an industrial problem. In Japan, more than 65 percent of the members of the boards of directors of the nation's leading companies have graduated from engineering and science programs, not graduate schools of business. In contrast, roughly two-thirds of the seats on boards of American companies are occupied by people trained in law, finance, or accounting.[2]

Whittaker[3] compares three studies on the proportion of chief executive officers of U.S. and Canadian companies having first degrees in engineering, science, or business (marketing, finance, accounting, etc.). Salgado's 1990 study of the *Business Week* top 1000 company CEOs[4] reported 25.6% had their first degree in engineering, 10.1% in science, and 46.7% in business. A 1986 survey of the Fortune 500 companies reported 24% had a first degree in engineering, 6% in science, and 33% in business.[5] These compare with a 1990 study of the top 200 Canadian executives[6] that showed 21.5% with a first degree in engineering, 9.5% in science, and 21.5% in business. (The U.S. business/engineering ratio is explained in part by the 1.4 U.S. ratio of business/engineering bachelor degrees awarded in 1959, a reasonable reference year for a CEO of average age 56 in 1990; the similar Canadian ratio was only 0.5.) The probability that the CEO of a major high-technology or resource-based company would have a first degree in engineering or science was roughly 60% in both countries; the probability for other companies was less (typically 20%), ranging down to 7% for U.S. financial organizations.

Lester Thurow has explained the importance of a technical background in today's executives:

> These nontechnical managers may understand the technologies being employed by their firms, but they don't have enough background to develop intuitions on which of the possible technologies now on the horizon are apt to further develop and which are apt to be discarded.
>
> As a result, incumbent managers have no way to judge the merits of revolutionary changes in production technology. So they procrastinate, waiting for it to become clear which technology is the best. By the time the answer is clear, foreign firms may have a two- to three-year lead in understanding and employing these new technologies.
>
> The problem is found not just among managers of manufacturing facilities. Those in the investment and marketing communities also don't know where to place their bets. The ignorance and resulting risk aversion of the industrial manager is reinforced by the ignorance and risk aversion of his investment banker, advertising manager and accountant.[7]

On the other hand, the engineering mindset can be a disadvantage in top-level politics (whether they be in government or a corporation). Columnist David Broder provides an example:

> [President Jimmy] Carter was a Naval Academy graduate and an engineer. His model of policy-making was rational, efficient, and introspective. Governing to him was an exercise in problem-solving. Come up with the right solution, check and recheck your calculations, and then act. Congress did not respond to that approach. And Carter did not react well when legislators tinkered with his solutions. Soon they were at odds.[8]

Should *You* Choose Management?

Why technologists switch. Badawy[9] has found in career goal interviews he has conducted with engineers and scientists that about 80% have indicated their career goal was to become a supervisor or manager within five years. He classifies the reasons they have given into six categories:

1. *Financial advancement.* Technologists believe that the managerial ladder offers "a bigger share of the organization's goodies" than does the technical path, even where a strong dual-ladder system (discussed later) exists.
2. *Authority, responsibility, and leadership.* Many believe that becoming a manager is the best way to make the right things happen, rather than submitting to the chaos that exists.
3. *Power, influence, status, and prestige.* Engineers that seek managerial positions tend to rank higher in McClelland's "need for power" and to find satisfaction in influencing events. As one increases in management level in most organizations, status symbols such as office size, secretarial support, and other "perks" increase in a clearly recognized way.
4. *Advancement, achievement, and recognition.* Advancement in management positions provides a clear recognition for achievement.
5. *Fear of technological obsolescence.* Some engineers and scientists find it extremely difficult to keep up with the complexity and rate of technological change, and they may see management as the only alternative. An engineer who wishes to stay at the forefront of technology needs to make his or her continuing technical

education a lifelong activity through self-discipline and some of the methods discussed in Chapter 16.

6. *Random circumstances.* An engineer may drift into management because a position has suddenly become open, or to prevent a disliked colleague from becoming the boss instead, or for a variety of other almost accidental reasons.

Making the right choice for you. Many of the reasons above can be the wrong reasons, and they can encourage an effective engineer to become an ineffective and unhappy manager. If the opportunity to move events and to achieve larger things through the leadership of other people gives you the most satisfaction, you will find this in a management career. If, instead, you find more satisfaction in what you accomplish and create personally, you may want to remain primarily a technologist.

Each professional needs to take personal charge of his or her career, and to determine its direction rather than letting events take their own course. By letting your managers know what your career goals are, preparing yourself for them, asking for transfers or reassignments that will enhance your chance of getting there, and seeking a position elsewhere if you cannot achieve it in your present organization, you can largely become the master of your own fate. If you really want to be a technical specialist, you should be aware that the movement into management for any period of time may be largely irreversible, since most engineers find it much more difficult to stay competitive in an advanced technology once they get into a managerial job.

The Dual Career Ladder

Many organizations attempt to provide a technical career ladder that is the equivalent of the management career ladder. Figure 17-1 provides an example of a dual career ladder; position titles will vary from organization to organization, but the typical dual ladder offers three or four levels of parallel position titles. In a survey of dual-ladder systems, companies gave the following reasons for implementing them (with percent responding):

- Retain the best professional and technical people (90%)
- Create a career path for those not interested in management (88%)
- Increase morale of the technical staff (67%)
- Create a more equitable nonmanagement compensation structure (61%)[10]

Unfortunately, in many organizations the technical ladder does not live up to its promise of equality with the managerial alternative. Badawy identifies five criteria that must be met for a dual-ladder system to work efficiently:

1. The technical and administrative ladders must be equally attractive to technologists in terms of salary scales and status symbols and other noneconomic rewards.

2. Neither ladder should be used as a dumping ground for individuals who are unsuccessful on the other ladder.

3. Criteria for promotion on the technical ladder must be rigorous and based on high technical competence and achievement.

4. Both ladders must have the full support of management.

5. The system must be fully accepted by the technical staff.[11]

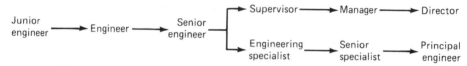

Figure 17-1. Example of dual-ladder system.

Allen and Katz reported[12] on career preferences of 2157 engineers and scientists responding from nine major U.S. organizations: two federal laboratories, three not-for-profit firms, and four industries (two aerospace, one electronics, and one food). Respondents were asked the extent (on a seven-point scale) they would like their career to be (a) a progression up the technical professional ladder; (b) a progression up the managerial ladder; or (c) "the opportunity to engage in those challenging and exciting research activities and projects with which you are most interested, irrespective of promotion." Of the 1495 indicating a preference for one of these three, 323 (21.6%) preferred the first (termed "technical orientation"), 488 (32.6%) the second ("managerial orientation"), and 684 (45.8%) the third ("project orientation"). With age, the project orientation tended to increase and the other two decrease somewhat. After age 40 there is general disagreement that high performance will lead to technical ladder promotion; only managers seem to believe that high performance leads to managerial promotion, and even their belief diminishes with age; only those with project orientation believe past age 40 that high performance leads to interesting project assignments—clearly, the existence of dual ladders has not solved the motivational needs of older engineers.

In a later analysis of these responses plus additional ones from two European industrial equipment manufacturers, Allen reports[13] that engineers and scientists with a Ph.D are more thoroughly indoctrinated in academic/scientific success criteria (publishing significant technical papers, developing new theoretical insights or innovative solutions to technical problems), retain these drives throughout their careers, and lean toward a technical ladder. Those stopping with a B.S. or M.S. degree are also "to some degree affected" by university values and are initially attracted to the technical ladder. "Shortly after that, however, reality begins to set in. They begin to understand that industry needs management as much as technology—that theories and publications don't put bread on the table and that commercially important projects are not necessarily those of the greatest scientific interest. This awakening occurs in the early to mid-thirties and results in a pronounced shift away from the technical ladder and increased interest in management."[14] Allen reports that a byproduct is a decoupling, with decreasing communication, between those on the technical and managerial ladders, hampering technology transfer and innovation.

Preparing for Managerial Responsibilities

Requirements. Effectiveness as a manager requires a combination of attributes, knowledge, and skills. Edwin Gee, while a senior vice president of DuPont, listed the following attributes of researchers that were indications they would become good managers:

1. They are able to identify a problem, analyze it, and synthesize a solution.
2. They are willing to accept and even seek responsibility.

3. They view their current assignment as the most important thing they have to do rather than as a step toward promotion.
4. They have good work habits, set personal goals, and plan ahead.
5. They are able to get results without upsetting people.
6. They have integrity.
7. In addition to technical talent, they have demonstrated at least some skill in such fields as marketing, finance, and employee relations.
8. Finally, the ability to make sound judgments is the key to both long-term potential and current readiness for promotion.[15]

Thamhain reports statistically significant correlation between personal desire to become a manager, actual promotion, and subsequent performance, but he adds, "personal desire alone is insufficient to gain a promotion. In the final analysis personal competence and organizational needs are the deciding factors. People who get promoted usually meet five key requirements:"[16]

1. Competence in current assignment and respect of colleagues
2. Demonstrated ability and willingness to take on greater responsibility (and good time management)
3. Preparation for new assignment through courses, seminars, on-the-job training, professional activities, and special assignments
4. Match of capabilities with current and long-range needs of the organization
5. Perceived aptitude for management

The skills that bring an engineer to prominence and result in promotion into a first management position are not necessarily the skills needed in the new management position. The engineer is a doer; an effective manager is a facilitator of other people's work. As we learned in Chapter 6, the new manager must learn to delegate, and the engineer's education and past experience have not prepared the new manager to do so. Indeed, the engineer and engineer-manager have quite different roles; these are contrasted by Amos and Sarchet in Table 17-2.

TABLE 17–2   Differences in Roles Between Engineers and Engineer Managers

Engineer's Roles	Engineer Manager's Roles
Originates projects	Evaluates projects
Creates, seeks new ideas	Provides facilities to help engineers
Works on specific programs	Does overall planning
Has limited responsibilities	Has responsibility for a department or group of people
Is specialized, is technically oriented	Is people-oriented, is responsible for and responsive to people
Obtains facts himself, is objective	Motivates others
Utilizes own skills	Utilizes skills of others to obtain goals
Has limited concern for finances	Has fiscal responsibilities

Source: John M. Amos and Bernard R. Sarchet, *Management for Engineers*, © Prentice-Hall, Inc., Englewood Cliffs, NJ, 1981, p. 16. Reprinted by permission.

TABLE 17–3 Stages in the Acquisition of Management and Business Skills

Stage	Description	Key Requirements
1. Preemployment	Normally completed within main education system/home environment	Personal discipline Pocket money economics, social effort, and communication Teamwork and leadership
2. Early years of employment	Premanagement stage when need to establish: Technical foundation Business foundation Technical leadership	Professional engineering development Small project leadership Understanding employer's business Basic communication skills (e.g., report writing, costing) Computer literacy Continuing personal development Goal setting/problem solving Understanding of industry and commerce
3. Supervisory or junior management (e.g., initial management position, team leadership)	First stage at which responsibility is assumed for other people and their work	Managing people Project management Industrial relations Management/business controls and financial systems Commerce and marketing Appraising and developing people Statutory obligations Communications skills (including presentations) Financial management
4. Middle or departmental management (e.g., group leadership)	Key stage when responsibility is assumed for a function or service, including junior management	In-depth understanding of all business functions (e.g., managing technology, finance, marketing, sales, legal, negotiating skills) and management of operations, information technology, and business systems Managing change Advanced communication skills Appraising and developing people
5. Senior management (e.g., departmental leadership)	The stage when fully accountable for a major business area or function Often part of the top executive team responsible for total operation of a unit, business, or organization	Business and strategic planning Management of human resources Management of financial resources Management of technology Decision making Appraising/developing managers Economics of engineering Communications skills (including use of media)
6. General management (e.g., direction of all aspects of business organization)	The level involving total accountability for the management, performance, and future direction of a whole unit, business, or organization	Managing managers/strategic management Creating business performance Strategic development—identifying discontinuities and managing consequences International diplomacy, politics, economics, and finance

Source: The Engineering Council, *Management and Business Skills for Engineers: Continuing Education and Training, A Statement*, 1988. Published with permission from The Engineering Council, London, United Kingdom.

The knowledge and skills required for carrying out management responsibilities vary considerably with the level of management. The [British] Engineering Council has outlined the key requirements by management stage (Table 17-3).

Formal degree programs. To prepare yourself for management, you will need to obtain *knowledge about management* and to develop *management skills,* which are two different things. The most common program for teaching knowledge about management is the master's degree in business administration (M.B.A.). In 1950 only 6 universities offered the M.B.A., but by 1980 almost 600 schools were granting more than 50,000 such degrees a year. Many of these programs were quite weak, and just over a quarter were accredited by the American Association of Collegiate Schools of Business (AACSB).[17]

There are also quite a few programs in engineering schools that combine engineering and management. As of 1990, I counted 80 engineering and 7 business schools in the United States and Canada offering programs called Engineering Management (78), Engineering Administration, or Management Engineering. Using a somewhat broader list of program titles, Kocaoglu[18] had verified 94 programs in the United States, 6 in Canada, and 21 elsewhere offering 166 degrees, 38 at the bachelor's, 99 the master's, and 29 the doctoral level. In addition, 18 engineering schools offered degree programs in construction or project management, primarily through their civil engineering departments.

The master's degree programs in engineering management (M.S.E.M.) typically require 30 to 36 semester hours of work beyond an acceptable engineering B.S. (Figure 17-2). This compares with requirements of from 42 to 60 semester hours for an engineer to obtain an M.B.A., largely because of the demand that engineers take remedial undergraduate courses in business subjects before tackling them at the master's level. The M.S.E.M. typically differs in that (1) courses are accelerated beyond the usual business first course, especially where quantitative capabilities are involved; (2) courses emphasize management of technical enterprises; and (3) faculty often have engineering backgrounds.

A third category of program is management of technology (MOT) or technology management, a growing group of programs largely being established in schools of business. For example, Weimer[19] lists 27 "universities and institutes [which] are among those offering degrees or individual courses and seminars in technology management" (18 U.S., 2 Canadian, and 7 European).

A commonly cited definition of MOT comes from the 1987 National Research Council report *Management of Technology: The Hidden Competitive Advantage:*

> Management of technology links engineering, science, and management disciplines to plan, develop, and implement <u>technological</u> capabilities to shape and accomplish the strategic and operational objective of the organization.[20]

The NRC report identified "four categories of technology-related issues" important in MOT. These can be conveniently paired into two classes, the *strategic* (long-term strategic and interfunctional policy issues) and the *operational* (current research, development, engineering, operations, and support services issues). The NRC report then lists 29 specific "MOT issues"; I have shown elsewhere[21] that only 4 of these issues fall into the strategic class, and the remaining 25 are operational issues addressed just as frequently in engineering management as in MOT educational programs. Supporting this, I have shown in a sur-

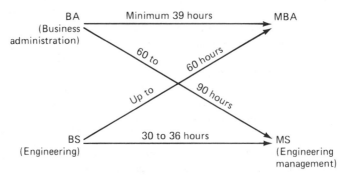

Figure 17-2. Comparison of M.B.A. and M.S.E.M. programs.
(From Daniel L. Babcock, "An Engineering Management Program Comes of Age," *Engineering Education,* November 1973, p. 103.)

vey[22] that the average required content in 33 engineering management and seven MOT programs were similar in the broad areas of organization and management, manufacturing, and decision making, but the MOT programs covered the strategic areas of technology versus business strategy, R&D strategy, entrepreneurship, innovation, and technology transfer much more intensively.

Nondegree coursework. Of course, even formal courses in management do not have to be part of a credit program. Noncredit short courses are offered by universities, by professional societies, by the American Management Association, by independent consultant/entrepreneurs, and through the educational programs of large employers. Courses may be full time for from one day to two weeks, or may be one or several times a week for some period, and may take place at your work site or elsewhere on your own time or your employer's time.

Noncredit courses permit the student and/or employer to select topics for their direct application to organizational needs, without the constraints of formal degree curricula. They are not as restrictive in admission requirements. They are usually less rigorous, without homework or tests. Younger engineers have a greater tendency to prefer degree programs; older professionals tend to favor noncredit courses that focus on topics of specific current need.

Experiential training. All the formal or informal coursework described above still only provides *knowledge about* management. Role playing, case studies, and management games attempt to develop understanding and enhance skills, but they are no substitute for real experience. Organizations that recognize the vital importance of developing future managers will have deliberate policies and programs to do so, and they will evaluate current managers partly on their success in developing future talent. Some of the methods used will be on-the-job training, coaching, selection of job assignments for their developmental value, job rotation, and temporary assignments to other areas. Often project assignments will give the new engineer an initial insight into the relation of his or her technical work to the total organization. McCall et al.[23] give examples of the "veritable encyclopaedia of executive education" that can be gleaned from appropriate assignments:

Assignments such as:	Can provide learning about:
Project/task forces	Giving up technical mastery; understanding other points of view
Line-to-staff switches	Coping with an ambiguous situation and understanding corporate strategies and cultures
Starting something from scratch	Identifying what is important and building a team
Turnaround jobs	Being tough, persuasive, and instrumental
Leaps in scope	Relying on other people and thinking like an executive

Unfortunately, too few organizations nurture the development of potential managers in this way. Often it is easier for managers to keep the engineering specialist working in a narrow area in which they have become proficient, rather than transfer them to something different and have to replace their special capability. Frequently, letting your manager know the direction you want to grow, with occasional reminders, will result in reassignment. You may wish to talk to the personnel office and to other managers about potential transfers and then tell your manager what you have learned. Your employer does not provide the only opportunity for growth—the local chapter of your professional society and local community and charitable organizations cry out for volunteers and give you an opportunity to try out your leadership skills without a direct impact on your career.

In the final analysis, each person must take responsibility for his or her own career. Let your immediate supervisor and others know, tactfully, what your interests and objectives are, and give them a chance to help you. Ask for an evaluation of what you need to do and what assignments you should look for to prepare yourself. Then if your present organization cannot or will not offer what you are looking for, you owe it to yourself to consider if another organization can.

INTERNATIONAL MANAGEMENT

Two groups of topics are discussed in the remainder of this chapter. First, trends in U.S. international trade, the nature of the multinational corporation, and their significance to the engineering career are considered. Second, the effect of cultural and economic differences on management is discussed, with special attention to management in Japan, Europe, and developing countries.

International Trade and the United States

In the past several decades, international trade and multinational management organizations have increased dramatically. From 1950 to about 1988 real U.S. gross domestic product (GDP) tripled, but real world GDP quadrupled and world trade grew seven fold;[24] GATT, NAFTA, and other trade agreements should assure that this continues. More and more American companies are finding that, even if they are not heavily involved in international sales themselves, they cannot escape the impact of international competition in their home markets. For the last decade the United States has experienced massive merchandise (commodity) trade deficits, amounting to $84.5 billion in 1992 (see Table 17-4) and $115.7 billion in 1993. The major bright star has been aerospace (principally commercial aircraft from

Boeing and McDonnell-Douglas), which reduced our annual deficit by $26 to $30 billion each year (at least from 1990 through 1993). Our next most important export commodities are farm products such as cereal grains (especially wheat) and tobacco ("?" in the table indicates that the corresponding import data were not available). The commodity creating our biggest deficit was (and remains) petroleum; the biggest deficit in manufacturing in 1992 was "road vehicles" (primarily automobiles), but this is decreasing since it is now significantly cheaper for Japanese and German companies to make their cars in the United States. Clothing (apparel and footwear) was the next largest contributor to our deficit, followed by telecommunications and sound recording and reproduction apparatus and equipment (including, one assumes, televisions and VCRs).

Our continuing merchandise trade deficit, balanced only in part by a positive U.S. trade balance in services, which has made the United States the world's biggest debtor, is something every American citizen needs to be concerned about. Engineers also need to be concerned as professionals about the movement of engineering design and manufacturing jobs out of the United States.

TABLE 17–4   U.S. Trade Imbalances, 1992, $ Billions

Commodity Group	Exports	Imports	Balance
Groups imbalanced by more than $2.5 Billion			
Transport equipment (except road)	38.527	8.252	30.275
Cereals and cereal preparations	12.237	?	12.237
Professional & scientific instruments	14.947	7.602	7.345
Tobacco & tobacco manufactures	6.177	?	6.177
Specialized machinery	17.250	11.814	5.436
Chemical materials & products	6.461	2.622	3.839
General industrial machinery	18.908	15.520	3.388
Power generation equipment	18.485	15.888	2.597
Furniture and bedding	2.600	5.503	−2.903
Nonferrous metals	5.073	8.513	−3.440
Photo equipment, watches, clocks	4.340	7.916	−3.576
Miscellaneous manufactures	23.324	28.543	−5.219
Nonmetallic mineral manufactures	4.885	10.172	−5.287
Office & ADP machines	30.997	36.377	−5.380
Iron and steel	3.868	9.320	−5.452
Footware	?	10.163	−10.163
Telecommunications & sound equipment	12.370	25.803	−13.433
Apparel	4.200	31.226	−27.026
Road vehicles	38.191	75.477	−37.286
Petroleum and related materials	6.095	50.361	−44.266
SUMS FOR IMBALANCED GROUPS	268.935	361.072	−92.137
ALL OTHER COMMODITY GROUPS	179.229	171.593	7.636
TOTAL ALL COMMODITY GROUPS	448.164	532.665	−84.501

Source: *Statistical Abstract of the United States 1994*, 114th ed., U.S. Bureau of the Census, Washington, DC, 1994, Table 1328, pp. 821–822.

? not available, thereby overstating balance.

International trade is very complex, as we can see in an example by Gilder:

> Intel Corp., for example, is a leading advocate of the U.S. crackdown on trade with Asia. But NEC [Japanese] is its fifth-largest customer; all of its other leading customers in the U.S. computer industry except IBM are utterly dependent on complementary devices produced in Japan; Mitsubishi Electric makes an increasing share of Intel's erasable programmable read only memories (EPROMs); South Korea's Samsung makes all of Intel's dynamic random access memories (DRAMs); Zymos, an American firm owned by the South Korean Daewoo, designs Intel's IBM clone chip sets; Daewoo in turn uses Intel microprocessors in computers made in Korea and designed and marketed in the U.S. by Leading Edge; Chips & Technologies, a leading force in bringing PC production back to the U.S. and to Intel microprocessors, has been heavily dependent on Japanese funds and factories from the beginning; and Intel itself packages chips in the Philippines and Malaysia, and buys capital equipment from Nikon, silicon wafers from Shin-Etsu and photomasks from Dai Nippon. Even Intel's leading customer, IBM, turned to Shimizu of Japan to construct its world-leading 648,000-square-foot chipmaking facility in Fishkill, N.Y.[25]

Two factors offer some hope of reversing the critical U.S. trade deficit in manufactured goods: the decrease in the artificially high postwar value of the dollar, and a more realistic response of American labor to wage rates and work rules. A *Business Week* cover story expands on this:

> Because of the dollar's fall—but also because of the new labor climate—American workers are now paid less than their German and Japanese counterparts. Energy and raw materials are cheap, and manufacturing in the U.S. guarantees access to the world's largest market while at the same time providing an export base. "It's the best country in the world now in terms of manufacturing costs," says Hans W. Decker, president of U.S. operations for West Germany's Siemens. Adds Hewlett-Packard President John A. Young: "The U.S. is certainly a lot more attractive a place to manufacture. Just look at all the Japanese firms setting up to manufacture here."[26]

Indeed, as early as 1988, numbers of Japanese firms had begun exporting parts to the United States, assembling them here, then reexporting the product to Europe in an attempt to avoid European tariffs and quotas on Japanese-made goods![27] Partly as a result, by the beginning of 1987 Japanese companies in the United States already employed 250,000 Americans. German labor enjoys such high benefits (six weeks vacation plus holidays, for example) that nonwage expenses (fringe benefits plus payroll taxes) in 1994 amounted to 46% of labor cost, versus 28% in the United States and 24% in Japan[28] (and, largely as a result, Mercedes built a plant in Alabama to produce cars as much for export as for sale in the United States).

Multinational Organizations

A multinational corporation is one with significant operations in more than one country. Jacoby[29] identifies six stages of multinationalization:

1. Exporting products to foreign countries
2. Establishing sales organizations abroad

3. Licensing patents and know-how to permit foreign firms to make and sell its products

4. Establishing foreign manufacturing facilities

5. Multinationalizing management from top to bottom

6. Multinationalizing ownership of corporate stock

An alternative to stage 3 or 4 that is becoming increasingly common is the formation of a *joint venture* between firms in two different countries to pursue some mutually desirable business undertaking. An example is NUMMI, the joint venture of General Motors and Toyota to produce automobiles at Fremont, CA, using Japanese management methods and designs and American labor and distribution. Reorganizing on a truly global basis can pay great dividends. Ford, for example, "is merging its large and culturally distinct European and North American auto operations and plans to later fold in Latin America and the Asia-Pacific region" in hopes "that by globalizing its product development, purchasing and supply, and other activities, it could save up to $3 billion a year"; similarly, IBM "is organizing its marketing and sales staffs into 14 industry groups, such as banking and retail, rather than by country."[30]

A number of foreign firms gain more than half their business outside of their headquarters country. Notable examples in 1990 were Philips (94% of sales outside of the Netherlands), ASEA Brown Boveri (85% outside Switzerland), L. M. Ericsson (81% outside Sweden), ICI (78% outside Britain), Sony (66% outside Japan), and Siemens (51% outside West Germany). American examples included engineering design and construction firms such as Foster Wheeler (72% international business), Brown and Root (65%), M. W. Kellogg (59%), and Bechtel Group (55%); manufacturers included IBM (59%), Dow Chemical (54%), Xerox (54%), and Hewlett-Packard (53%).[31] Facing saturated markets in the United States for products such as power turbines and jet engines, General Electric is targeting China, India, and Mexico for its growth; from 1988 to 1993 it increased its percentage sales from international activities from 30% to more than 40%, doubling its contribution to the U.S. balance of trade to $6 billion in the process.[32]

A wide variety of factors makes international management more complex than single country management. Robinson[33] categorizes the important variables in the international system into six groups, based on differences in (1) national sovereignty, (2) national economic conditions, (3) national values and institutions, (4) timing of national industrial revolutions, (5) geographical distances, and (6) areas and population.

Pricing policies are a strategic tool of multinationals. One example is *transfer pricing*. For example, an automobile firm may produce an engine in country A, a transmission in country B, and assemble these into a car in country C. The price at which components are transferred to the assembly process determines where profits are accumulated and taxed and the level of import duties, yet often there is no competitive price for such components. Obviously, the countries losing revenue in this process may press for "more realistic" transfer prices. Multinationals also have engaged in "penetration pricing" to gain market share in a new market by unusually low prices. Japanese and other Pacific rim countries have been accused of "dumping" goods in the United States and Europe at prices far below their own internal markets, leading to legislation and quotas limiting such imports. In mid-1994,

North Dakota farmers were accusing Canada of dumping durum wheat in the United States; Canadians replied that U.S. subsidies of wheat shipments to Europe left a shortage of wheat needed by U.S. spaghetti makers, which they were filling.

Significance for the Engineering Career

American engineers have been in demand internationally for a century and a half, especially since World War II. Many engineers have found they could earn up to twice as much abroad in places like Saudi Arabia, returning after as little as 15 years with a comfortable retirement. Engineers in truly global concerns like the large design and construction firms previously cited or manufacturers with half their sales overseas often find at least one international tour necessary to rise very far in the firm. Again, many engineers find an international tour satisfies their desire for adventure, or returns them to the culture their family originated from, or adds an invaluable experience in the education and growth of their children.

There are negative factors as well. Companies who are not fully committed to global operations still send engineers and managers abroad but, when they return, the expatriates find they have been "out of sight—out of mind" and have lost momentum in their career. Two-career families may find that the career of the expatriate's spouse is "on hold" during the overseas period, and families with children in high school or college (or with elderly parents) may find being overseas disruptive. With today's weak dollar, the traveler must also be sure that compensation is sufficient for the cost of international living—the dollar may go further in Australia, for example, but a modest house in a Tokyo suburb may rent for $11,000 a month.

The young engineer needs to consider carefully what an international assignment will do for (or the harm it might do to) his or her overall career or progress within a specific firm, and the impact it might have on personal goals and family life.

Japanese Management Styles

Ever since the Japanese began their ascendancy in the 1970s, an ocean of printing ink has been consumed in books and articles about Japanese management and cultural practices alleged to be at the root of their success. Among the practices most frequently mentioned, with some discussion of their implications, are:

1. Recruitment of employees directly out of secondary school or college, which (**a**) makes acceptance into a prestigious university critical, and (**b**) limits the opportunity for midcareer transfer to another company (although the recent success of the *heddo-hantaa* (from the American "headhunter") in arranging midcareer movement of people with needed skills is moderating this.

2. The fabled "lifetime employment" of Japanese firms only applied to full-time male employees of the larger firms; at its peak in the mid 1970s it only applied to half of Japan's workers, and is estimated to be only half that now; nearly a fifth of the work force is part time (threefold that of 20 years earlier) with no security (a practice that has recently increased in the United States); further cutbacks are accommodated by bringing work in-house from subcontractors or foreign plants.[34]

Still, in the recent recession the nominal Japanese unemployment rate of 2.7% (December 1993) included neither women nor older workers who had given up hope nor the estimated "window tribe" of 1.4 million office workers who sit by the window all day with nothing to do, reducing overall productivity.

3. Promotions infrequent and based (along with salary) on seniority and teamwork, although Honda and other firms are now beginning to base bonuses on performance rather than seniority.

4. Retirement or restricted employment for older (55-year-old) workers, except for the few rising to top management; with a large aging population, this practice will give Japan serious problems.

5. *Michi,* or "the way" (to heaven or to corporate success), which emphasizes mastery of even a small task—a Zen philosophy of self-improvement to do the present job better, contrasted with the Western (and Confucian) emphasis on the next promotion.

6. *Giri,* a sense of duty of honor or obligation to observe community (read, company) customs; for the modern-day Japanese "salaryman" this may involve compulsory parties and unwanted assignments overseas or away from family, yet require loyalty to the company.[35] Respected Japanese managers and technicians are *moretsu shain* (fanatical workers) and *yoi kigyo senshi* (good corporate soldiers), and thousands become victims of *karoshi* (death from overwork), leaving "*karoshi* widows" behind.[36]

7. From the company in return, an almost womblike concern for the (full-time) worker, including family support, housing, recreational activities, social gatherings and cultural events, and festivities to commemorate promotion, internal transfer, and retirement, a security that has diminished in the recent recession.

8. Ceremonies and rituals (such as group calisthenics and the company song) to foster love of company and group identification.

9. Rotation between functional departments to ensure broad exposure to the company (at the cost of the skills honed in specialization).

10. *Omikoshi* management, "the practice of having the middle [and lower] levels in the corporate organization . . . plan new projects on their own initiative, obtain top management approval for it, and then carry it out on their own . . . so that they carry the company like a group of people carrying the small portable shrine (*omikoshi*) that is a traditional feature of Shinto festivals."[37]

11. The *ringi* system of having people sign on the bottom-up proposals, after which responsibility for implementation is borne by the group rather than a specific individual.

These practices, although many now apply only to only a minority of Japanese workers, are not without their problems. Odaka cites four disadvantages whose "drawbacks have become increasingly obvious in the postwar years":

1. Encouraging employee dependency and suppressing individual creativity

2. Discriminatory employment and impediments to the formation of a free horizontal labor market

3. Harmful effects of the escalator system [in which those first on the employment "escalator" rise before later arrivals] and middle management promotion gridlock
4. Work that has no joy and seemingly has no meaning[38]

Japanese management is moderating some of the sharpest differences from methods of Western management—the gradual, but slow, improvement in opportunities for women is a case in point. The Japanese postwar baby boomers, or *dankai sedai,* now in their forties, are more affluent and better educated (especially in Western culture); as they gain ascendancy they are insisting on a balance between Western ideas and Japanese values. On the other side, leading American corporations (which Ouchi[39] terms "Theory Z" corporations) have selectively adopted the most useful and transferable of the lessons from Japanese management.

Japanese production plants have several characteristics that set them aside from American competitors. Some of these have already been mentioned in earlier chapters: the almost fanatic emphasis on quality, the just-in-time approach that almost eliminates inventory on the plant floor, and the quality control circles and other devices through which the production employee directly works to improve the product. Others are a high priority on plant cleanliness, regarded as a prerequisite to quality; operating equipment below its ultimate capacity to extend its life and precision; automation and warning systems that require fewer workers; and a mutual trust (and unified language and ethnicity) that requires fewer management review levels.

Any of the management methods being successfully employed in Japan or other countries can be adapted to U.S. needs, and many of them were originally developed in the United States. As Drucker puts it: "What we can learn from foreign management is not what to do. What we can learn is to do it."[40]

Europe and the Common Market

The United States has long been the world's most attractive single market, with its quarter-billion affluent people in a single market free of internal barriers, and it has therefore been a prime target of European and Pacific rim manufacturers of every sort. On January 1, 1993, Western Europe was transformed into a market free of internal barriers that equals or exceeds that of the United States, with profound implications for international trade. The *Wall Street Journal* summarized these implications in early 1988:

> The EC [European Community] may never become the United States of Europe its founders dreamed of in the 1950s. The Europe of 1992 will continue to differ from the United States of 1988. The 12 nations will still speak nine different languages, spend twelve different currencies in different ways, and be governed by different laws.
>
> But as the EC enacts some 300 new rules aimed at dismantling virtually all intra-European trade barriers by the end of 1992, nearly every major European country is rushing to adjust to a future market of 320 million customers. Although nobody can predict all the winners and losers, it's clear that bigger, more powerful companies will arise in Europe and then move into world markets. At the same time, multinational companies, be they European, American, or Japanese, should find it simpler to manufacture and sell in Europe.[41]

As of the EC opening in January 1993, some 94% of the needed single-market directives had been passed, although some of the tough ones (articles of incorporation, trademarks, corporate income tax, uniform labor rules) had not. By the end of 1994, real progress had been made in merging smaller national companies into larger multinationals that could be competitive not only throughout Europe, but worldwide. U.S. multinationals such as IBM and Ford that have established integrated manufacturing and marketing systems in Europe were finding new opportunities in the barrier-free EC, and there will be many opportunities for internationally minded American engineers and engineer managers to assist companies in this growth. Companies that do not establish a presence in Europe may not only find themselves no longer competitive in that market, but also at a disadvantage in competing in the United States and the rest of the world with the new, larger multinational companies being established in Europe.

Nonetheless, progress has not been smooth, and has been held back by a recession throughout Western Europe since before the EC began, as governments have postponed changes to protect jobs. In the early 1980s, unemployment was less than 4% in most West European nations; by late 1993 it was over 10% (almost 20% for those under the age of 25). Strong unions and a strong currency have given Germany one of the highest labor costs in the world; the other leading European economies also have higher labor costs than does the United States. Part of the opposition to the EC is national pride and culture—the British and French, for example, fear the domination of German banks if a common currency is adopted (and the British loss of the traditional pint of beer if metric system adoption becomes total!).

The fall of the Iron Curtain and the disintegration of the Soviet Union are causing great changes we are only beginning to understand. The reunification of Germany in 1990 was launched with great optimism:

> At the time of reunification, the German Chancellor figured a combination of seed capital and West German knowhow would quickly transform East Germany's economy, which on paper, at least, looked surprisingly strong. After all, these were fellow Germans: With a little leadership, they would show the same drive that created West Germany's *Wirtschaftswunder,* the economic miracle of the 1950s. That was the dream. The sad reality is a manufacturing base far more inefficient than ever imagined, a heavy dependence on the old Soviet Union for trade—now vanished—and environmental pollution on a massive scale. But West German leaders were also to blame. They exported their high wages, generous benefits, and stifling regulations to the east—just when this brand of capitalism was beginning to flag under global competition.[42]

Instead, by late 1993, 70% of eastern Germany's productive capacity was shut down, and the German government was transferring $120 billion a year from the (somewhat) prosperous west, ⅔ of which was used to support otherwise unemployable workers who clung to the communist illusion that the state would always attend to their needs. Naturally, the remaining "guest workers" attracted to West Germany from southern Europe and Turkey (which amounted to 1.4 million by 1970) were now resented as taking jobs from Germans.

Other industrial countries of Eastern Europe (such as Poland, the Czech Republic, and Hungary) have the same problems of low productivity, low quality, and pollution as eastern Germany, but without the artificially high wage scale. Western European, Japanese, and American companies are rushing to acquire companies in these countries and to build or upgrade factories to take advantage of the affordable skilled labor and growing market demand in these areas. These same countries seek membership in the European Community, and even in NATO, which was created in part to defend *against* their armies, and must now redefine its purpose.

Russia and the other countries of the former Soviet Union have even greater problems, since they have no memory of free market life; the industrial nations of the West are feeling their way towards economic integration with them. Further, we in the West are just now relearning about the centuries-old ethnic hatreds kept bottled up by Serbian and Russian control in the former Yugoslavia and Soviet Union, and we cannot predict what terrible conflicts may follow on the rape of Bosnia.

Suffice it to say that conditions in the European Community, Eastern Europe, and the former Soviet Union are evolving in ways that are hard to predict, but their changes will have major effects on the world economy.

International Trade Agreements

North American Free Trade Agreement (NAFTA). The United States and Canada have long been touted as having the world's longest undefended border. What few U.S. citizens realize is that Canada is our largest trading partner (Japan is second), as are we theirs: in 1993 the United States exported $100.2 billion in merchandise to Canada, and imported $110.9 billion (U.S.) in return. While Canada has about 10% the population of the United States, with proportionate GNP (Table 17-5), we have many similarities. Both countries were settled on the Eastern seaboard about the same time from England and other European countries; both grew westward, and both have been a welcome melting pot for immigrants of all cultures. While our governments have taken different viewpoints in matters like health care and the Vietnam War, we have fought side by side in two world wars, and our border has been wide open for commerce, migration, and employment.

Mexico is quite another matter. Our 2000-mile border is garrisoned not by an army (which is unnecessary for the United States and pointless for Mexico), but by a Border Patrol whose function is to try to reduce the flow of illegal workers northward. The one-party (PRI) government had been in power since the early 1930s without challenge, and each President hand-picked a successor. The PRI developed a system of government-owned, -controlled, or -protected industries for the domestic market, and it protected them through

TABLE 17–5   Statistics of NAFTA Countries Compared

	Mexico	U.S.	Canada
Population (1991, millions)	87*	250*	26*
18-year-olds (1992, millions)	2.0	3.2	0.4
Gross domestic product (1992, billion $U.S.)	334	5951	552
Hourly labor cost (1992, $U.S. per hour)	2.5	16#	17#
Average annual productivity increase, 1988–92	6.3%	3.0%	NA

Sources: *Business Week* articles: April 19, 1993, pp. 84–86; (*) November 12, 1990, p. 103; (#) May 31, 1993, p. 48.

NA: Not available

high tariffs and import regulations, so they did not have to be efficient or competitive outside Mexico. As a result, Mexico became continually poorer and more dependent on the United States.

About 1965 the innovation of industrial parks called *maquiladora* began on the U.S. border. Although foreigners could not own industries elsewhere, they could own 100% of a *maquila* (a plant in a maquiladora), and could import parts and supplies duty-free to assemble into products that were returned to the United States (with duty only on the value added in Mexico). As a result the border cities boomed. When I visited Cuidad Juarez (next to El Paso, TX) in 1954 it seemed a sleepy town of a few hundred thousand whose most visible economic activity seemed to be cheap handcrafts and prostitution; when I taught graduate engineers there in 1971 it boasted a million people, many in sprawling slums, and I watched its workers (mostly women) in the maquilas assembling circuit boards for U.S. televisions, wire harnesses for U.S. refrigerators, and even entering bales of grocery "cents-off" coupons into computer memory so that U.S. banks could reimburse the grocers. Practically all new jobs created in Mexico in the 1980s were in the maquiladoras, and by 1990 they employed 500,000 and created the bulk of foreign-exchange earnings.[43]

Next, U.S. auto makers uncovered the potential of young, semiskilled Mexican workers:

> By the late 1980s, Buick Centuries and Mercury Tracers made in Mexico were winning quality awards; Mercedes-Benz and Nissan Motor Co. were setting up plants, and [Mexican President] Salinas' chief asset, his vast labor pool, was suddenly enhanced by the new mantra "quality." The carmakers were the main showcases for Mexican quality. But other manufacturers are taking the time to train young Mexicans—and reaping rewards. The turnaround began [about 1986], when Mexico joined the General Agreement of Tariffs and Trade [GATT] and tore down its trade barriers. Suddenly, companies that were producing overpriced goods for Mexico's closed economy had to upgrade operations to world-quality levels. . . . And they delivered.[44]

The sum of exports and imports between Mexico and the United States increased from $29 billion in 1986 (in Mexico's favor) to $58 billion in 1990 (more balanced). Presidents George Bush and Carlos Salinas de Gortari began trade discussions in late 1990 that later involved Canada; these tripartite talks were successfully formalized in August 1992 as the North American Free Trade Agreement; Bush signed it, but he lost the election. It remained for President Clinton to push it through Congress. It was opposed by labor in both the United States and Canada, but it finally passed the U.S. Congress in 1993 by a narrow margin, with greater support by Republicans than Democrats.

By the end of 1994, it appeared NAFTA would be a success. Retiring Mexican President Salinas had "fixe[d] salaries through a complex business-labor agreement that's known as *el pacto,* maintaining industrial labor costs (1992) at about ⅙ those in the United States,[45] well below those in Hong Kong, Korea, Singapore, and Taiwan. Many Japanese and American plants that might have been located in these four Asian countries (or Thailand or China) began to appear in Mexico. *Business Week* could report that "the 'great sucking sound' [of jobs going south feared by U.S. auto workers] turned out to be Detroit autos going south, not jobs," and bilateral trade increased from a 1993 total of $82 billion to an estimated $95 billion for 1994.[46] Other Central and South American countries were crying for membership in NAFTA, which seemed to have the potential to profoundly change not just the North American, but the entire continental economy.

However, Salinas had been maintaining the Mexican peso at an artificial 3.5 to the U.S. dollar (about $.29); on December 19, 1994, soon after taking office, new President Ernesto Zedillo had to abandon this fiction, and the peso plunged to $.18 within a week. Despite bold efforts by Presidents Zedillo and Clinton, it stood at about $.17 as this chapter went to press (early May 1995), and the stability of NAFTA seemed in serious trouble.

GATT. The General Agreement on Tariff and Trades (GATT) is a much more comprehensive agreement, currently involving 124 nations. It has been in effect for almost 50 years and has been modified from time to time. The latest change, called the "Uruguay round" from the country where negotiations began, took several years of negotiation for agreement. Approval by the United States was critical but remained in doubt until President Clinton called Congress back into session after the November 1994 election, and it then passed comfortably. This modification cuts import duties and lowers other trade barriers well below where they were previously. It also "includes basic rules protecting patents, copyrights and other intellectual property for the first time in [GATTs] 47-year history," and requires the 124 countries to "pledge to prosecute sales of knock-off drugs, purloined software and pirated videos."[47] American "high-tech industries, those that are our most globally competitive and make world-leading products" should especially benefit from this provision. At this writing it is too early to predict quantitatively the impact of these changes, but they almost certainly will lead to significant increases in world trade. Over the last decade U.S. manufacturers have regained their competitiveness, and they should be a net beneficiary of this expanded trade.

Management in Developing Countries

Background. There is certainly a "world of variety" in the problems of managing activities in developing countries, which we can define for our purposes as the entire world except for the United States, Canada, Europe, Japan, Australia, New Zealand, Russia, and a few other countries such as Israel and the most industrialized parts of Brazil and South Africa. The Pacific rim countries of South Korea, Taiwan, Hong Kong, and Singapore are rapidly developing in a manner similar to Japan's development and will also not be considered further. I discuss here a few general concerns with operations in developing countries, then follow with some specific comments on Middle Eastern countries as an unusual example.

Characteristics of developing countries. Third-world countries differ widely, but the following characteristics will be found in many of them:

1. A shortage of capital, often with financial resources controlled by a few families.
2. Government planning of the economy, and frequently, government operation of utilities and major industries.
3. A shortage of skilled workers, professionals, and support services.
4. A high level of government control of foreign subsidiaries in terms of approvals to establish or modify business activities; import controls; currency exchange

rates; control of repatriation of profits and capital; requirement of partial or majority local ownership of foreign subsidiaries; and taxation policies, including tax incentives.

5. Cultural differences influencing attitudes toward industrial work: Hindu, Buddhist, Moslem, or even Catholic beliefs may not be conducive to the economic motivations offered to industry.

6. Different preferences regarding leadership style. Hofstede[48] measured the "power distance" between the superordinate and subordinate (a measure of the extent of autocratic leadership style). For example:

   - Five South American countries and Mexico had power distance indexes from 63 to 81, indicating an authoritarian style (Argentina was 49).
   - Five Romance-language European countries (Italy, France, Belgium, Spain, Portugal) had indexes from 50 to 68.
   - Eight South and Southeast Asian countries ranged from 54 (Japan) to 77 (India).
   - The Philippines, both Asian and Spanish in influence, had the highest index (94).
   - The United States had an index of 40.
   - Nine West and North European countries had indexes of 18 to 39, indicating a less authoritative style in these developed countries.

7. In many Asian and South American countries the family is paramount, and enterprises tend to be small, with management confined to family members.

8. While engineers from developed countries tend to specify the advanced technology they are familiar with for use in developing countries (and the host countries often encourage this), there is seldom a local ability to operate and, especially, repair such technology. Often a more *appropriate technology* is possible, which may be more labor intensive (not necessarily a penalty where labor is cheap and unemployment high), that can be used and maintained more easily.

9. Without long experience in an industrial economy, nationals (including government officials) may take a much more relaxed attitude about getting things done on time.

Ethical considerations. Because cultures vary so much, engineering and managerial work in developing countries may involve ethical decisions that would not present themselves in the United States. For example, bribery of public officials may be a way of life and may be necessary to get permits issued or spare parts released from customs. Minor officials may be paid so little that to maintain a decent livelihood, they count on supplements that would clearly be illegal in the United States. Should the American plant manager provide such "grease" (or baksheesh or whatever the local term is)? On a higher level, should a U.S. sales manager pay millions of dollars in bribes to top officials in a foreign government to facilitate sales of jet aircraft to the national airline, an illegal act under U.S. law, or abandon the sale to less fastidious competitors?

Plant operation presents its own questions. Should a U.S.-owned plant just maintain levels of pollution control and plant safety consistent with local requirements, or should the higher levels required in U.S. plants be maintained, at higher cost? To what extent should an American plant manager insist on equal opportunity regardless of race, religion, and sex in a society that distinctly discourages such equality? Like many of the issues in professional ethics discussed in Chapter 18, actual cases of such problems often do not have simple solutions.

A Middle Eastern example. Since the countries of the Middle East enjoy half the world's petroleum reserves, they have been able to forge ahead with economic development at an unprecedented rate, and thousands of American, European, and other engineers have become acquainted with this very different culture. Actually, the Middle East is not a single culture but varies from the state capitalism of Iraq to the fanatical theocracy of Iran to the ultraconservative monarchy of Saudi Arabia. The discussion following relates to Saudi Arabia and is based on discussions with engineering employees of the Arabian American Oil Company (ARAMCO) from America, Europe, Saudi Arabia, and other Moslem countries who were students in classes taught by the author in Saudi Arabia in 1983 and 1985. The first American oil company engineers literally waded ashore and set up a tent camp in 1935 to begin exploration. Following World War II, facilities and production were increased at a record rate until a capacity of 11 million barrels per day (Mbd) was reached in the early 1980s; at the peak of $40 a barrel, this represented an astounding potential $160 *billion* a year in corporate (and, due to expropriation, national) income. Projects to create this capacity were immense and exciting engineering achievements. (In the oil glut of the mid-1980s, production fell at one time to 3 Mbd at $18 a barrel, reducing national income to ⅛ of the level noted above, thus eliminating most new projects.) At one time or another during the development of ARAMCO capacity, engineers faced the following types of problems:

1. Transportation and utilities, which one takes for granted, had to be developed from scratch by ARAMCO or the country or both: major water ports, roads, trucking capacity, railroads, and electric power; even fresh water was unavailable in acceptable quantities, and mammoth facilities to create it from seawater were built.

2. Climatic effects: common summer temperatures of 120°F limited human output; wind and sand on the desert and offshore weather and corrosion for oil wells in the Arabian (Persian) Gulf resulted in abnormal (and expensive) material specifications.

3. Remoteness from the U.S. (Texas) ARAMCO design headquarters led to shutdowns from parts shortages—reduced by locating immense quantities of parts inventory "in Kingdom" and dedicating a Boeing 747 to two weekly round trips to carry cargo and passengers from Texas to Dhahran (Saudi Arabia).

4. Communication timing problems due to the nine-hour time difference and the four-day "weekend" (Thursday and Friday in Dhahran, Saturday and Sunday in the West).

5. Islamic law and culture: reduction in efficiency during the fasting month of Ramadan; time and facilities for prayer five times daily; restrictions on women's work and prohibition of their driving; intolerance of other religions, cultures, and lifestyles.

6. Political and governmental factors: boycotts on products and parts from Israel and specified companies elsewhere; priority on developing Saudi personnel for management and later engineering positions and for local subcontracting; customs and other bureaucratic delays.

7. Shortage of skilled labor, professionals, local suppliers and contractors.

8. Expatriate problems in hiring (lead time, premium pay, and restrictions on nationality, religion, and sex); impact of extended vacation schedules on management continuity; and integrating the multinational work force with their different pay scales, languages, cultures, values, motivations, and work methods.

These same problems will not evidence themselves in another part of the third world; there will, instead, be a different set. The engineer and manager who seeks to operate in an unfamiliar environment must be very careful to become familiar with the local culture, politics, and people and learn to operate under these constraints; those who can do this effectively are of great value to their organization, and many find that they truly enjoy working and succeeding in exotic environments.

DISCUSSION QUESTIONS

17-1. Comment on the effect of the organizational changes of the 1990s on the expected progress of engineers into management.

17-2. Why do you think U.S. companies have a smaller percentage of engineers and scientists in top management and on industrial boards of directors then do Japan and many West European countries?

17-3. Condense the six reasons "why technologists switch" in the text into a simpler statement. State whether you think these reasons are valid.

17-4. The dual career ladder concept was developed to offer technical professionals an equal alternative to a management career, but it often proves less successful than hoped. Why do you think this is so?

17-5. Gee has listed attributes of *researchers* that indicate a potential for management. What would you include in a similar list for engineers involved in design or manufacturing?

17-6. You are an experienced engineering manager in a growing organization, have just selected several capable engineers for promotion to their first management positions, and are responsible for guiding their growth into their new responsibilities. What are some of the skills your new managers will need to acquire, and what advice will you give them about the new problems they will face?

17-7. As owner of a company making roller conveyor systems, you must appoint a new manager for your customer service group, which is made up equally of engineers, technicians, and clerical personnel. One candidate is an engineer who has designed many of your systems but has had no management responsibility; the other is a business administration graduate now heading your public relations effort. Discuss the pros and cons of appointing each of these.

17-8. In what ways are master's programs in business administration (M.B.A.) or in engineering management deficient as preparation for a management position?

17-9. What are the likely consequences if the United States *does not* reduce its present excess of imports over exports?

17-10. To what extent do you feel that U.S. companies should adopt the management practices of Japanese companies discussed in the text?

17-11. Congratulations! You have served as an R&D manager so effectively that you have been appointed Director of R&D for your company's affiliate in Tokyo. What differences will you expect to find in your new assignment (other than language), and how might this affect your life and your management style?

17-12. What is the significance of the elimination of trade barriers between European countries to American companies?

17-13. What have been the effects of (**a**) the NAFTA treaty approved in 1993 and (**b**) the GATT modifications approved by the U.S. Congress in late 1994 on the U.S. economy?

17-14. Select a specific developing country and comment on which of the nine "characteristics of developing countries" seem to apply. What additional characteristics would you add?

17-15. Give an example of an "appropriate technology" that might be preferable in a specific developing country situation instead of the technology used to solve the same problem in the United States.

NOTES

1. "Up and Out Is No Longer In for Researchers at Some Japanese Companies," *Wall Street Journal,* April 23, 1991, p. 1.

2. Lawrence T. Grayson, "Education for Leadership," *Engineering Education News,* February 1989, p. 3.

3. John Whittaker, "Engineers as Chief Executive Officers [CEO's]: A Canadian Perspective," *Proceedings of the Joint American Society for Engineering Management/National Council for Systems Engineering Annual Conference,* Chattanooga, TN, October 1991, pp. ASEM 253–255.

4. Elizabeth B. Salgado and Dundar F. Kocaoglu, *CEOs with Engineering/Science Background,* paper presented at ORSA/TIMS Joint National Meeting, Philadelphia, October 1990, reported in Whittaker, "Engineers as CEOs," p. ASEM-253.

5. Maggie McComas, "Atop the Fortune 500: A Survey of the CEOs," *Fortune Magazine,* April 28, 1986, pp. 26–31, reported in Whittaker, "Engineers as CEOs," p. ASEM-253.

6. Wayne Gooding, "Canada's Corporate Elite," *Financial Post Magazine,* Special Issue, December 1990, reported in Whittaker, "Engineers as CEOs," p. ASEM-254.

7. Lester C. Thurow, "The Task at Hand," *Wall Street Journal,* June 12, 1987, p. 46D.

8. David Broder, "Clinton's Performance Won't Mimic Carter's," *St. Louis Post-Dispatch,* November 17, 1992, editorial page.

9. Michael K. Badawy, *Developing Managerial Skills in Engineers and Scientists* (New York: Van Nostrand Reinhold Company, Inc., 1982), pp. 37–42.

10. Survey by Hewitt Associates, reported in "Dual-Career Ladders: New Evidence That They Really Work," *Engineering Department Management & Administration Report,* March 1994, pp. 5–6.

11. Badawy, *Developing Managerial Skills,* p. 45.

12. Thomas J. Allen and Ralph Katz, "The Dual Ladder: Motivational Solution or Managerial Delusion," *R&D Management,* 16:2, 1986, pp. 185–197.

13. Thomas J. Allen, "Age, Education, and the Technical Ladder," *IEEE Trans. on Engineering Management,* August 1992, pp. 237–245.

14. Allen, "Age, Education," p. 244.

15. Edwin A. Gee and Chaplin Tyler, *Managing Innovation* (New York: John Wiley & Sons, Inc., 1976), p. 172.

16. Hans J. Thamhain, "Determining Aptitudes for Engineering Managers," *Proceedings of the International Engineering Management Conference, Institute of Electrical and Electronic Engineers,* Santa Clara, CA, October, 21–24, 1990, pp. 5–10.

17. Badawy, *Developing Managerial Skills,* pp. 109–110.
18. Dundar F. Kocaoglu, "Research and Educational Characteristics of the Engineering Management Discipline," *IEEE Trans. on Engineering Management,* August 1990, pp. 172–176.
19. William A. Weimer, "Education for Technology Management," *Research•Technology Management,* May–June 1991, p. 42.
20. National Research Council, *Management of Technology: The Hidden Competitive Advantage* (Washington, DC: National Academy Press, 1987).
21. Daniel L. Babcock, "Engineering and Technology Management University Degree Programs: Comparisons and Contrasts," in T. A. Khalil and B. A. Bayraktar, eds., *Management of Technology II* (Norcross, GA: Institute of Industrial Engineers, 1990).
22. Daniel L. Babcock, "Are We Training Tomorrow's Engineers to Meet Yesterday's Needs?" *Annual Conference Proceedings, American Society for Engineering Education,* New Orleans, LA, June 16–19, 1991, pp. 83–88.
23. M. W. McCall, M. M. Lombardo, and A. M. Morrison, *The Lessons of Experience: How Successful Executives Develop on the Job* (Lexington, MA: Lexington Books, 1988), summarized in Alan W. Pearson, "Management Development for Scientists and Engineers, *Research•Technology Management,* January–February 1993, p. 45.
24. National Academy of Engineering, *The Technological Dimensions of International Competitiveness* (Washington, DC: National Academy Press, 1988), p. 17.
25. George Gilder, "Trade Gap Is Inevitable—and Good," *Wall Street Journal,* January 15, 1988, editorial page.
26. "Made in U.S.A.," *Business Week,* February 29, 1988, p. 61.
27. "Is Japan Using the U.S. as a Back Door to Europe?" *Business Week,* November 14, 1988, p. 57.
28. Robert J. Samuelson, "The Useless 'Jobs Summit,'" *Newsweek,* March 14, 1994, p. 53.
29. Neil H. Jacoby, "The Multinational Corporation," *The Center Magazine,* 3, May 1970, pp, 37–55.
30. "Borderless Management: Companies Strive to Become Truly Stateless," *Business Week,* May 23, 1944, pp. 24–26.
31. Data from *Business Week,* May 14, 1990, and *Engineering News Record,* May 24, 1990, summarized in *Graduating Engineer,* September 1990, p. 52.
32. "GE's Brave New World," *Business Week,* November 8, 1993, pp. 64–70.
33. Richard D. Robinson, *Internationalization of Business: An Introduction* (Hinsdale, IL: CBS College Publishing, 1984).
34. "Many Japanese Find Their 'Lifetime' Jobs Can Be Short-Lived," *Wall Street Journal,* October 8, 1992, pp. A1, A8.
35. Linda S. Dillon, "Can Japanese Methods Be Applied in the Western Workplace?" *Quality Progress,* October 1990, pp. 27–30.
36. Frank H. Squires, "Karoshi Widows," *Quality,* November 1990, p. 66.
37. Kunio Odaka, *Japanese Management: A Forward-Looking Analysis* (Tokyo: Asian Productivity Organization, 1986), p. 68.
38. Odaka, *Japanese Management,* p. 57.
39. William G. Ouchi, *Theory Z—How American Business Can Meet the Japanese Challenge* (Reading, MA: Addison-Wesley Publishing Company, Inc., 1981).
40. Peter F. Drucker, "Learning from Foreign Management," *Wall Street Journal,* June 4, 1980, editorial page.
41. Philip Rezvin, "Getting Together: Despite Difficulties, Moves to Strengthen Common Market Gain," *Wall Street Journal,* February 23, 1988, p. 1.
42. "Germany: Is Reunification Failing?" *Business Week,* November 15, 1993, pp. 48–50.

43. Peter F. Drucker, "Mexico's Ugly Duckling—the Maquiladora," *Wall Street Journal,* October 4, 1990, editorial page.

44. "The Mexican Worker," *Business Week,* April 19, 1993, pp. 84–91.

45. "The Mexican Worker," pp. 85–86.

46. "Celebrate—and Expand—Free Trade," *Business Week,* December 26, 1994, p. 206.

47. "GATT Aids Makers of Drugs, Software," *St. Louis Post-Dispatch,* December 2, 1994, p. 13D.

48. Geert Hofstede, "Hierarchical Power Distance in Forty Countries," in C. T. Lammers and D. J. Hickson, eds., *Organizations Alike and Unlike* (London: Routledge & Kegan Paul Ltd., 1975), p. 105, quoted in Anant R. Negandhi, *International Management* (Newton, MA: Allyn and Bacon, 1987), pp. 322–323.

CHAPTER ⬤ 18

Special Topics in Engineering Management

PREVIEW

In this closing chapter we consider a series of topics important to engineers and engineering managers. First is the changing position of women and of minorities (especially blacks and Hispanics) in engineering and management. Particular attention is paid to the reasons for the disproportionately small numbers of each entering engineering schools and graduating from them, and the barriers they face as engineering professionals and managers. A second topic in the chapter concerns how we make the best use of the scarcest resource we will have as professionals—our own time.

Fully half the chapter is devoted to the subject of professional ethics and conduct, beginning with some definitions and a look at engineering codes of ethics. Ethical problems that face the engineer in consulting and contracting and in industrial employment are discussed, ending with case studies of the (Kansas City) Hyatt Regency and the *Challenger* space shuttle disasters from the point of view of engineering ethics. The chapter and the book close with a brief look at some of the changes that may be predicted for the future.

WOMEN AND MINORITIES IN ENGINEERING AND MANAGEMENT

Thirty years ago the ranks of American engineers, especially engineering managers, were almost completely white and male. Although this situation has changed dramatically and continues to change, there still exist attitudes and perceptions that can make career opportunities different for women or minorities than for the traditional white male, and they need to be understood. Also, employers who expect to be successful in the 1990s and beyond must learn how to employ the full range of its workers effectively, since it has been estimated that 80% of those being added to the work force in the near future will be women, minorities, or immigrants.

Women as Engineering Students

Although four-year degree programs in engineering for men began in 1817 (at the U.S. military academy) and grew rapidly, it was not until 1892 that the first woman received a U.S. engineering degree (Elmina Wilson, in civil engineering from Iowa State). Careers of the few that followed in the next several decades typically ended with marriage. By 1920, when women got the right to vote, only 90 women had received engineering degrees (from 20 U.S. schools). During World War II many engineering schools set up engineering training programs for women, but both before that war (during the Great Depression) and immediately following it women were pressured to become homemakers so men could have the jobs. During the 62-year period from the first degree in 1892 to 1953, only one engineering school averaged over one degree to a woman per year (Purdue, with 103; the next highest was the Colorado, with 47; the presence of Lillian Gilbreth [see Chapter 2] on the Purdue faculty 1935–50 and as advisor to women students there long after that surely contributed to Purdue's record). Many engineering schools and professional engineering societies did not admit women (Tau Beta Pi awarded them a "Badge" rather than membership), leading to the formation of the Society of Women Engineers about 1949.[1]

Even in 1970 women made up less than 1% of engineering bachelor's degrees. The number then began to climb, reaching almost 10% by 1980 and 15% by 1985, but had only reached 15.7% in 1992 (Table 18-1) and 16.1% in 1993.[2] Graduate engineering degrees for women lagged this increase, but reached 15.5% of master's and 9.9% of doctoral engineering degrees by 1992. The representation of women among engineering technology graduates is even smaller, totaling 11.8% of two-year associate degrees and 7.9% of B.E.T. degrees (1991–92). Vetter summarized some of the reasons fewer women choose engineering and physical science:

- The most important reason by far that women choose to study and work in engineering is that they are good in math and science, according to a recent pilot study

TABLE 18–1   Women Earning Engineering B.S. Degrees

Year	Number	Percent of Graduates
1970	358	0.8
1975	878	2.3
1980	5,631	9.7
1985	11,516	15.0
1991	10,016	15.7
1992	9,972	15.7

Source: Engineering Workforce Commission data.

of about 500 women and men engineers by the Society of Women Engineers. Only 32% of the men named that as the most important reason, compared with 44% of the women. But men are far more likely than women to believe they are good in math and science, even when their grades are the same, or lower. This is not an accident, nor does it result from genetic differences. Its origin is societal conditioning.

- Most K–8 teachers, almost all women, suffer from inadequate preparation in science so they fear teaching [it]. . . . A woman teacher's fear of or feeling of helplessness with science or mathematics says to the girls that it is natural for them not to like or be good at these subjects.
- Teachers of all ages and both sexes discriminate in the classroom. . . . have lower expectations for girls than for boys in science and mathematics. . . . call on and praise boys more than girls, let boys interrupt girls, but praise girls for being polite and waiting their turn.
- A recent study by the American Association of University Women found that pre-teen girls are self-confident, and generally equal to boys in mathematics, but that as they enter adolescence, their confidence erodes, as society, its commercials, and value systems, tells them they are inferior to boys and men in intellectual abilities, particularly those requiring mechanical or mathematical skills, and in leadership qualities.
- Parents also discriminate against their daughters in failing to give them the most elementary training in the use of tools to build or repair mechanical things.
- As a group, parents have lower educational aspirations for their daughters than for their sons, and to no one's surprise, women high school graduates have lower educational aspirations than their male classmates.
- Although men are more confident than women of their abilities to learn mathematics, women earn better grades. . . . In high school, students [male or female] with higher math achievement scores will tend to take chemistry. . . . Physics is another story. Regardless of mathematical excellence, women are much less likely to take physics.
- Among doctoral scientists employed in academic institutions women earn less than men at every rank, in every type of institution and in every field. It is also more difficult for women to find employment in academic institutions . . . to achieve tenure when they do receive academic appointments, and to advance in rank.[3]

Eleanor Baum, America's first female Dean of Engineering (at Cooper Union), says women go into engineering from a desire to do satisfying and interesting work at a good salary, and to have a profession that is transferable around the world—only 1% do so to "meet men." The single most important factor in their retention is knowing a faculty member who cares, and the lack of women as role models among science and engineering faculty is a factor in fewer women who begin engineering graduating in that field.[4]

The Minority Engineering Student

The percentage of African Americans among U.S. engineering freshmen rose from 2.8% in 1972 to 6.1% in 1979 to 8.0% in 1989–90, compared with their 12% share in the population. Unfortunately, their dropout rate is high, and only about 3 in every 10 black engineering freshmen persevere to earn their B.S. degree (Table 18-2). Hispanics made up 5.7% of en-

TABLE 18–2 Engineering Graduation Rates by Racial and Ethnic Group

Racial or Ethnic Group	1983–90 Engineering Graduates	
	% of 1980–86 Freshmen	% of 1981–87 Sophomores
African Americans	30.6%	50.0%
Hispanics	44.8	70.2
American Indians	35.6	59.1
Nonminorities	68.4	* 87.4

Source: *Research Letter: Minority Graduation Rates*, National Action Council for Minorities in Engineering, Inc., New York, NY, December 1991.

gineering freshmen (1980–90), with a graduation rate of about 45%. American Indians represent about their population share (0.5%) of engineering freshmen, but only a third graduate. The graduation rate of entering freshmen, excluding these three minorities, was just over two-thirds.

The low percentages of blacks, Hispanics, and American Indians entering engineering colleges and their low retention rates can be traced to common problems. Family backgrounds, neighborhood environments, lack of role models (the three totaled only 6.4% of U.S. full-time engineering faculty in Fall 1992[5]), and the quality and offerings of neighborhood schools all contribute. *Stand and Deliver,* the film based on the real-life experiences of mathematics teacher Jaime Escalante in Garfield High in East Los Angeles, shows what a difference inspired (and well-trained) teachers can make. [In the movie James Olmos, playing Escalante, tells kids: "Tough guys don't do math. Tough guys fry chicken for a living."] Unfortunately, there are very few teachers well trained in math and science who are teaching in schools with high percentages of minorities.

Several engineering schools have been able to improve the retention rates for minorities through carefully conceived minority engineering programs (MEPs). The University of Missouri—Rolla, for example, runs a mathematics, science, and English review program for minorities emphasizing study habits in the summer preceding freshman entry; over half of the freshmen entering following this MEP achieve an engineering B.S. The MEP model published by the National Action Council for Minorities in Engineering (NACME) emphasizes three structural elements: (1) a formal orientation course for new freshmen featuring orientation to the university and its engineering program, study skills, and motivation through career awareness; (2) clustering MEP students in common sections of their classes to reduce ethnic isolation and encourage group study; and (3) providing a student study center for MEP student use. The California State University MEP achieved three-year retention rates of 79% for black and 88% for Mexican-American MEP students, versus the 31% and 45%, respectively, shown in Table 18-2.[6]

A number of organizations exist to offer help to specific minority groups in engineering. These include the following:

- American Indian Science and Engineering Society (AISES)
- League of United Latin American Citizens (LULAC), offering about 300 engineering scholarships a year to Hispanic-American students

- Mexican American Engineering Society (MAES)
- National Action Council for Minorities in Engineering (NACME)
- National Consortium for Graduate Degrees for Minorities in Engineering (GEM)
- National Society of Black Engineers (NSBE)
- Society of Hispanic Professional Engineers (SHPE)

Career Problems of the Woman Engineer

Women who persevere and graduate as engineers find that entry-level job opportunities and salaries are comparable with those of men. (This is true for other mathematics-based professions as well: in 1991 starting salaries for women averaged 102% of those for men in engineering, 101% in computer science, 99% in accounting, and 98% in mathematics and statistics; women fared poorer in the sciences [93–94% in biology and health sciences and 90% in chemistry] and in fields where they represented a majority [89% in elementary education and 87% in psychology].)[7]

Women comprised about 8.6% of the engineering work force in 1993, compared with 0.5% in 1970 and 1.9% in 1980. The Society of Women Engineers commissioned the American Association of Engineering Societies to launch a full-scale survey of the profession in 1992. Active members of 22 major engineering societies were sampled in a stratified random sample, with 1917 respondents (58.2%), about equally divided by sex:

> The study reveals striking differences, as well as similarities, in how women view the profession and their engineering careers. . . . Along with a lack of parity in women's compensation—a gap that increases as they advance—the study confirmed that women participate less in management, even when compared to men within matched age groups. Three times as many women as men say they are aware of some degree of job discrimination. But on the whole, both men and women express satisfaction with their careers and value their education highly.[8]

About 61% of the women engineers (mean age 33.9) in this survey were married, and 91% of these had a working spouse; 81% of the men (mean age 42.7) were married, and 62% had a working spouse. Interestingly, 56% of the women (and but 6% of the men) were married to engineers. Unemployment was higher for women engineers than men (2.5% versus 1.2%) but still far below average unemployment.

Attitudes about women in management have changed: In a 1965 *Harvard Business Review* survey, 54% of men *and 50% of women* thought women rarely expected or desired positions of authority; in a follow-up survey of 1900 executives in 1985, only 9% of men and 4% of women still thought women didn't want top jobs.[9] In practice, however, many women have found they bump against a "glass ceiling" as they approach executive levels and find men being promoted past them and achieving higher salaries. Between 1981 and 1991, women increased from 41 to 46% of the overall labor force, and from 27 to 41% of total managers. However, they increased only from 1 to 3% of senior executives (who report directly to the CEO), and actually decreased (from 0.2 to 0.1%) among CEOs of the top 1000 companies.[10] One missing factor is mentoring, a practice of senior professionals in showing promising juniors "the ropes" and guiding them up the corporate ladder. Few men seem comfortable in taking on female protégés, in part because of the sexual innuendoes that can result.[11]

Some of the limitations on women's careers stem from conscious choices, often involving family and children. Women can be more expensive to employ: One multinational corporation found "that the rate of turnover in management positions is 2½ times higher among top-performing women than it is among men."[12] Further, women are not as likely to accept the travel and relocation expected of "upwardly mobile" men. As reported in the *Wall Street Journal:*

> Female executives are stay-at-homes. Only one in 20 corporate women travels for business, says Damin Aviation. Among male executives, by contrast, the rate is one in three. While one in six men prefers a day trip to an overnight stay, almost half the women want to go and come back in the same day.[13]

Schwartz distinguishes between two types of women: the career-primary woman and the career-and-family woman:

> Like many men, some women put their careers first. They are ready to make the same trade-offs traditionally made by the men who seek leadership positions. They make a career decision to put in extra hours, to make sacrifices in their personal lives, to make the most of every opportunity for professional development. For women of course, this decision also requires that they remain single or at least childless or, if they do have children, that they be satisfied to have others raise them. Some 90% of executive men but only 35% of executive women have children by the age of 40. . . .
>
> The majority of women are, however, what I call career-and-family women, women who want to pursue serious careers while participating actively in the rearing of children.[14]

Jacobs reports of the latter group:

> Women are adjusting their expectations accordingly, looking for more balanced lives that include fulfillment outside the corporate arena, as well as in it. The least happy women we interviewed several years ago were the so-called "corporate brides," married to the company to the exclusion of spouses, family, and friendships.
>
> Now we're seeing more women question the myths and expectations that drove them so hard. A lot of women are recognizing that you can have it all, but not all at once. . . . To paraphrase Lily Tomlin, women have finally discovered that the problem with the rat race is even if you win, you're still a rat.[15]

There is no ethical reason why the burden of home and family should fall more heavily on the woman professional than on the man, except for giving birth to children, but in much of our society it does. Employers who have a serious interest in encouraging women professionals can provide child-care facilities at cost, and can experiment with more liberal parental leaves and part-time job situations. Nor is there a "right" choice as to the relative emphasis an engineer (male or female) should put on his or her career relative to other life goals. Each of us must make choices, and the rate of career progression often depends on these choices.

Minority Engineers

Minority engineers face some of the same problems in engineering practice as do women, and there are some differences. The problem of finding mentors has similarities, since it is easiest for executives to mentor people like themselves. Moreover, senior/junior relationships are often enhanced outside the workplace—in the community, country club, or church—and when minorities move in different social groups than their managers, they are at a disadvantage. Black engineers, for example, tend to have higher unemployment (about 4%) than whites (1.2%), and report lower salaries ($35,700 versus $41,000 in 1986).[16] Racial intolerance still exists:

> As minority professionals have become more generally included in every kind of American enterprise . . . they work alongside some colleagues who are still intolerant of racial and cultural differences. Although a majority of minority professionals have learned to function relatively well in these settings, it has meant that many of them have been compelled to "commute psychologically" between the world of work and their home base.[17]

Minorities can enhance their career opportunity through further study. Blacks, Hispanics, and American Indians together made up about 6000 of the 65,000 U.S. engineering B.S. graduates in the 1993–94 academic year; these three "underrepresented minorities" constituted just 4.4% of U.S. doctoral degrees in engineering in 1993. The National Consortium for Graduate Degrees for Minorities in Engineering (GEM) offers fellowships for M.S. study to reverse this imbalance. Pointing out that employment opportunities for minorities have improved greatly in the past generation offers little consolation for the person still facing discrimination. It does, however, provide hope of continuing progress in the future.

MANAGING YOUR TIME

Work Smarter, Not Longer

Time is a very democratic resource: "The prince and the pauper" both have exactly the same amount to spend in a day. Yet the engineering manager (and the engineer as well) never seems to have enough. Amos and Sarchet explain the problem:

> All new engineer managers constantly hear that to get ahead takes "hard work," which implies long hours, and that as he [or she] is promoted, he will assume greater responsibilities and have a greater span of management. However, each promotion brings more authority, which allows more delegation to get the work done through others. When the engineer puts in long hours at his office, does not take regular vacations, and spends little time with his family, he also fails to have time to develop the creative aspects of engineering management that are important to his success.[18]

Amos and Sarchet follow this with a self-test (Table 18-3) to help the individual engineer who fits this pattern recognize it, and then they provide some good ideas on planning time:

> Being busy is simple for the engineer manager, but being effective is difficult. Planning activities is a necessary requirement because he does not have time to do all the things that his conscience or imagination tells him he needs to do, but he must decide what *to do* and what *not to do*. . . . The solution is not working long hours. Instead, it is setting priorities. Otherwise, he will constantly put second things first by default. Then he will be in trouble.[19]

Time Wasters and What to Do About Them

Every writer on managing time seems to have a laundry list of activities that waste time. Especially applicable is one by LeBoeuf,[20] who surveyed 50 engineering managers in a number of countries and found their "top 10" ranked as follows:

1. Inadequate, inaccurate, or delayed information
2. Ineffective delegation
3. Telephone interruptions
4. Meetings
5. Unclear communication
6. Crises
7. Leaving tasks unfinished
8. Indecision and procrastination
9. Drop-in visitors
10. Lack of self-discipline

No manager can completely avoid these problems, but the problems can be minimized through good time-management practices. Let's tackle each of these in turn:

1. *Information.* The effective manager thinks through the decisions that will require external information well ahead of time, sends out requests that define clearly what information will be needed, why, and when, and then gets on with other matters in the interim.
2. *Delegation.* As we have seen in Chapter 6, new engineering managers often find it difficult to trust others for matters they used to handle personally. Nonetheless, they will progress no further until they learn not only to assign jobs, but to delegate authority and still exact accountability for results.

TABLE 18–3 Evaluation of Personal Work Habits

Answer each question either "yes" or "no"
1. Are you a self-appointed witness of all the organization's activities?
2. Do you assume everyone's responsibilities?
3. Do you always have a messy desk?
4. Is there a disarray of important papers and memos?
5. Do you have lunch at your desk, not take breaks, or not leave the office because things will go wrong?
6. Do you take home a briefcase full of work every evening?
7. Are you the last to leave the office?
8. Have you missed taking your vacation?
9. Do you only take emergency trips?
10. Are you unable to meet deadlines?
11. Do you not have time for your family?
12. Are you always volunteering to do a job?

If more than three questions were answered "yes," you have very poor work habits and your time is being used ineffectively.

Source: John M. Amos and Bernard R. Sarchet, *Management for Engineers*, © Prentice-Hall, Inc., Englewood Cliffs, NJ, 1981, p. 319. Reprinted by permission.

3. *Telephone interruptions.* The telephone can be a tyrant if you permit it. Train your secretary (if you have one) to screen your calls, put through what is urgent (or those from your boss), ask if he or she can help others, and give you the rest at a specified time when you can return a number of calls together. Make effective use of answering machines and electronic mail to avoid the "telephone tennis" of always catching the other person out. When you do get on the phone, give it your full attention and learn how to probe for the essentials of the call and bring it to a courteous yet prompt close.

4. *Meetings.* Think through what meetings you must attend, and which you can ask a subordinate to represent you in. For meetings that you initiate, ask first if the meeting is really necessary; if it is, use the techniques of preparation, meeting conduct, and follow-up discussed in Chapter 6 to make the meeting effective and efficient.

5. *Unclear communication.* Practice the techniques of oral and written communication discussed in Chapter 16.

6. *Crises.* Crises will occur and must be handled, and you must leave a degree of freedom in your schedule to handle the unexpected. Before charging off to fix the problem yourself, however, ask yourself who among your subordinates and colleagues can shoulder some of the unexpected burden.

9. *Drop-in visitors.* Learn the difficult dividing line between being available to your subordinates (and your colleagues, since you will need their help another time, and your boss, since helping him or her is your job) and spending excessive time on nonessentials. If you can make the contact in *their* office, it is much easier to get away!

7, 8, 10. Consider some of the scheduling approaches that follow.

Tools of Time Management

List goals and set priorities. The Swiss economist and sociologist Vilfredo Pareto is credited with the observation that in many collections of items, 80% of the value or importance is represented in only 20% of the items. Lakein[21] emphasizes the need to (1) list possible long-, intermediate-, and short-term goals; (2) set priorities; (3) schedule the most important goals; and (4) follow through as scheduled.

Categorize your goals as either A (highest priority), B (lower priority), or C (desirable but postponable), and the time that you estimate each will require. Schedule your most important A task first; if it is overwhelming and looks like it needs a large block of time that you do not have at the moment, ask yourself what part of it can be tackled *now* to prepare for the rest. Avoid C tasks: they are beguilingly easy, but they do not help solve the real problems for which you are being paid.

Make a daily action list. List things you plan to do that day, with assigned priorities (A or B). Maintain a tickler file of important deadlines. I find it useful to combine the two in a computer "Action" file—updating it regularly without having to copy over future items—and print out a hard copy to carry with me. Those who carry a notebook or smaller computer will find this even easier.

Make a time log. Periodically, list what you *actually* do with your time, minute by minute, for a week or two—or have your secretary keep it for you. Summarize your activities by category and by ABC priority. Most people who do this find that they are spending much more time than they expected on matters that are trivial or could be delegated to someone else, and much less on their A items; knowing this, they can make some needed changes in the way they work.

Handle each piece of mail once if possible. Keep a large wastebasket by your desk, and use it freely. Identify categories of items that you do not want to see, and get yourself off mailing lists and unnecessary magazine subscriptions. If a subordinate can help, mark the item "Please recommend a solution by [date]" or, better yet, "Please handle" and get it in your out basket. Set aside a "C drawer" in your desk for other less important matters; if they do not come up again, move the stack farther away and eventually dump it. (Lakein suggests this even for a C item that your boss mentioned; if he or she mentions it again, it becomes a B; if your boss storms into your office, it becomes an A, but this will not happen often.) The remaining items are your A's and B's; schedule them and start on them.

Consider your energy cycle and your environment. You should be aware that your energy level is not a constant, but varies from hour to hour. When project managers were asked to rate their perceived energy level hour by hour, they reported a peak in the late morning, a low around lunchtime, and a second peak in the afternoon. But people differ: Some do their best work in the morning, some in the afternoon, and a wise manager learns when he or she functions best and tries to schedule difficult decisions or confrontations for that period. Days matter, too: The same project managers perceived their energy to be highest on Tuesday through Thursday and lowest on Sunday.[22]

The environment in which work is done is also very important. The "bullpen," or sea of desks in which some engineers work, helps foster communication, which can be either desirable or time wasting depending on whether your work of the moment involves coordination or independent creativity. Noise, poor lighting, uncomfortable seating, and inadequate space also inhibit your best work. The "open door" of the manager's office almost rules out those blocks of time needed for the tough problems during the day, forcing them into evenings and weekends. Some executives have gone so far as to acquire a "second office" known to only a few, where they can disappear for periods unless matters of real urgency surface.

PROFESSIONAL ETHICS AND CONDUCT

Some Definitions

Koestenbaum describes the importance of being ethical in business:

> Little else is as distinctively human as our ethical conscience and our moral sense. To be ethical means to live by the stern demands of reason and not to be governed or swayed by the seduction of the emotions. To be ethical is to be just, consistent, and predictable. It is preeminently our capacity to act ethically and our possession of a moral sense, which set us aside from the animals.[23]

Gluck distinguishes between morality and ethics as follows:

> *Morality* is concerned with *conduct* and *motives,* right and wrong, and good and bad *charac-*
> *ter. Ethics* is the philosophical study of morality; it is *moral philosophy.* When we do moral phi-
> losophy, that is, when we practice the philosopher's craft, we are subjecting the questions *of*
> morality to other critical and analytical questions *about* morality.
>
> "Morals" is a set of rules of conduct and standards of evaluation that a culture uses to
> guide its individual and collective behavior and direct its judgments. Codes of professional
> moral conduct ["codes of ethics"] are specialized subsets of these rules and standards.[24]

Study of *moral philosophy* used to be a formal part of every liberal education, preparing the graduate for ethical analysis of future problems occurring in one's life and career.
This study is regrettably uncommon in the crowded curricula of the twentieth century.

Ethicists, after lifetimes of thought, are unable to agree on a simple definition of
ethics. Instead, their definitions fall into several categories:

1. *Utilitarian* ethics is a goal-based approach in which we seek to obey those rules or
   choose those acts which will result in the greatest good for the greatest number of
   people. This involves value judgments in the weighting of different "goods," and
   raises the question of the rights of minorities.
2. *Ethical egoism* is a goal-based theory of "rational" self-interest. Adam Smith developed the theory that if everyone acts in his own self-interest, the "invisible
   hand" of the marketplace will transform this into social good. Economist Milton
   Friedman and novelist Ayn Rand [*Atlas Shrugged* and *The Fountainhead*] are
   more recent proponents of this philosophy.
3. *Duty-based,* or *deontological* or *normative,* ethics asserts that there are moral imperatives that we must obey, regardless of the consequences. This view stems
   from the *categorical imperatives* of the German philosopher Immanuel Kant, who
   believed that to steal, lie, or break promises is universally immoral, regardless of
   the consequences.
4. *Rights-based* ethical theories are based on the belief that there are certain fundamental human rights, and that moral obligations arise in the context of these rights.
   The English philosopher John Locke (1632–1704) believed that these rights included life, liberty, and property; his writings inspired the framers of our Declara-
   tion of Independence to declare it "self-evident" that all men are endowed with
   "certain unalienable rights" including those of "life, liberty, and the pursuit of
   happiness."
5. *Environmental* ethics, a development of the last 40 years, broadens the moral
   community to whom we owe ethical responsibility to include animals, plants, and
   even inanimate objects. Environmental ethics may be either goal-based (utilitarian)
   or duty-based (deontological). They should be of particular interest to civil and
   chemical engineers.

No single one of these views provides us with a simple, reliable guide to resolving the
ethical problems we are sure to encounter in our lives and careers. In the end, we must fall
back on our personal set of *values.* Table 18-4 provides an extensive "compilation of val-

TABLE 18–4 Compilation of Values

Four categories of values are highlighted in the following listing: individual, professional, societal, and human. The values in any category are not mutually exclusive, nor is the listing complete.

Individual Values	Civic consciousness	Prestige	Privacy
Curiosity	Collegiality	Pride in work	Progress
Endurance	Communication	Problem-solving ability	Public service
Family	Compassion	Professionalism	Social justice
Flexibility	Competence	Prudence	Societal harmony
Friendship	Conformity	Rationality	Survival of society
Hard work	Conscientiousness	Realism	Tradition
Honor	Cooperation	Recognition of accom-	
Independence	Courtesy	plishments	**Human Values**
Initiative	Creativity	Self-education	Autonomy
Intellectual stimulation	Curiosity	Selflessness	Beauty
Intelligence	Decisiveness	Service to others	Beneficence
Leisure	Devotion to principle	Tolerance	Bravery
Optimism	Duty	Trustworthiness	Fairness
Personal liberty	Economy		Faith
Personal morality	Effectiveness	**Societal Values**	Freedom
Personal power	Efficiency	Capitalism	Friendship
Personal security	Fair play	Centralization	Happiness
Privacy	Flexibility	Change	Health
Property	Forthrightness	Competition	Hope
Quality of life	Freedom of inquiry	Culture	Human dignity
Self-advancement	Honesty	Democracy	Humility
Self-control	Idealism	Education	Idealism
Self-fulfillment	Imagination	Equality	Justice
Self-reliance	Informedness	Equality of opportunity	Love
Self-respect	Initiative	Freedom of religion	Morality
Self-worth	Innovativeness	Freedom of thought	Pleasure
Strength	Integrity	Governance by law	Prevention of evil
Success	Leadership	Improvement of society	Progress
Wealth	Literacy	Individual rights	Promotion of goodness
Wit	Loyalty	Individualism	Prudence
	Obedience	Liberty	Reason
Professional Values	Openness	National prosperity	Reverence for life
Ability to analyze	Patience	National pride	Self-sacrifice
Ability to synthesize	Perseverance	Order	Truth

Source: Heinz C. Luegenbiehl and Don L. Dekker, "The Role of Values in the Teaching of Design," *Engineering Education*, January 1987, p. 245.

ues," divided into four categories: individual, professional, societal, and human values. In the final analysis, each of us must determine what values are most important.

Engineering Codes of Ethics

The profession's first code of ethics was written by Isham Randolph and adopted by the American Association of Engineers in 1918.[25] Soon thereafter, five engineering societies collaborated in preparing a uniform code for engineers, but the American Society of Mechanical Engineers was the only one to adopt it (in 1922).[26] Oldenquist and Slowter trace the history of more recent attempts to achieve a universal code:

> By 1947, ECPD [the Engineers' Council for Professional Development] developed Canons of Ethics which eventually had acceptance by a significant number, but not a preponderance, of engineering societies; these canons were updated in 1963, and again some acceptance was obtained; more recently, a major effort by ECPD produced a Code of Ethics for Engineers in 1974 which contained three levels of specificity: Fundamental Principles; Fundamental Canons; and Suggested Guidelines.
>
> Although a number of societies participated in the preparation of this three-level code and it seemed to offer opportunity for universal acceptance of at least the Fundamental Principals, it has not secured the support of a majority of the professional societies, and the goal of a universal code continues to elude the profession.[27]

Oldenquist and Slowter have analyzed current codes, and they identify 20 basic concepts that pervade them, divided into three groups: (1) the public interest; (2) truth, honesty, and fairness; and (3) professional performance (see Table 18-5).[28] They add that "mixed with these ethical principles [in the existing codes] are a number of rules and customs concerning business practices and political conventions which, at least in the eyes of much of the public, seem more designed to protect the interests of engineers than to serve the general good." A prime example was the provision of the former Section 11(c) of the National Society of Professional Engineers (NSPE) code, which held that the Engineer "shall not solicit or submit engineering proposals on the basis of competitive bidding. . . . defined as the formal or informal submission, or receipt, of verbal or written estimates or cost . . . whereby the prospective client may compare engineering services on a price basis" before selecting one engineer or engineering organization for negotiations. This section was deleted by order of the U.S. District Court, as affirmed by the U.S. Supreme Court on April 25, 1978. At the same time, provisions of Section 3(a) limiting advertising by engineers to the identification of name, address, telephone number, and fields of practice were replaced by statements emphasizing the ethical consideration of avoiding misrepresentation of fact.

Ethical Problems in Consulting and Construction

Significance. Engineering codes of ethics have traditionally been geared toward the consulting engineer in private practice, especially the civil engineer. Much of the literature on ethical problems stems from the NSPE, whose membership is weighted heavily toward civil engineers and private consultants since these are the engineers most motivated to become registered PEs. The engineer in private practice and in public works management has a special set of ethical problems, some of which are discussed below.

Political contributions. Private consultants must rely on public and private clients for work. In an article entitled "I Gave Up Ethics—To Eat,"[29] an anonymous engineer who went into consulting in the public works field after a good education and "a long apprenticeship with an old, established engineering firm" describes how he went two full years without work until he was learned about "political engineering." He was introduced to "The Reverend," who "was in pretty good with the public works department and other state agencies."

TABLE 18–5 Core Concepts in Engineering Ethics

I. The public interest
  A. Paramount responsibility to the public health, safety, and welfare, including that of future generations
  B. Call attention to threats to the public health, safety, and welfare, and act to eliminate them
  C. Work through professional societies to encourage and support engineers who follow these concepts
  D. Apply knowledge, skill, and imagination to enhance human welfare and the quality of life for all
  E. Work only with those who follow these concepts
II. Qualities of truth, honesty, and fairness
  A. Be honest and impartial
  B. Advise employer, client, or public of all consequences of work
  C. Maintain confidences; act as faithful agent or trustee
  D. Avoid conflicts of interest
  E. Give fair and equitable treatment to all others
  F. Base decisions and actions on merit, competence, and knowledge, and without bias because of race, religion, sex, age, or national origin
  G. Neither pay nor accept bribes, gifts, or gratuities
  H. Be objective and truthful in discussions, reports, and actions
III. Professional performance
  A. Competence for work undertaken
  B. Strive to improve competence, and assist others in so doing
  C. Extend public and professional knowledge of technical projects and their results
  D. Accept responsibility for actions and give appropriate credit to others

Source: Andrew G. Oldenquist and Edward E. Slowter, "Proposed: A Single Code of Ethics for All Engineers," *Professional Engineer*, May 1979, p. 9.

On a 10% commission of his fees to The Reverend, and through a practice of large campaign contributions (to both parties, just in case), our engineer began to prosper. The editors of the compendium in which this appeared conclude of this author: "His story may not be true for all branches of private practice nor in all parts of the country, but his story is authentic for his field of practice and also his state. We checked it."

Distribution of public services. Price describes at some length[30] the 1971 case of *Andrew Hawkins* v. *Town of Shaw, MS*.[31] Shaw was a segregated town of 1500 black inhabitants (including Hawkins) and 1000 whites. Ninety-eight percent of the homes fronting on unpaved streets and 97% of those not served by sanitary sewers were occupied by blacks. White neighborhoods had mercury vapor street lights and 6-inch water mains; black neighborhoods had bare bulbs and 2-inch mains. The 5th Circuit Court of Appeals found that this violated the equal protection clause of the 14th Amendment (the same clause that was applied to segregated schools in the famous *Brown* v. *Board of Education* case in 1954). Public works improvements in this case were from the city's general tax funds; improvements reimbursed by special assessments on the property owners benefited would not be subject to this equal protection clause. This case is just one example of the political environment in which the city engineer or public works director may work.

Construction safety. If you seek maximum safety, I suggest you work in a nuclear plant or for an explosives manufacturer, for in these occupations safety engineers examine hazards exhaustively and large sums are spent to mitigate them. Avoid unskilled construction jobs such as digging trenches! At least 100 fatal accidents occur in the United States each year from trench cave-ins, most of which could be prevented by prudent engineering and construction supervision. Large contractors, driven by liability insurance costs, OSHA (Occupational Safety and Health Administration) regulations, and desire to avoid bad publicity, have few such accidents; small contractors are tempted to cut corners and ignore regulations and have a disproportionately greater percentage of accidents.[32] The ethical implication is clear, not only for the contractor's engineer, but also for the client's engineer, who has an ethical obligation "to hold paramount the safety" of the public and to specify and insist that construction be carried out in a safe manner.

High-rise construction offers many hazards also, and with it, obligations for the ethical design or construction engineer. OSHA's largest fine, $5.11 million, was imposed against contractors for "serious disregard for basic fundamental engineering practices, a factor directly related to the cause of the collapse . . . of an apartment complex under construction in Bridgeport, Conn., where 28 workers were killed."[33] A rod connected to a hydraulic jack being used to lift a concrete floor slab into place slipped, "triggering a chain reaction that brought both wings of the 13-story structure down in a matter of seconds." Quotations in this article from John Pendergrass, OSHA chief, make it clear that engineers and contractors in this construction failed to meet their ethical obligations to worker safety:

> We found obvious design deficiencies in a lifting system that could have been easily detected with an engineering analysis, but this was not done even after the system failed on two occasions prior to [this building] collapse. Furthermore, we found a pattern of sloppy construction practices throughout the project and an overall sense of employer complacency for essential work place safety considerations.[34]

Construction case study: the Hyatt Regency disaster. Engineers in construction bear ethical obligations to the public that uses the structures they design that are at least as important as the obligations to workers that build these structures. The Hyatt Regency Hotel in Kansas City, MO, was designed with a high-roofed atrium over the lobby area, crossed by three pedestrian bridges, or "skywalks," of which the one at the fourth-floor level was directly above the one at the second-floor level. These two skywalks, each about 8 feet (2.4 meters) wide and 120 feet (36.6 meters) long, were suspended from the ceiling by three pairs of steel rods about 30 feet (9.1 meters) apart. On July 17, 1981, not too long after the hotel opened, a tea dance was held, and the skywalks as well as the lobby floor were crowded with people. A structural support for the fourth-floor skywalk gave way, causing it to crash into the lobby, taking the second-floor walk with it, killing 114 people and injuring more than 200. The original design of the hotel specified the support rods for these skywalks be made from 1.75-inch (4.44-cm) diameter high-strength steel; the actual construction used 1.25-inch (3.18-cm) diameter mild (A36) steel rods. Moreover, the six support rods were intended to be continuous from the ceiling to the second floor, threaded so that separate nuts would support the two skywalks; in a field change the rods were terminated at the fourth floor, with additional rods slightly offset beginning there and going down to the second floor, adding some torsional stress. A structural engineering consulting firm, GCE International Inc., was retained by the architects and "sealed" the drawings; no one at GCE did a detailed analysis of the support of the skywalks despite their promise to do so, maintaining instead that this was the customary responsibility of the steel fabricators providing the actual steel components.[35]

In February 1984 the Missouri Board for Architects, Professional Engineers and Land Surveyors (Missouri's licensing agency) filed a complaint against Daniel M. Duncan and Jack D. Gillum, principals in GCE, and a 26-day hearing was held before Administrative Hearing Commissioner James B. Deutsch. In a carefully written decision, Judge Deutsch supported the complaint, and the Missouri Board stripped Duncan and Gillum of their Professional Engineer licenses and GCE of its engineering license, an action that was upheld in appeals to the City of St. Louis Circuit Court in December 1986 and the Missouri Court of Appeals on January 26, 1988. In discussing the case later, before a civil engineering class at the University of Missouri–Rolla, Judge Deutsch identified three sources of professional responsibility:

1. The statutory responsibility "to protect the public interest" and the regulatory law amplifying this
2. Contractual responsibility (in this case, to perform all structural engineering and analysis on the design)
3. The customs and practices of the engineering profession, including the common law requirement to act as a "reasonable man" would and its modern extension to act as a reasonably skilled and prudent professional should[36]

In the Hyatt affair, Deutsch told the class, none of these three were observed. The engineer who imprinted his PE seal on the applicable drawing did not design it, supervise it, or even check it. Deutsch rejected the defense that "everybody does it this way," quoting a prior opinion that an unsafe practice does not become acceptable even if uniformly adopted. As a result, Professional Engineers can have no doubt of their personal responsibility for designs that bear their seals. (Following the Hyatt action, liability insurance for Missouri structural engineers soared to 9.3% of billings, three times the level of 1983.)

Ethical Problems in Industrial Practice

Significance. The examples of ethical problems in the section above refer to engineers in private consulting practice or in public works, to which much of engineering codes of ethics have been directed. But as Wilcox points out:

> First, engineers are, for the most part, employees of large corporations. They face the dilemmas of many professionals whose work is situated in the context of complex organizational structures. How are engineers as professionals to understand themselves as employees of institutions which do not have the same ethical codes as the engineers?[37]

Four examples of problems engineers may face in corporate employment appear in the following paragraphs, culminating in two case studies: a whistle-blowing case involving General Electric, and what is perhaps the best-known recent ethical failure of engineering managers, which led to the *Challenger* disaster.

Environmental responsibilities. Engineers and engineering managers are intimately involved in decisions about effluent discharge from their plants and the effect it has on the air we breathe and the water we drink. For example, oil- or coal-burning electric power plants are one of the largest sources of sulfur dioxide in our atmosphere, a prime cause of smog and acid rain. Reducing the sulfur dioxide requires using much more expensive low-sulfur fuel or construction of very expensive "scrubbing" equipment. These costs *may* be recoverable by increasing electrical rates, but achieving rate increases requires lengthy and uncertain pleadings before regulatory agencies, and any costs not recovered fall heavily on stockholders.

The decision may be even more difficult in unregulated industries. Ideally, one might want a chemical plant or a metal smelter to return water to our streams as clean as it was received. To accomplish this, however, requires substantial expenditure. If industrial competitors, whether in the United States or abroad, ignore their environmental responsibilities, it may not be economically feasible for a single firm to modify an old plant to meet them completely, and the alternative may be closing the plant and laying off all the workers. In such situations the ethical answer is not always clear-cut.

Conflict of interest. There are a number of situations where conflict of interest may arise in the work of an engineering employee of a corporation. Three of the more obvious are:

1. *Gifts.* Engineers typically provide the specifications for technical products and recommend which among alternative offerings should be accepted. Salespersons (including sales engineers) for suppliers attempt to build goodwill and capture business from potential customers/clients. Part of the custom in this relationship is the offering of tokens of appreciation (gifts). Each person (and each organization) needs to establish what is an acceptable gift, and at what level a gift carries a connotation of undue influence. Many organizations establish an approximate dollar limit. Federal government employees may have almost a zero limit, down to not accepting a free meal. One common rule on gifts is to offer and to accept only consumables: accept a bottle (not a case) of spirits but not a champagne cooler; accept a (small) box of golf balls but not a golf club; accept an "extra ticket" of modest value but not a major league season pass. No hard and fast rule can be provided—each engineer must develop a feeling for what is proper for the industry, his/her employer, and the sensitivity involved in the job held with that employer.

2. *Moonlighting.* Most engineers would agree that they should not compete with their employer in bidding on a project. However, if the employer declines to bid, may the engineer offer to do the job privately on nights and weekends? If that brings the client back to the engineer rather than his or her employer on the next offering, must the employer again be offered first opportunity?

3. *Inside information.* Engineers who have a significant percentage ownership in a small firm are limited in their right to buy and sell large blocks of stock based on "inside information" of favorable and unfavorable information that will affect stock purchases. Is the engineer who owns an insignificant amount ($10,000 stock in a billion-dollar corporation) similarly constrained? Was the author wrong in 1954 when, impressed as a young Air Force officer by the F101 "Voodoo" aircraft under test, he bought 10 shares of McDonnell Aircraft (and sold them a year later at a 90% increase)?

Postemployment limitations. What are reasonable limitations to the contracts required by some employers of engineers and other professionals that they will not work for a competitor for a specified time after leaving their current employer? Certainly, it is reasonable to require an engineer working on an unusually important confidential development not to move to another firm and use this proprietary information (i.e., trade secrets) to create a product that competes directly with the original employer. On the other hand, knowledge and skill in a specific specialty is all that a knowledge worker has to offer a potential new employer, and to deny an engineer the right to use these skills (as opposed to proprietary information) elsewhere would also be unprofessional. Where the ethical divider between these extremes lies must fall to individual consciences and, failing that, the courts to decide.

Whistle-blowing. Engineers and other professionals may well come across an example where their employer is doing something that they believe is dishonest, illegal, or damaging to the public, and the employer is unwilling to change. Here, the engineer's professional ethics may come in conflict with loyalty to the employer (and "business ethics," if these are perceived differently). DeGeorge provides some guidelines for such occasions:

> Whistle blowing is morally justifiable when there is impending danger and a concerned employee "has made his moral concern known" to his immediate superior who has subsequently failed to act. When this happens, advises DeGeorge, a concerned employee should take his or her complaint upward through company channels, if necessary, to top management. After all internal efforts have failed, public disclosure is justifiable. For whistle blowing to be obligatory as well as justifiable, two more conditions must be met: first, that the employee have documentation or other hard evidence (else his chances are slim); and second, that he "must have good reason to believe that by going public he will be able to bring about the necessary changes."[38]

An evening graduate student of mine interrupted at this point in class with his own experience. He had been hired to develop a self-cleaning oven for home use at a time when several competitors were developing similar ovens, and to include a window in the oven door so that the housewife could "see what's cooking." After building a prototype he measured the temperature on the outside of the oven and found it to be 300°F (about 150°C), quite enough to burn the hand of a toddler crawling around the kitchen. My student pointed this out to his boss, and to his boss's boss, but both regretted that nothing could be done because "management" was anxious to beat out the competition on this new product. Only when my student took the matter still higher, and pointed out that he would be ethically obligated to notify the Consumer Product Safety Commission if the company would not act, was development slowed to incorporate the safer design to which housewives are now accustomed.

My student did not lose his job as a result of this event (although he no longer worked for that employer at the time he was in my class), but most whistle-blowers are not so fortunate. Westin[39] documents the experiences of 10 whistle-blowers in American corporations. All 10 were terminated, and only one was reinstated. Two secured partial damages in court, but the other seven have been unable to obtain reinstatement, damages, or vindication of their professional reputation.

Whistle-blowing case study: General Electric Company (GE). Occasionally the good guys win! In July 1992, GE pleaded guilty to fraud involving diversion of U.S. funds by high-level company employees and an Israeli general in connection with Israel's acquisition of GE jet engines. This resulted from the action of GE employee Chester Walsh, who was based in Tel Aviv and gathered evidence for more than four years before suing GE in Cincinnati in November 1990 under the federal False Claims Act. GE paid $9.5 million criminal and $59.5 million civil fines and penalties, of which latter Walsh was entitled to up to 25% under the Act. The U.S. Department of Justice (under President Bush), charged with assisting and aiding whistle-blowers, joined GE in contending the employee had "manipulated" the False Claims Act for his personal gains, and "asked for a smaller award in order to send a 'strong message' that whistle blowers shouldn't get 'large windfalls' for preparing their cases 'in a dilatory fashion.'" GE maintained it would have put a stop to the fraud if Walsh had immediately reported it to the company. Judge Carl B. Rubin took GE to task, citing the case of another GE employee, Alaric Fine, who was reassigned after he reported his suspicions to the company, as "an example of what happened to GE employees who in fact called attention to irregularities to the persons committing them," and awarded Walsh $13.4 million.[40] Walsh no longer works for GE, but then, he no longer needs to!

The *Challenger* disaster. On January 24, 1985, Roger Boisjoly, Senior Scientist at Morton Thiokol Inc. (MTI), watched the launch of Flight 51-C of the space shuttle program and remained to inspect the solid rocket boosters after their recovery from the Atlantic Ocean. These immense boosters are too large to transport, so they are manufactured in cylin-

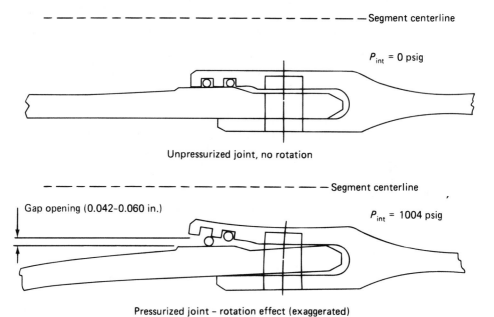

Unpressurized joint, no rotation

Pressurized joint – rotation effect (exaggerated)

Figure 18-1. Cross section of *Challenger* booster flange.
(From Russell J. Boisjoly and Ellen Foster Curtis, "Roger Boisjoly and the *Challenger* Disaster: A Case Study in Engineering Management, Corporate Loyalty, and Ethics," *Proceedings of the Eighth Annual Meeting, American Society for Engineering Management,* St. Louis, MO, October 11–13, 1987, p. 10.)

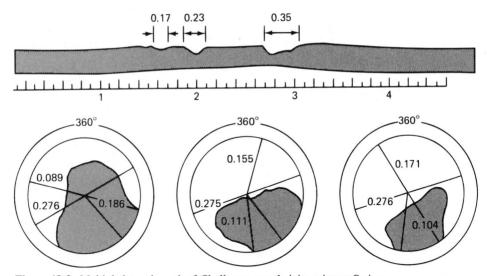

Figure 18-2. Multiple burn-through of *Challenger* nozzle joint primary O-ring.
(From Russell J. Boisjoly and Ellen Foster Curtis, "Roger Boisjoly and the *Challenger* Disaster: A Case Study in Engineering Management, Corporate Loyalty, and Ethics," *Proceedings of the Eighth Annual Meeting, American Society for Engineering Management,* St. Louis, MO, October 11–13, 1987, p. 8.)

drical sections and fastened together with "field joints" before launch. At each joint, the straight terminal ring of one segment (the "tang") slid into a clevis ring (with a Y-shaped cross section) on the mating segment, and this joint was sealed with two O-rings (see Figure 18-1). Boisjoly, then "considered the leading expert in the United States on O-rings and rocket joint seals," was dismayed to find "that both the primary and secondary O-ring seals on a field joint had been compromised by hot combustion gases (i.e., hot gas blow-by had occurred) which had also eroded part of the primary O-ring," although the temperature of the field joint at launch was believed to be a comfortable 53°F (12°C) (see Figure 18-2).[41]

Since rocket motor pressurization following ignition causes some rotation in the field joint, opening the annulus sealed by the O-rings, Boisjoly sponsored a series of subscale laboratory tests in March 1985 of the effect of temperature on O-ring resiliency. In these tests O-rings were squeezed, the pressure removed, and the time for the O-ring to regain shape measured. At 100°F (38°C) recovery was immediate, at 75°F (24°C) it took 2.4 seconds, but at 50°F (10°C) the seal had not recovered even after 10 minutes (600 seconds). In the ensuing months, Boisjoly emphasized in the strongest terms the need to redesign the field joint. On August 20, 1985, Robert K. Lund, MTI Vice President, Engineering, announced formation of a Seal Erosion Task Team, but little progress was made on solving the problem despite further blow-by on a flight on October 30, 1985, when the field joint temperature was estimated at a balmy 75°F (24°C).

The stage is now set for the eve of the *Challenger* tragedy:

At 10 a.m. on January 27, 1986, Arnie Thompson [MTI Supervisor of Rocket Motor Cases] received a phone call from Thiokol's Manager of Project Engineering at MSFC [Marshall Space Flight Center], relaying the concerns of NASA's Larry Wear, also at MSFC, about the 18°F (-8°C) temperature forecast for the launch of Flight 51-L, the *Challenger,* scheduled for the next day. This phone call precipitated a series of meetings within Morton Thiokol; at the Marshall Space Flight Center; and at the Kennedy Space Center [KSC] that culminated in a three-way telecon, involving three teams of engineers and managers, that began at 8:15 p.m. E.S.T.[42]

Present on the telephone were 14 managers and engineers at Thiokol's Wasatch (UT) Division Management Information Center, 15 at MFSC, and 5 at KSC. Boisjoly and Thompson began by detailing the flight and laboratory experience outlined above. Lund presented the final chart, recommending against launch unless the O-ring seal temperature exceeded 53°F (12°C); Joe Kilminster, MTI Vice President, Space Booster Programs, supported his engineers and would not recommend launch below 53°F.

George Hardy, Deputy Director of Science and Engineering at MSFC, was "appalled at that recommendation," but would not recommend to launch if the contractor was against it. Lawrence Mulloy, Manager of Booster Projects at KSC, also strenuously objected, saying "My God, Thiokol, when do you want me to launch? Next April?" Boisjoly continued to object to launch, but finally Kilminster asked for a five-minute caucus of Thiokol people. Jerry Mason, MTI Senior Vice President of Wasatch Operations, began the caucus by saying that "a management decision was necessary" (influenced, very likely, by the fact that MTI was at that time negotiating a billion-dollar follow-on contract with NASA). Thompson and Boisjoly re-reviewed their reasons for not launching but quit when it was obvious that no one was listening; Mason then turned to Bob Lund and in a memorable statement, asked Lund to "take off his engineering hat and put on his management hat." At that point Lund, Mason, Kilminster, and Calvin Wiggins (MTI Vice President and General Manager of the Space Flight Division) held a brief discussion and voted unanimously to recommend *Challenger*'s launch.[43] The following day, about 73 seconds into launch, the *Challenger* exploded in a ball of flame on the television screens of the entire world.

In discussing this "management" decision, Florman concludes:

> The four so-called Thiokol "managers" are, in fact, engineers. Mason has a degree in aeronautical engineering, Lund in mechanical; Wiggins has a degree in chemistry, and Kilminster a master's in mechanical engineering on top of an undergraduate degree in mathematics. The two NASA "officials," Hardy and Mulloy, who urged that Thiokol approve the launch, are also engineers, as are the key NASA people above them. These men were educated as engineers and had worked as engineers, eventually moving into positions of executive responsibility. They did not thereupon cease being engineers, any more than a doctor who becomes director of a hospital stops being a doctor.
>
> Were these engineer-executives under pressure to meet a launching schedule? Of course. But pressure is inherent in engineering. . . . Pressure goes with the job like the proverbial heat in the kitchen. It may help explain, but it cannot excuse, an engineering mistake.[44]

Boisjoly's testimony before the Rogers Commission regarding the foregoing events led to increasing friction with MTI management. "Although given the title of Seal Coordinator for the redesign effort, he was isolated from NASA and the seal redesign effort. His design information had been changed without his knowledge and presented without his feedback."[45] As Boisjoly later concluded: "The research on [whistle-blowing] leads to two conclusions. First, all whistle-blowers attempt to achieve problem resolution through their organizational chain of command; and, second, they are all punished by the organization after whistle-blowing outside the organization."[46] He cited as "timeless" the advice of Adolph J. Ackerman in a June 1967 IEEE article:

> Engineers have a responsibility that goes far beyond the building of machines and systems. We cannot leave it to the technical illiterates, or even to literate and overloaded technical administrators to decide what is safe and for the public good. We must tell what we know, first through normal administrative channels, but when these fail, through whatever avenues we can find. Many claim that it is disloyal to protest. Sometimes the penalty—disapproval, loss of status, even vilification, can be severe.[47]

Boisjoly understands the last sentence well. His position at MTI became untenable, and he requested extended sick leave on July 21, 1986, with a case of posttraumatic syndrome. More than two years later, when the redesigned shuttle put America back in space, it was clear that Boisjoly would never return to work.

In a later analysis of testimony before the Rogers Commission on the correlation between temperature and O-ring erosion, Lighthall[48] quotes testimony of participants Boisjoly ("I couldn't quantify it"), MTI engineer Jerry Burn ("it is speculation"), NASA's Mulloy ("I can't get a correlation"), and NASA's Hardy ("obviously not conclusive"), then shows by statistical analysis of data available before the flight a better than 99.5% probability of just such correlation. He concludes "that none of the participants had ever learned, or had long since forgotten, elementary ideas and methods of statistical analysis and inference," a conclusion of obvious significance for engineering education.

Company Responses to Ethical Problems

Corporate managers are faced with many problems with ethical content, both collectively on behalf of the company and individually. Companies have varying responsibilities to a variety of stakeholders, and these responsibilities may conflict with each other. A typical enterprise's major stakeholders include:

- Owners (usually stockholders)
- Customers
- Employees
- Suppliers
- Competitors
- Governments
- The local community
- Society as a whole

Corporations are, indeed, concerned with the ethical conduct of their employees. Indeed, a Conference Board survey of 300 major companies revealed that:

- Three-fourths had written codes of conduct.
- Sixty-one percent had terminated employees in the past five years for ethical conduct violations.
- The support of the CEO "by word and example" was the most important factor in achieving compliance.[49]

Regrettably, companies with codes of ethics are actually cited by federal agencies *more often* than are those that lack such standards. This is because "the codes are really dealing with infractions against the corporation, rather than illegalities on behalf of the corporations,"[50] reports Marilynn Mathews in a survey of ethical codes of 202 Fortune 500 companies. She found that more than three-fourths of these codes stressed relations with the U.S. government (87%), customer/supplier relations (86%), political contributions (85%), conflicts of interest (75%), and honest records (75%). On the other hand, more than three-fourths *did not* mention personal character matters (94%), product safety (91%), environmental affairs (87%), product quality (79%), and civic and community affairs (75%).

Certainly there are companies that make exceptional efforts to assure that their organizational values, which can only be an extension of the values of top managers, are impressed on members of the organization. As a single example, Johnson and Johnson has a one-page statement of values known as its Credo, which is a unifying force in a company otherwise divided into 150 somewhat autonomous business units. Between 1975 and 1978 more than 1200 J&J managers attended two-day seminars of 25 people to challenge the Credo and assure that it was still up to date. Either CEO James E. Burke or President David R. Clare presided at each of these many meetings.[51] J&J is not alone: books such as *In Search of Excellence*[52] identify many other companies whose top managers go to great lengths to emphasize the values important to the organization.

Summary: Making Ethical Decisions

This half chapter has shown that engineers, managers, and other professionals have many occasions in which corporate, professional, and personal objectives and values may conflict. In some of these there will be real conflicts between positions that each have ethical justification. The engineer, or any other person, can then only fall back on his or her personal values. Questions such as the following can lead to a solution you can adopt with self-respect and live with:

- Does the action I am considering make good sense?
- Does this action fit my best concept of a dedicated professional engineer?
- Will my action unnecessarily harm others? Is there some way that I can compensate them?
- Would my action stand up to close public scrutiny? Would I have difficulty explaining it to a reporter? to a judge and jury? to my colleagues? to my own family?
- Am I hiding behind a superior's judgment or wish, or can I justify it based on my own values?

FUTURE CONSIDERATIONS IN ENGINEERING AND MANAGEMENT

Future directions in engineering and management are, of course, inseparable from future trends of society as a whole. Some of the driving forces affecting the beginning of the twenty-first century will include (1) the continuing computer-based information revolution;

(2) an increasing technological sophistication of society; (3) international and political considerations in a shrinking, interdependent world; (4) demographic considerations; and (5) interactions of food, energy, materials, and the environment. Let us look at each briefly, then speculate about the effects on the engineering career.

The Computer Revolution

We are still in the early phases of the computer and information revolution, which is just as significant to world history as was the industrial revolution that preceded it. Let us speculate on its effects on some major parts of society:

* *The virtual organization.* The availability of integrated management information and decision-making systems, together with user-friendly software and increased computer literacy among managers, has decreased the number and levels of middle managers and staff professionals needed to analyze and relay information in industrial organizations, yet increased the empowerment and responsibility of non-managers. As members throughout an organization more and more share common purposes and a common information system with trusted suppliers and customers, boundaries between them will become more fluid and they will become interdependent members of the new "virtual organization." Ultimately, the lessons learned in reducing hierarchy, breaking down functional boundaries, and empowering individuals in profit-making industry will find real (but perhaps more limited) application in government, military organizations, and academia.
* *The virtual product.* The traditional practice of manufacturing standard goods for inventory and ultimate sale will more and more disappear. Designs and specifications for families of "virtual products" will exist in computer memory, and they will be transmitted to computer-integrated manufacturing systems when needed and as specified by the customer. The need for repetitive labor in manufacturing will continue to decline, and the availability of inexpensive labor will become less critical than the availability of knowledge workers in determining industrial locations.
* *The home.* The computer and information revolution has already begun to affect the way we live, and we can expect that it will do so increasingly. Computer systems in the home can be expected to handle home security, climate control, and optimization of energy use. Computer, television, and telephone will blend into a total system that connects us with the rest of the world for communication (oral, visual, and written), education, and interactive banking, shopping, and entertainment.

Education and the Technological Society

To function effectively in this technological society, at least in the developed nations, will require a more general understanding of technology. Education to prepare for this world will require a good understanding of mathematics and science and widespread computer literacy. Such a change can only be gradual, since elementary and secondary school teachers are in general (at least in the United States) poorly prepared in such areas. However, we can speculate on the age when something close to an engineering education will be the standard for an educated person—especially the policy makers who control society.

The Shrinking World

The former Communist bloc has shattered, and it would certainly appear that, like Humpty Dumpty, it cannot be put together again. Nonetheless, the transition from a totalitarian but inefficient communist economy largely controlled from Moscow to more democratic and independent market economies throughout Eastern Europe and the former Soviet states will be difficult, and probably not without conflict. To the "developed countries" of Western Europe, the United States, Canada, and Japan, we have already added the Four Tigers of Asia (South Korea, Taiwan, Singapore, and Hong Kong); other countries of Southeast Asia (Thailand, for one), Latin and South America (Mexico and Brazil, in particular) can be expected to follow. Russia and India, next in population of scientists and engineers after the United States, and China, with the world's greatest population, will add unpredictably but powerfully to the world economy as they develop.

We can expect a continuing increase in the size and ubiquity of the great multinational companies, to the point that the determination of "what country they belong to" will become more and more uncertain. Clearly, the engineering manager of the future will need a much greater world view in designing, manufacturing, and marketing products.

Demographics

The increasing demand for knowledge workers will hasten the integration of women and minorities in all levels of our economy, especially in engineering and in all levels of management and government. New approaches in child care, part-time work, and work at home may accompany this. The increase in elder citizens will begin to blur the age of retirement, with some workers retiring early and beginning a second (possibly part-time or entrepreneurial) career, while others continue with their employer in at least a limited way into their seventies. While medical science (and engineering) continues to find cures that keep us alive longer, the increasing burden of the aged on the working population will expand the ethical decisions that medicine must face.

Food, Energy, Materials, and the Environment

While birthrates can be expected to decrease as countries develop, world population will place increasing demands on static or decreasing farmlands and especially on diminishing natural resources of fossil energy and minerals. As industrialization continues to spread and the rain forests disappear, world problems of air, water, and thermal pollution can only increase. We can expect continuing pressure for energy conservation, and perhaps a grudging reemphasis in the United States on nuclear power to avoid the air and water pollution from fossil fuels, unless fusion energy is proven both feasible and economic. All of these problems present difficult decisions in public policy at local, national, and world levels. They also offer endless challenges to be met by the engineer and the engineering manager.

Changes in the Engineering Career

As a result of the decline of middle management in the industrial organization, engineering (and other) specialists can expect to spend a longer portion of their career working as professionals before being considered for management positions, but they will need some understanding of other specialties and of management's concerns while working as a team member.

Engineers will find their careers less stable than in previous generations. Bahrami explains:

> Many firms have reexamined their employment policies—initiating early retirement programs and other incentives to reduce the size of their workforce. As pointed out in other studies, the critical tradeoff in this context is between corporate flexibility and individual security. Many corporations rely on temporary workers, specialized vendors, and consultants in order to flexibly deal with unique contingencies. Additionally, this trend points to a fundamental shift in the foundation of employer-employee relationship, away from the traditional patriarchal orientation toward what may be characterized as a peer-to-peer relationship. This sentiment is echoed in the following comment which encapsulates the implicit relationship between Apple Computer and its employees: "You own your own careers; we provide you with the opportunity."[53]

Engineers will therefore have to manage their career progress more than ever before. Because of rapidly changing technology, they will need to emphasize lifetime learning, not only in formal coursework, but by spending much more time following changes in their specialty, their industry, and general business trends. Design engineers will find that the evolution of computer-aided design makes them more efficient and productive as long as they continue to master new tools, but more vulnerable if they do not, since fewer designers may be needed. And, like production workers, they will find they are competing in a global market, and that modern communication permits engineering to be integrated across the globe. Texas Instruments has been designing integrated circuits in India since 1986 (taking advantage of the world's third largest pool of scientific and engineering talent, after the United States and Russia, who uniformly speak English); Sun Microsystems has hired Russian scientists for software and microprocessor research; CrossComm Corp. has their communications software written by Poles at the University of Gdansk.[54]

In conclusion, the engineer who remains alert to changing products, processes, technologies, and opportunities and who manages the progress of his or her career should find the beginning of the twenty-first century exciting and rewarding. Conversely, the engineer who simply reports to work and follows directions is likely to become, in the expressive British term, redundant, well before it is time for retirement.

DISCUSSION QUESTIONS

18-1. What is the percentage of women in the freshman class and in the last graduating class in the school of engineering at your university? Compare it with national trends.

18-2. Repeat question 18-1 for each minority grouping, and compare this with the percentage of minorities in the population and high schools from which your school primarily draws students.

18-3. What is the percentage of women and minority Americans among the graduate assistants and the faculty in your school of engineering? If it is low, what might be done to increase it?

18-4. For a company with which you are familiar, identify (**a**) any inhibitions that exist to equal opportunity for female and black professionals, and (**b**) any intentional programs to increase such opportunity.

18-5. Record your expenditure of time about every hour in 10-minute segments. At the end of the week, categorize your use of time for its contribution to A-, B-, and C-quality objectives. Report briefly on your findings.

18-6. Identify the top several "time wasters" in your life (from the "top 10" in the chapter and any others you identify). What might you do to reduce their effect?

18-7. Estimate how your personal energy level varies throughout the day. How might you use this information in planning your activities?

18-8. Which of the five categories of definitions ethicists have developed for "ethics" do you feel best apply to your activities as an engineer?

18-9. Obtain a copy of the code of ethics subscribed to by your engineering or other professional society. How much of its content is clearly based on ethical principals? Of what does the rest consist?

18-10. Identify the most important real or potential ethical problems facing you in your present position or one for which you are preparing. Select one and describe your position.

18-11. Prepare a brief report on a specific case from the public print or your own experience on an ethical problem faced by an engineer or engineering manager, and how it was resolved.

18-12. Describe the formal or informal code of ethical conduct established by your employer or by a company with which you are familiar.

18-13. Select a specific aspect expected in tomorrow's world (from the end of this chapter or from other sources) and discuss its impact on engineers and engineering managers.

NOTES

1. William K. LeBold and Dona J. LeBold, "An Historical Perspective on Women Engineers and a Futuristic Look," *Annual Conference Proceedings, American Society for Engineering Education,* University of Illinois, Urbana-Champaign, June 20–24, 1993, pp. 1259–1263.

2. "Opportunity Barometer," *Graduating Engineer,* February 1995, p. 9.

3. Betty M. Vetter, "Women in Science III—Ferment: Yes . . . Progress: Maybe . . . Change: Slow," *Mosaic,* (National Science Foundation), Fall 1992 (final issue), pp. 34–41.

4. Eleanor Baum, "Women in Engineering," Engineering Education in the Twenty-first Century Lecture Series, University of Missouri—Rolla, October 9, 1990.

5. Linda J. Zimbler, "Faculty and Instructional Staff: Who Are They and What Do They Do?" U.S. Dept. of Education National Center for Education Statistics Survey Report NCES 94-346, October 1994, p. 14.

6. Raymond B. Landis, "The Case for Minority Engineering Programs," *Engineering Education,* May 1988, pp. 756–761.

7. College Placement Council data reported in Vetter, "Women in Science III," pp. 39–40.

8. Sheila Humphreys, "Men, Women, and Engineering," *ASEE Prism,* January 1994, pp. 20–23.

9. C. D. Sutton and K. K. Moore, "Executive Women—20 Years Later," *Harvard Business Review,* September–October 1985, pp. 42–43.

10. "Corporate Women," *Business Week,* June 8, 1992.

11. Selwin Feinstein, "Women and Minority Workers in Business Find a Mentor Can Be a Rare Commodity," *Wall Street Journal,* November 10, 1987, p. 33.

12. Felice N. Schwartz, "Management Women and the New Facts of Life," *Harvard Business Review,* January–February 1989, p. 65.

13. *Wall Street Journal,* March 15, 1988, p. A1.

14. Schwartz, "Management Women," pp. 69–70.

15. Nehama Jacobs, "The Problems Female Managers Face," *Graduating Engineer,* February 1989, pp. 64–69.

16. Melissa J. Lane, "The Current Status of Women and Minorities in Engineering and Science," *Engineering Education,* May 1988, p. 753 (from National Science Foundation data).

17. Helen Gouldner, "The Social Context of Minorities in Engineering," Appendix C in National Research Council, *Engineering Employment Characteristics* (Washington, DC: National Academy Press, 1985), pp. 59–62.

18. John M. Amos and Bernard R. Sarchet, *Management for Engineers* (Englewood Cliffs, NJ: Prentice-Hall, Inc., 1981), p. 316.

19. Amos and Sarchet, *Management for Engineers,* p. 321.

20. Michael LeBoeuf, "Managing Time Means Managing Yourself," *Business Horizons,* February 1980, p. 41.

21. Alan Lakein, *How to Get Control of Your Time and Your Life* (New York: Peter H. Wyden, Inc., 1973).

22. Harold Kerzner, *Project Management: A Systems Approach to Planning, Scheduling, and Controlling,* 4th ed. (New York: Van Nostrand Reinhold Company, Inc., 1992), pp. 367–368.

23. Peter Koestenbaum, *The Heart of Business: Ethics, Power, and Philosophy* (Dallas, TX: Saybrook, 1987), p. 75.

24. Samuel E. Gluck, "Ethical Engineering," in John E. Ullmann, ed., *Handbook of Engineering Management* (New York: John Wiley & Sons, Inc., 1986), p. 176.

25. For Randolph's code and a variety of more recent ones, see Robert J. Baum and Albert Flores, *Ethical Problems in Engineering* (Troy, NY: Rensselaer Polytechnic Institute, Center for the Study of the Human Dimensions of Science and Technology, 1978), pp. 33–52.

26. Fenton Bagley, "Ethics, Unethical Engineers, and ASME," *Mechanical Engineering,* July 1977, p. 42.

27. Andrew G. Oldenquist and Edward E. Slowter, "Proposed: A Single Code of Ethics for All Engineers," in James H. Schaub and Karl Pavlovic, eds., *Engineering Professionalism and Ethics* (New York: John Wiley & Sons, Inc., 1983), pp. 446–447.

28. Oldenquist and Slowter, "Proposed: A Single Code," p. 449.

29. Anonymous, "I Gave Up Ethics—To Eat," in Schaub and Pavlovic, *Engineering Professionalism and Ethics,* pp. 233–238.

30. Willard Price, "Values in Public Works Decision Making: The Distribution of Services," in Daniel L. Babcock and Carol A. Smith, eds., *Values and the Public Works Professional* (Rolla, MO: University of Missouri—Rolla, 1980), pp. 23–35.

31. *Andrew Hawkins et al.,* Plaintiffs-Appellants v. *Town of Shaw,* MS, Defendants-Appellees, U.S. Court of Appeals, 5th Circuit, 23 January 1971, 437 Fed. Rep., 2d, 1286.

32. Manuel M. Davenport, "Ethical Issues in Excavation Safety," *Business and Professional Ethics,* quarterly newsletter of the Center for the Study of the Human Dimensions of Science and Technology (Troy, NY: Rensselaer Polytechnic Institute, n.d.), p. 11.

33. "Builders Fined $5 Million in Fatal Collapse," *St. Louis Post-Dispatch,* October 23, 1987, p. 1F.

34. "Builders Fined $5 Million," p. 1F.

35. Paul Munger (Professor of Civil Engineering, University of Missouri—Rolla and chairman (1981–84), Missouri Board for Architects, Professional Engineers and Land Surveyors), personal communication.

36. James B. Deutsch, talk before civil engineering class, University of Missouri—Rolla, November 3, 1988.

37. John R. Wilcox, "The Teaching of Engineering Ethics," *Chemical Engineering Progress,* May 1983, pp. 15–20.

38. Richard T. DeGeorge, *Business Ethics* (New York: Macmillan, 1982), quoted in Gluck, "Ethical Engineering," pp. 185–186.

39. Alan F. Westin, *Whistle-Blowing! Loyalty and Dissent in the Corporation* (New York: McGraw-Hill Book Company, 1981), pp. 132–133.

40. Amal Kumar Naj, "Federal Judge Awards Ex–GE Staffer Record Amount in Whistle-Blower Case," *Wall Street Journal,* December 7, 1992, p. A5.

41. Russell J. Boisjoly and Ellen Foster Curtis, "Roger Boisjoly and the *Challenger* Disaster: A Case Study in Engineering Management, Corporate Loyalty, and Ethics," *Proceedings of the Eighth Annual Meeting, American Society for Engineering Management,* St. Louis, MO, October 11–13, 1987, pp. 6–7. [Russell Boisjoly (like Curtis, associate professor in the College of Management Science, University of Lowell) is Roger's brother.]

42. Boisjoly and Curtis, "Roger Boisjoly," p. 9.

43. Boisjoly and Curtis, "Roger Boisjoly," p. 10.

44. Samuel C. Florman, *The Civilized Engineer* (New York: St. Martin's Press, 1987), p. 163.

45. Boisjoly and Curtis, "Roger Boisjoly," p. 11.

46. "Boisjoly Says Self-Respect, Ethics Guided Actions," *ASME News,* March 1988, p. 1, a summary of Roger M. Boisjoly, "Ethical Decisions—Morton Thiokol and the Space Shuttle *Challenger* Disaster," presentation to the American Society of Mechanical Engineers Winter Annual Meeting, December 1987, ASME paper 87-WA/TS-4.

47. As quoted in "Boisjoly Says," *ASME News,* p. 1.

48. Frederick F. Lighthall, "Engineering Management, Engineering Reasoning, and Engineering Education: Lessons from the Space Shuttle *Challenger,*" *Proceedings of the International Engineering Management Conference,* Santa Clara, CA, October 21–24, 1990, pp. 369–377.

49. Bill Kester, "Much Ado About Ethics in Business," *St.Louis Post-Dispatch,* October 14, 1987, business section.

50. Marilynn C. Mathews, quoted in Rick Wartzman, "Nature or Nurture? Study Blames Ethical Lapses on Corporate Goals," *Wall Street Journal,* October 9, 1987, p. 21.

51. John A. Byrne, "Businesses are Signing Up for Ethics 101," *Business Week,* February 15, 1988, pp. 56–57.

52. Thomas J. Peters and Robert H. Waterman Jr., *In Search of Excellence: Lessons from America's Best-Run Companies* (New York: Harper & Row, Publishers, Inc., 1982).

53. Homa Bahrami, "The Emerging Flexible Organization: Perspectives from Silicon Valley," *California Management Review,* Summer 1992, pp. 33–52.

54. "What's Wrong: Why the Industrialized Nations Are Stalled," *Business Week,* August 2, 1993, pp. 42–59.

Index